Monographiae Biologicae

Volume 92

Series editor
Henri J. Dumont, Ghent University, Department of Biology, Ghent, Belgium

Aims and Scope

The Monographiae Biologicae provide a forum for top-level, rounded-off monographs dealing with the biogeography of continents or major parts of continents, and the ecology of well individualized ecosystems such as islands, island groups, mountains or mountain chains. Aquatic ecosystems may include marine environments such as coastal ecosystems (mangroves, coral reefs) but also pelagic, abyssal and benthic ecosystems, and freshwater environments such as major river basins, lakes, and groups of lakes. In-depth, state-of-the-art taxonomic treatments of major groups of animals (including protists), plants and fungi are also elegible for publication, as well as studies on the comparative ecology of major biomes. Volumes in the series may include single-author monographs, but also multi-author, edited volumes.

More information about this series at http://www.springer.com/series/6304

Victor R. Alekseev • Bernadette Pinel-Alloul
Editors

Dormancy in Aquatic Organisms. Theory, Human Use and Modeling

Springer

Editors
Victor R. Alekseev
Zoological Institute of Russian
Academy of Sciences
St. Petersburg, Russia

Bernadette Pinel-Alloul
Département de sciences biologiques
Université de Montréal
Montréal, QC, Canada

ISSN 0077-0639 ISSN 2215-1729 (electronic)
Monographiae Biologicae
ISBN 978-3-030-21212-4 ISBN 978-3-030-21213-1 (eBook)
https://doi.org/10.1007/978-3-030-21213-1

This Springer imprint is published by the registered company Springer Nature Switzerland AG.
The registered company address is: Gewerbestrasse 11, 6330 Cham, Switzerland

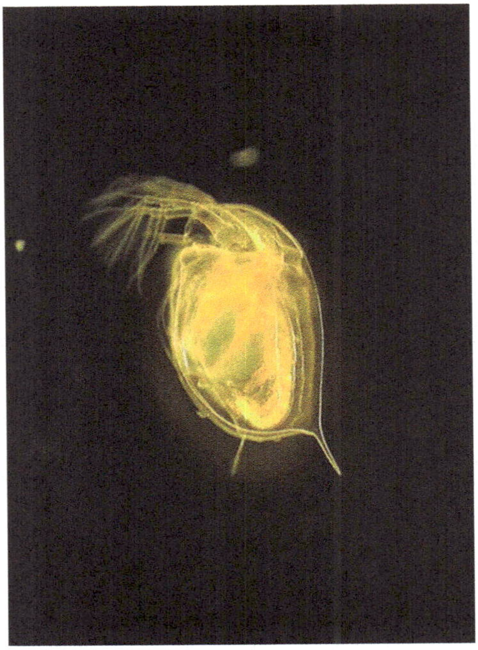

Ad memoriam Professor Oscar Ravera

Preface to the Second Edition

This book is dedicated to Professor Oscar Ravera, the outstanding Italian scientist who organized the 2003 Palazzo International Conference on Diapause in Aquatic Invertebrates. Many of this edition's authors, as well as those of the first edition (entitled *Diapause in Aquatic Invertebrates*), met each other at this conference. Chapter 1 features a fascinating biography of Dr. Ravera written by Nicoletta Ricardi, and it describes his significant impact in many fields of freshwater biology, including dormancy.

The theoretical sections of this volume include several groups of aquatic organisms which are under-studied in terms of dormancy's role in the life cycle. This does not mean, however, that its role is trivial. The first group of organisms is tardigrades (Chapter 3 by Roberto Bertolani and coauthors). Members of this group are known as the best survivors of extremely harsh conditions (radiation, high pressure, etc.). The second group, also absent in the first issue, is fish (Chapter 4 by Dmitry L. Lajus and Victor R. Alekseev). Resting stages in marine zooplankton are covered in Chap. 5 (Genuario Belmonte and Fernando Rubino). Marine ecosystems were not covered in the previous edition.

Part 2 includes chapters devoted to the use of dormancy phenomena in science and potential human applications. It opens with Chap. 6, by Victor R. Alekseev and coauthors, which describes experiments on dormancy in space. Such work aims to advance new technologies capable of supporting human life outside our planet, such as long-term space flights, living in orbital stations, etc. Survival of dormant stages in harsh human-engineered conditions is described in Chap. 7 (Egor Zadereev, Tatiana S. Lopatina, and Natalia Oskina). Chap. 8 covers birds' participation in distribution of aquatic invertebrate dormant stages (Miroslaw Slusarczyk, Barbara Pietrzak, and Bernadette Pinel-Alloul). The chapter, in fact, describes the second part of a mechanism described in the first edition (Panov and Caceres, 2007) regarding biological invasion by alien species, after crossing zoogeographical barriers, with human participation.

A chapter from the first edition, devoted to seasonal diphenism in *Daphnia* from temperate environments, has been revised and is presented (Chapter 9 by Kęstutis Arbačiauskas). The role of chemical signaling in crustacean dormancy induction

(Chapter 10 by Egor Zadereev and Tatiana S. Lopatina) has also been significantly updated relative to the first edition. Chapter 11 covers new medical approaches aiming to exploit dormancy in human practice (Mei-Hiu Tseng, Jiang-Shiou Hwang, and Victor R. Alekseev).

Part 3 is comprised of several examples of using modeling in relation to dormancy phenomenon, and it opens with a theoretical analysis of studies of biological information, including seasonal information (Chapter 12 by Victor R. Alekseev). A detailed description of a *Daphnia longispina* population dynamics model, that includes several aspects of dormancy's role in facilitating optimal life cycles, is presented in Chap. 13 (Tamara Kazanzeva and Victor R. Alekseev).

The role of biotic and abiotic factors in cyclopid summer diapause is described in Chap. 14 (Bernadette Pinel-Alloul and Victor R. Alekseev), including application and testing of a principal model. Chapter 15 features a cyclopid plankton population dynamics model based on photoperiod's role in dormancy induction (Victor R. Alekseev and coauthors). Final Chap. 16 examines aspects of dormant egg production by the rotifer *Brachionus calyciflorus*, including biotic and abiotic factors, such as infection by parasites (Alexander K. Gorbunov and coauthors).

St. Petersburg, Russia — Victor R. Alekseev
Montréal, QC, Canada — Bernadette Pinel-Alloul

Acknowledgments

This work was partly supported by the RFBR grant 17-04-00027 and the Russian Government program on Biodiversity #AAAA-A19-119020690091-0. Victor R. Alekseev kindly appreciates the significant help of Olga Chaban for organizing and preparing the work for submission to the publisher.

Contents

Contributors

Victor R. Alekseev Zoological Institute of Russian Academy of Sciences, St. Petersburg, Russia

Tiziana Altiero Department of Education and Humanities, University of Modena and Reggio Emilia, Modena, Italy

Kęstutis Arbačiauskas Nature Research Centre, Vilnius, Lithuania

Genuario Belmonte Department of Biological and Environmental Sciences and Technologies, University of the Salento, Lecce, Italy

Roberto Bertolani Department of Education and Humanities, University of Modena and Reggio Emilia, Modena, Italy

Boris V. Fiks Medicine University of Lubeck, Lübeck, Germany

Alexander K. Gorbunov Astrakhan State Nature Biosphere Reserve, Astrakhan, Russia

Roberto Guidetti Department of Life Sciences, University of Modena and Reggio Emilia, Modena, Italy

Susanna H. Hakobyan Institute of Hydroecology and Ichthyology of Scientific Center of Zoology and Hydroecology of NAS of RA, Yerevan, Armenia

Jiang-Shiou Hwang Institute of Marine Biology, National Taiwan Ocean University, Keelung, Taiwan

Karen G. Jenderedjian Project "Promotion of Eco-corridors in the Southern Caucasus", Sevan Natural Reservation, Yerevan, Armenia

Tamara I. Kazantseva Zoological Institute of the Russian Academy of Science, University Emb., St. Petersburg, Russia

Dmitry L. Lajus Saint Petersburg State University, St. Petersburg, Russia

Margarita A. Levinskikh Institute of Biomedical Problems of RAS, Moscow, Russia

Tatiana S. Lopatina Institute of Biophysics, Krasnoyarsk Research Center SB RAS, Akademgorodok, Krasnoyarsk, Russia

Diane R. Nelson Department of Biological Sciences, East Tennessee State University, Johnson City, TN, USA

Natalia D. Novikova Institute of Biomedical Problems of RAS, Moscow, Russia

Natalia Oskina Institute of Biophysics, Krasnoyarsk Research Center SB RAS, Akademgorodok, Krasnoyarsk, Russia
Siberian Federal University, Krasnoyarsk, Russia

Barbara Pietrzak Faculty of Biology, Department of Hydrobiology, University of Warsaw at Biological and Chemical Research Centre, Warszawa, Poland

Bernadette Pinel-Alloul GRIL, Département de sciences biologiques, Université de Montréal, Montréal, QC, Canada

Lorena Rebecchi Department of Life Sciences, University of Modena and Reggio Emilia, Modena, Italy

Nicoletta Riccardi CNR-IRSA, Verbania, Italy

Fernando Rubino Water Research Institute, Unit Talassografico "A. Cerruti", National Research Council, Taranto, Italy

Miroslaw Slusarczyk Faculty of Biology, Department of Hydrobiology, University of Warsaw at Biological and Chemical Research Centre, Warszawa, Poland

Vladimir N. Sychev Institute of Biomedical Problems of RAS, Moscow, Russia

Mei-Hiu Tseng Medical College, National Taiwan University, Taipei, Taiwan

Elena B. Vinogradova Laboratory of Experimental Entomology, Zoological Institute RAS, University emb. 1, St. Petersburg, Russia

Egor Zadereev Institute of Biophysics, Krasnoyarsk Research Center SB RAS, Akademgorodok, Krasnoyarsk, Russia
Siberian Federal University, Krasnoyarsk, Russia

Part I
Strategies and Mechanisms of Diapause in Aquatic Organisms

Chapter 1
Professor Oscar Ravera: Long Life in Science

Nicoletta Riccardi

Abstract The long scientific way of Dr. Oscar Ravera started soon after the Second World War and lasted for more than 60 years. A vast number of his research projects covered practically all main fields in limnology of the twentieth century.

Keywords Aquatic ecology · Limnology · Bioindication · Radioactivity · Sediment dating · Environmental ethics

Professor Oscar Ravera getting an award in limnology

N. Riccardi (✉)
CNR-IRSA, Verbania, Italy
e-mail: nicoletta.riccardi@irsa.cnr.it

V. R. Alekseev, B. Pinel-Alloul (eds.), *Dormancy in Aquatic Organisms. Theory, Human Use and Modeling*, Monographiae Biologicae 92,
https://doi.org/10.1007/978-3-030-21213-1_1

Just after graduation in Natural Sciences (1949), he started his research career at the Istituto Italiano di Idrobiologia under the direction of Edgardo Baldi. Here, he spent 8 years focusing on basic research in limnology, mainly dealing with zooplankton and zoobenthos demography. During this period, he also took the opportunity to perform a research on the relative stability of benthic populations at the Biological Station of Lunz am See (Austria) under the direction of Prof. Franz Ruttner. In 1956, he moved to Sicily to run the Micropaleontological Service of the mining company Edison Co., which he directed until 1959. He returned to research in 1960 with a position as a hydrobiologist at CNEN (National Committee on Nuclear Energy) and then as a Senior Scientist at the Joint Research Center (JRC), Commission of the European Communities (CEC), Ispra, where he spent the following 30 years of his scientific career. Qualified as university lecturer in hydrobiology (University of Milan 1962) and as a full professor in "biogeography" (University of Catania 1976), he taught ecology at the Universities of L'Aquila, Venice, and Urbino. After retirement in 1997, he decided to continue his research as a private scientist at the Institute where he had started over 40 years earlier. Here, he worked until 2012, when age forced him to permanently discontinue a career that spanned 62 years. During his stay at the CNR Institute of Ecosystem Studies (previously Istituto Italiano di Idrobiologia), he continued his research on bioindicators and populations dynamics of freshwater mollusks and zooplankton.

During his long career, he leaded or collaborated to several CEC, OECD, UNESCO, INTECOL, NATO, and EEA projects. He was a founding member and President of the Italian Ecological Society (S.It.E.), a member of many Italian and foreign scientific societies, and national representative of SIL and European Ecological Federation (EEF). In addition to his scientific research, Ravera has had broad involvement as adviser to the Commission of the European Communities (CEC, Brussels) in formulating environmental research programs (1971–1990). He was a member of the Technical Bureau of the OECD International Program on Eutrophication (1972–1980); a member of the Technical-Scientific Committee of the International Commission for the protection of Italian-Swiss waters (1972–1987); the scientific coordinator of the UNESCO International Expert Group on the Lagoon of Venice (1997–1999); a member of the "Special Program Panel on Eco-Sciences" of the NATO Brussels (1979–1983); and several others. In addition, he collaborated with the International Expert Commission, coordinated by Prof. Ph. Bourdeau, on the environmental impact of the mobile flood barriers on the Venice Lagoon (1998). Thanks to his long-term experience on bioindicators and bioaccumulators, he has been for many years a coordinator of the working group of the SIL. He published more than 200 scientific papers and 3 books on general ecology, radioecology, and environmental ethics.

The 30 years of scientific activity of the research group that O. Ravera coordinated at the Joint Research Center (JRC) of the Commission of the European Communities (CEC) and his activity as an adviser to the CEC in Brussels have made the JRC one of the most important points of reference in Europe in the field of aquatic ecology.

His research interests have covered a very wide range of basic and applied ecology topics, which, starting from radioecology, have addressed crucial issues such as eutrophication, acidification, heavy metal contamination, the use of bioindicators for monitoring fresh and marine waters, and the assessment of human impacts on the ecosystem and its different components. This broad spectrum of interests is reflected by its most significant achievements, among which the following deserve particular mention:

- He was the first to measure fallout ^{54}Mn in European lakes and to use fallout radioactivity for dating lake sediments. His discovery of the great accumulation by freshwater mussels of ^{54}Mn from fallout (100,000 times the concentration in the lake water) was one of the most convincing demonstrations for ameliorating the CEC legislation on the radiocontamination risks (Ravera 1964).
- Using freshwater snails and mussels as models, i) he traced the distribution and fate and turnover rate of fall-out radioisotopes and of phosphorus in biotic and abiotic compartments of the lake ecosystem; ii) he evidenced the variations of X-ray effects at various life stages from germ cells to adults (Ravera 1967).
- He was a pioneer in the use of zooplankton remains as a proxy for trophic change in lakes, and he developed a new approach for the analysis of plankton community structure based on combined assessment of species/stage density with their relative biomass.
- The same methods used in the study of radiocontamination were transferred to the study of heavy metal pollution, producing important results in the assessment of the fate and the biological effects of low concentrations of toxic metals.

He numbers among his major achievements the promotion of ecology in Italian universities, the training of many Italian and foreign students, and helping to raise the consciousness of the general public as regards ecological concepts and the principles of environmental ethics (Ravera 1989).

His passion for research, continued actively even after retirement, is evidenced by the organization of two workshops of great scientific importance and lasting impact: one dealing with bioindicators ("Biological Monitoring," Verbania Pallanza, Italy, September 4–5, 2000) and one with dormancy/diapause in freshwater animals ("Diapause In Aquatic Invertebrates," Verbania Pallanza, Italy, October 21–23, 2003) (Alekseev et al. 2004). These workshops gave rise to collaborations and produced valuable scientific publications. For example, the diapause workshop promoted the publication of the first monograph on "Dormancy in aquatic invertebrates" published by Springer in 2007. As a pebble in the pond can generate a wave that is slow to die out, a bit of merit for the production of the current updated monograph goes back to the workshop of the now distant 2003.

The public recognition of his scientific merits, their amount of influence, success and originality, arrived in 2001, when he was awarded by the Global 500 Roll of Honor of the UNEP.

His enthusiasm for science, and ecology in particular, kept him active for over 60 years, only illness and death could defeat his optimistic and perennially young spirit. He "slipped away to the next room" just after turning 90, but even if out of sight, he will never be out of mind, *just around the corner*.

References

Alekseev V, Giussani G, Ravera O, Riccardi N (eds) (2004) Diapause in aquatic invertebrates. J Limnol 63: 97p

Ravera O (1964) Distribution of Mn-54 from fall-out in populations of freshwater Lamellibranchs (Unio mancus var. elongatulus, Pfeiffer). Verh Internat Verein Limnol 15:885–892

Ravera O (1967) The effects of X-rays on the demographic characteristics of Physa acuta (Gastropoda, Basommatophora). Malacologia 5:95–109

Ravera O (1989) Degradation, Protection and recovery of lake ecosystems. https://doi.org/10.1016/B978-0-08-040816-3.50025-9

Chapter 2
Introduction to Dormancy in Aquatic Invertebrates: Mechanism of Induction and Termination, Hormonal and Molecular-Genetic Basis

Victor R. Alekseev and Elena B. Vinogradova

Abstract Dormancy is a profound and ancient adaptation found in a wide spectrum of plants and animals of all habitats. In diapause, the switch between active and dormant states is driven by hormonal mechanism that usually includes a photoperiodic pacemaker. Temperature, food limitation, and some other stress factors as well are shown as driven by diapause induction in aquatic invertebrates. In the last decade, diapause studies from a wide variety of topics have demonstrated that diapause switch mechanisms may be developed to create novel applications in biotechnology. Resting eggs accumulated in the surface lake sediments represent a "bank" of zooplankton species that assures their persistence in a community, in spite of periodic harsh conditions.

Studies on the vertical distribution of resting eggs in sediment cores yield useful information to opening important perspectives for paleolimnological climate reconstruction and paleoecology. Cultivation of live food, like rotifers, Daphnia, Artemia, or marine copepods, is an expanding application of practical use of diapause in modern aquaculture. Biotechnologies can now be imagined for maintaining ecosystems outside the Earth's biosphere. Resting stages provide at least two properties highly suitable for such ecosystems. They can be easily transported in space for a long time without special care as compared with an active ecosystem. In addition, storage of seeds and diapausing animals will provide a reserve in case of an unpredictable destruction of the active part of an ecosystem caused, for example, by a meteorite strike.

V. R. Alekseev (✉)
Zoological Institute of Russian Academy of Sciences, St. Petersburg, Russia
e-mail: alekseev@zin.ru

E. B. Vinogradova
Laboratory of Experimental Entomology, Zoological Institute RAS, University emb. 1, St. Petersburg, Russia

V. R. Alekseev, B. Pinel-Alloul (eds.), *Dormancy in Aquatic Organisms. Theory, Human Use and Modeling*, Monographiae Biologicae 92,
https://doi.org/10.1007/978-3-030-21213-1_2

The term alien species takes on a new meaning when one considers another aspect of space biology. By enlarging the distribution area of the species, colonization of new environments could be a safeguard against its extinction. Thus, it would also be important to develop technologies to guard against invasions of other species via ship ballast waters and similar means.

We also suggest that molecular-genetic insights of diapause in invertebrates provide new ways of looking at carcinogenesis. Tumor cells may have parallels in postdiapause embryonic cells.

Keywords Dormancy · Signal factors · Diapause in invertebrates · Environmental factors · Photoperiod · Terminology

2.1 Introduction

In both continental and oceanic waters, aquatic organisms exhibit conditions that are characterized by wide ranges of heterogeneity. This heterogeneity is mainly determined by the annual cycle of solar radiation and is also influenced by wind and water current activity and directions (Odum 1963). The two main strategies organisms use to adapt to environmental heterogeneity are resistant and tolerant strategies. The first is based on responses to stress either at an individual or group level, actively overcoming environmental stress. Migration is an example of resistant strategies (Lampert 2003). Tolerant strategies are characterized by minimizing vital functions or/and forming resistant (dormant) stages.

At high and mid-altitudes, and at different times in the tropics, invertebrates adapt to seasonal conditions by adjusting their life cycles to these periodic fluctuations in external factors. Periods of reproduction and population growth may either shift to certain periods of the year, or be delayed during unfavorable periods. As a result, life cycles can be seen as having alternating periods of active reproduction and population growth with periods where these processes are delayed or stopped. The latter has been called dormancy. The best-defined types of dormancy according to the cause of the arrest in development are diapause and quiescence. Quiescence is driven directly by the dynamics of environmental factors, whereas diapause is determined by a predictive mechanism combining environmental signals and an internal hormonal mechanism (Danilevsky 1961). The regulatory hormonal system acts at the organism and the single-cell level. It always includes both inhibition and stimulation of vital processes (Bertalanfy 1969).

A wide range of aquatic organisms exhibit diapause, or similar dormancy states (Table 2.1). It is interesting that, practically in all plant and animal phyla studied, photoperiod plays a leading role in the seasonal biological clock of organisms. Professor Alexander Danilevsky, the founder of a school of photoperiodism in invertebrates, argues that the mechanism of photoperiodic response is based on a common principle in all organisms. His ideas are supported by a comparison of photoperiodic responses (PPR) in the induction of insect, crustacean, rotifer, and even plant diapause (Alekseev 1990; Pourriot and Clement 1973; Tyshchenko 1977).

Table 2.1 Diapause in aquatic organisms (Alekseev 1990)

Types, classes	Species	Diapausing stages	Suppressed function
Diatomea	*Coicinodiscus concinnus*	Aukospora	Development
Spongia	*Halichondria panicea*	Gemmules	Development
Coelenterata	*Aurelia aurita*	Plannules	Embryogenesis
Turbellaria	*Hemaniella retunuova*	Eggs	Embryogenesis
Nemertini	*Prostoma graescense*	Cysts	Growth
Rotatoria	*Notommata copeus*	Eggs	Embryogenesis
Polychaeta	*Dinophilus teaniatus*	Cysts	Growth
Oligochaeta	*Aelosoma hemprichii*	Cysts	Growth
Crustacea	*Daphnia pulex*	Eggs	Embryogenesis
Insecta	*Culex pipiens*	Eggs	Embryogenesis
Mollusca	*Sepia officinalis*	Larvae	Maturation
Bryozoa	*Lophopodella carteri*	Statoblasts	Development
Echinodermata	*Stichopus japonicus*	Adults	Growth and breeding
Pisces	*Nothobranchius gardneri*	Eggs	Hatching

These aquatic organisms share many similarities, not only in common principles but also in details of diapause induction, termination, and life cycle organization (Alekseev 1990). This suggests a monophyletic origin of this ancient adaptation and a similarity in the molecular basis and genetics of diapause mechanisms among these organisms.

The adaptive function of diapause applies both to biorhythms and defensive responses (Danilevsky 1961). Biorhythms determine synchronization of the life cycle with environmental seasonal rhythms. Defensive traits allow an organism to endure the actions of the complex set of suppressive or even lethal factors occurring during an unfavorable period.

The defensive function of diapause provides an organism unspecific resistance to a wide complex of unfavorable actions. It is based on reducing the metabolic rate. In the case of anabiosis, the metabolic rate approaches zero (Winberg 1936). Secondly, a variety of protective structures can be formed. Typically, their size and characteristics, both morphological and physiological, help maintain a good viability level until the end of the harsh environment (Fryer 1996). Examples are gemmules, shelves of resistant eggs, cysts, cocoons, statoblasts, and ephippia (Makrushin and Stepanova 2003). Diapausing eggs may even maintain viability after passing through the digestive system of a predator, while normal eggs of the same species are digested (Hairston and Olds 1987).

Diapausing eggs contain nutritive storage substances. Given the depressed metabolism, the quantity of storage substances can suggest the duration of diapause (Zaffagnini 1987).

Consequently, the postdiapause (exephippial) generation may grow faster and mature earlier and has greater fecundity than the individuals from parthenogenetic females (Arbačiauskas 1998, 2001). In addition, the exephippial individuals also have significantly higher levels of metabolites than individuals from parthenogenetic eggs (Arbačiauskas and Lampert 2003). As a result, the daphnids hatching from diapausing eggs are better adapted to optimal environmental conditions, while those from parthenogenetic eggs are better adapted to an unpredictable environment (Arbačiauskas 2004).

Physiological functions are suppressed in diapause. In larval and adult diapauses, there is a reduced rate of oxygen consumption (Fig. 2.1). In addition, other functions (nutrition, mobility, reproductive activity) are also suppressed (Alekseev 1998; Hirche 1996). Defensive functions may be met from a combination of physiological and behavioral responses, including such things as vertical and horizontal migrations and searching for shelters (Alekseev 2004). Even though the combination of behavioral, constitutional, and metabolic elements is usually species- or even population-specific, there are many common features in the range of each type of diapause.

The duration of diapause is determined by the combined effect of four general groups of factors. These are genetic characteristics, the result of selective mechanisms, environmental constraints, and the organism's physiology (Hairston and Cáceres 1996). However, typically the decrease of oxygen consumption is most important in determining diapause duration. The relative importance of other factors varies according to the situation.

During the period of adverse environmental conditions, diapause preserves an adequate number of viable individuals from a species population to assure its permanence in the community (Danilevsky 1961). And yet, to achieve this aim, the duration of the diapause can neither be too short nor too long. It cannot be shorter than the period of unfavorable conditions; neither can it be so long that the mortality in the passive individual increases greatly from such things as predation and bacterial and fungal infections (Alekseev 1990; De Stasio 1990; Marcus et al. 1994).

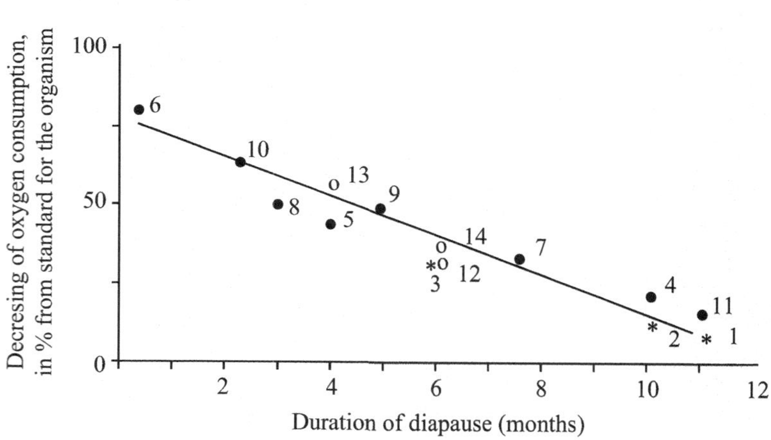

Fig. 2.1 The correlation between duration of diapause and decrease of oxygen consumption for crustaceans. 1–3, species with embryonic diapause; 4–11, species with larval diapause; 12–14, species with abult diapause. (Alekseev and Starobogatov 1996)

2.2 Ecological Stimulants of Diapause in Aquatic Organisms

What conditions argue evolutionarily for diapause? What are the factors which suggest diapause is necessary in an organism's life cycle? There are at least two demands, that diapause be both important for life and occur in a periodic fashion. Such factors would guarantee the appearance of diapause and likely lead to its natural selection in populations (Tsukerzis and Shashtokas 1977).

Seasonality occurs from the subtropics to the arctic. These meteorological rhythms lead to cyclical pulses of the main factors important for life. This is equally true in the atmosphere, on land, and in both continental and oceanic waters. For the overwhelming majority of aquatic organisms, the warm season of the year is the most favorable. In high latitudes, it is often the only period in which life functions can thrive. The appearance of favorable temperatures is determined by one of the most stable processes, the revolving of the Earth. Therefore, seasonal fluctuations of temperature should be regarded as one of the most important factors triggering development of diapause among aquatic invertebrates.

Periodic trophic conditions are also essential regulators for aquatic animals. The significance of trophic factors in the induction of diapause has been experimentally supported many times (Makrushin 1968; Stross 1969; Stuart and Banta 1931; Woltereck 1911). For many crustaceans, especially those who live near the bottom of the deepest part of highly trophic basins, the winter and summer decline of oxygen concentration likely triggers diapause and leads to the interruption of active development (Alekseev et al. 1999; Marcus 1996).

Another periodic factor is predation. Outside of tropical waters, the pressure of predators, especially juvenile fish, is a likely trigger. Some authors consider it to be the main cause of the appearance of diapause in these habitats (Gliwicz and Rowan 1984; Nilssen 1978).

Two main strategies appear to have evolved to help aquatic invertebrates avoid fish predation. One is migration into oxygen-depleted zones to find a refuge. This has been shown for cyclopoid copepods (Ulomsky 1953; Fryer and Smyly 1954; Sarvala 1979). Diapause in cyclopoid copepods reduces basic metabolism so they are able to remain in refuge for months until juvenile fish leave the plankton for nearshore areas (Alekseev et al. 1999, 2001).

Another strategy to avoid fish predation is to form resting eggs that are stable enough to survive passage through fish digestive systems (Hairston and Olds 1987). If these eggs can still reactivate later, at least part of the population will survive until the beginning of the next season. This adaptation has been found in some clones of large-sized *Daphnia magna* living in shallow ponds (Slusarczyk 1995; Pijanowska 1997).

In some water basins, diapause may also lead to a decrease of the crustaceans' survival or even lead to their extinction from certain ecosystems (Alekseev 1990). In the diapause state, along with reduced metabolism, their behavioral reactions and speed are also reduced.

A special factor that selects organisms for their ability to form diapausing stages is the drying of temporary basins. Field data and experimental analyses show that dormancy is present in different orders of crustaceans inhabiting temporary basins (Alekseev 1984; Champeau 1970; Monchenko 2003).

In temperate and polar areas, dormancy seems to be a common strategy for surviving low temperatures. Earlier authors have argued that this proves dormancy originated in these regions. It is clear, however, that other forms of seasonality in tropical regions are equally unique. One periodic influence is monsoon rainfall (Williams 1980), supported by studies of oceanic calanoid copepods from tropical upwelling environments (Herman et al. 1981; Owen 1981; Thiriot 1978). The periodic lack of biogenic income from the deeper layers is similar to the situation in basins that dry seasonally. Because the gradient of trophic conditions is huge near algal blooms and regions outside of the upwelling zone, the crustaceans, which occur outside of the bloom or that experience the end of a bloom, feel the same as if their environment has "disappeared." In such situations, some calanoid copepods, for example, *Calanus hyperboreus*, demonstrate the ability to form diapausing stages. These stages sink to a depth of 400 m where low-oxygen concentrations occur. Here, they stay until new local upwelling lifts them up to the surface. When this happens, they again start new generations that will find themselves in a temporary flowering "oasis" of food.

In the middle of the last century, it was postulated that the ability of invertebrates to form diapausing stages decreased from high latitudes to the equator. It was speculated that this would occur because of the reduction of hydrometeorological seasonality (Danilevsky 1961). More recently, investigations indicate that this idea should be revised. It is now evident that diapause appears wherever annual or seasonal rhythmic fluctuations of important life factors take place (Alekseev 1990, 2004).

2.3 Induction of Diapause

For well over a century, it has been clear that sexual reproduction and the appearance of diapause eggs (ephippia) can be triggered by changes in ecological factors such as temperature, food source, and population density (Weismann 1880).

Banta and Brawn (1929) showed that female parthenogenesis could account for up to 800 generations of *Daphnia pulex*. However, by manipulating temperature, quantity of food, and population density, males could be easily induced. Berg (1934) suggested that the appearance of males was a function of the physiological state of females (stress theory). This state could be achieved once a background state of unfavorable factors was present. This occurred with a range of many factors, and thus was not specific (Mortimer 1936). Diapause induction occurred under the influence of isolated factors such as starvation (Stuart and Banta 1931), low temperature (Grosvener and Smith 1913), food quality (Scharfenberg 1914), diseases of females (Woltereck 1911), pH (Munuswamy et al. 1992), predator's smell (Slusarczyk 1995), alarm signals (Pijanowska 1997), or the presence of some insecticidal hormones (Olmstead and LeBlanc 2003).

The first experiment conducted by Fries (1964) on *D. magna* showed that photoperiod, together with temperature, triggered gamogenesis. Then, Stross (1965, 1969) determined that photoperiod is an essential factor for daphniid life cycles to synchronize with seasonal rhythms of the environment. Other investigations also showed that this factor is very significant in the transition of Cladocera to sexual reproduction and induction of their embryonic diapause (Burner and Halcrow 1977; Kubersky 1977; Shan 1974; Shan and Frey 1968).

Photoperiod also plays a leading role in the induction of embryonic diapause in Calanoid copepods. In the marine environment, Marcus' (1982, 1986) works proved that the ratio of day to night determines the percentage of diapause eggs in sacs of the copepod *Labidocera aestiva*. These calanoids at times completely stop laying subitaneous eggs under the influence of this factor.

Hairston and Olds (1987) showed the role of photoperiodism in causing the freshwater copepod *Diaptomus sanguineus* to switch from laying subitaneous eggs to producing diapause eggs. These authors also showed that photoperiodic responses of populations from neighboring basins were different. This difference might be caused by differences in trophic conditions and population densities, which, together with photoperiod, participate in diapause induction. Also, maternal effects might influence the diapause signal in these populations, as they do for some cladocerans (Alekseev and Lampert 2001).

In general, it should be assumed that embryonic diapause induction depends essentially on seasonal fluctuations of night and day lengths. Other factors may also operate to induce this type of diapause but very often they work along with photoperiod.

Investigations of larval diapause in cyclopoid copepods show that similar factors trigger diapause. Some causal factors triggering a change from active development to dormancy are lower water temperatures (Nilssen and Elgmork 1977; Ulomsky 1953), food depletion (Nilssen 1978), oxygen depletion in the hypolimnion (Einsle 1967), H2S contamination, and foraging by fishes (Gliwicz and Rowan 1984; Nilssen 1978). Further experiments proved the essential role of seasonal fluctuations of day and night lengths in induction of larval diapause for freshwater cyclopoids (Alekseev 1984; Cooley 1971; Stross 1966).

Experimental research on some decapod crustaceans also proved the essential role of photoperiod in the induction of adult diapause (Aiken 1969, 1981).

All of the above examples are facultative types of diapause. Obligatory diapause is genetically fixed and does not require an inductive mechanism. In these cases, reactivation conditions become the most important consideration (Danilevsky 1961). Monocyclic development of many species may be overcome by environmental factors. A good example of this is the experimental manipulation causing a doubling of the number of molts and missing of diapause in the crayfish *Astacus astacus* from Sweden (Westin and Gydemo 1986).

In summary, the following are the main inductive factors for invertebrate diapause: photoperiod, trophic conditions, temperature, and population density. All may usually act in combination but sometimes independently. Effect of density may be manifested either through food supply or by signal action of chemical and/or behavioral agents.

A further distinction can be made in all these inductive factors, between signals (among which photoperiod is the most important) and ecologically important conditions (e.g. trophic conditions). An important intermediate factor is temperature. In a certain range of conditions, it acts as a signal; then, when its values exceed tolerance levels, as an ecological condition. The effect of density is analogous. The only difference is that the action of signal agents (metabolite accumulation, hormone secretion, and/or tactile contacts between specimens) operates within tolerance limits (compared with temperature) and frequently acts in conjunction with trophic conditions. Among all these factors, photoperiod is the most stable and is closely correlated with the seasonal transformation in amount of solar energy.

2.4 Diapause as a Photoperiodic Response

There are two main groups of photoperiodic responses (PPR) in Crustacea, aquatic insects, rotifers, as well as many other aquatic organism studied: qualitative and quantitative PPR (Danilevsky 1961; Pourriot and Clement 1973; Alekseev 1990; Nelson et al. 2010; Alekseev et al. 2007). Quantitative responses are changes of some measurable characteristic (e.g. body length, fecundity), which occur under the influence of an increase or decrease in day length. During qualitative PPR, organisms change between alternative states. These might include development with or without diapause, parthenogenetic reproduction, or gamogenesis (Fig. 2.2).

Quantitative PPRs are more known among crustaceans with adult diapause. One such example—decrease of time intervals between molts—was found for the isopod *Armadillidium vulgare* (Jassem et al. 1982).

As another example, *Daphnia pulicaria*, the first results obtained on the life cycle parameters affected by quantitative PPR were quite impressive (Alekseev and

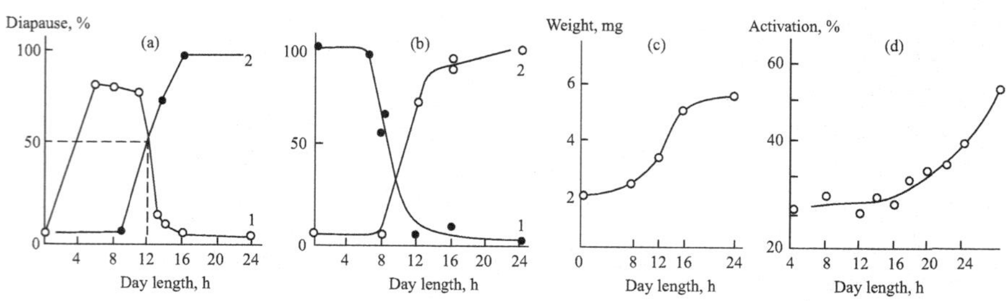

Fig. 2.2 Quantitative (A, B) and qualitative (C, D) photoperiodic responses (PPR) in Crustacea. A, Long- and short-day diapause induction in *Diacyclops navus* and *Diacyclops* sp. (After Watson and Smallman 1971 with changes) B, Diapause induction in two alternative lines of *Macrocyclops albidus*. (After Alekseev 1984) C, Growth of body mass in a phyllopod *Streptocephalus torvicornis*. (Modified from Alekseev 1990) D, Activation of females in crevettes *Palaemones varians* after winter diapause. (Modified from Bouchon et al. 1985)

Lampert 2004). In this species, photoperiod influenced the maturation time, the first clutch size, and even the survival of the parthenogenetic offspring.

Reserve materials, stored by diapausing organisms at the end of their previous season, are available and used at the beginning of the next season. Thus, even when food is limited, they serve as the basis for successful reproduction of cladocerans (Stross and Chisholm 1975). This conveniently reduces the divergence between the time of reactivation and that of the mass development of food organisms (especially phytoplankton microalgae).

PPRs may be divided into long-day and short-day ones according to the length of the photoperiod inducing diapause (Fig. 2.2a). These types are evidently alternative. With one species, a long day serves as a signal for transition to active development, whereas for another, it may be a signal to enter diapause. Most crustaceans develop throughout the warm period of the year and thus short-day PPRs are the most widespread.

PPRs may be presented as graphs, where day length (hours) is plotted along the *X* axis and the number of diapausing specimens (%) or some other quantity is plotted along the *Y* axis. PPR threshold (50% response) is a very important statistic of the PPR. The range of the zone of reaction, which to some degree is due to population or clone polymorphism, is ecologically significant. Species with a narrower zone of reaction have less variation in their response to day length.

The PPR threshold and the width of the zone of reaction regularly change under the influence of other factors (e.g., temperature), which participate together with day length in the realization of photoperiodic responses. All of this information permits phenological predictions (Alekseev 1990).

When different species are compared, it may be necessary to identify species with the greatest plasticity in their response to photoperiod. This information may be useful for understanding the acclimatization and range expansion of introduced (alien) species (Alekseev 1986; Panov et al. 2004).

Appearance patterns of PPRs in aquatic organisms can be subdivided into gradual and threshold responses (Alekseev 1990). The latter appear in a narrow photoperiodic range, while the former occur throughout the whole range of photoperiods. Generally, threshold responses are characteristic for diapause, but they also involve other phenomena such as crustacean vertical migrations or mortality in different seasons (Alekseev 2004). Besides absolute (scalar) values of day and night lengths, the direction of the vector of photoperiod change (increase or decrease of day length) may also be essential for realization of PPRs by crustaceans (Juchault et al. 1980).

Danilevsky (1961) considered PPR types to be a very fundamental adaptation because it determines the whole complex of environmental conditions for the species, such as temperature, food presence, and predation risk. Most surprising are cases where contrasting responses to day length exist within one species, as noted for some insects (Shull 1943) and the littoral cyclopoid *Macrocyclops albidus* (Alekseev 1990). Oddly, no reproductive isolation was found between the groups with long-day and short-day responses (Alekseev and Lampert 2004; Alekseev et al. 2006a, b; Alekseev and Sychev 2006).

2.5 Effects of Population Density

What are the possible signals for changes in population density? Some of these signals appear to be food cessation, worsening of food quality, increase in concentration of the signal (e.g., increase in metabolites, pheromones, or hormones), or the number of tactile interactions. Ecological significance of these factors appears to vary widely.

The trophic conditions promoting transition to gamogenesis may reach significant levels rather easily even if population density of the consumer is relatively low (Stross 1971). This occurs through changes in structure or quality in food organism communities. On the contrary, necessary contents of metabolites or number of tactile interactions may be achieved only under high, particularly constant, concentrations of organisms. There is a suggestion that this limitation in high metabolite concentration may be removed when crustaceans become more sensitive to factors like metabolites and/or tactile interactions (Slusarczyk 1995).

About a century earlier, the common belief was that the transition to gamogenesis was determined by worsening of trophic conditions. At that time, some researchers suggested otherwise that change in the food quality was the most important (Scharfenberg 1914). Others considered that change in the type of reproduction was triggered by starvation (Stuart and Banta 1931). Later, Von Denn considered that food quality determined appearance and accumulation of fat in bodies of *Moina macrocopa* females and this, in turn, caused formation of males and ephippia.

These experiments, however, rarely took into account photoperiodic conditions, and this is believed to seriously compromise the results. The situation changed when gamogenesis induction studies carefully tracked photoperiod, density, and temperature together (Cáceres 1997; Carvalho and Wolf 1989; Kubersky 1977; Stross 1965, 1969).

The most detailed data on the combined effects of these factors on Cladocera gamogenesis induction was obtained by Stross (Stross 1965, 1969). Using threshold densities, the effect of photoperiod (facultative, but rather essential from the ecological point of view) is expressed as the change necessary to trigger a change of reproductive mode.

In permanent basins, where populations are not large because crustaceans are eaten by predators (presumably fishes), dynamics of food conditions are likely the most important factor. However, in temporal basins where consumers do not limit the quantity of crustaceans, density becomes the most important factor. While moving from south to north, the signal effect of photoperiod noticeably increased. Starting at 20 hours of light, all females of *Daphnia middendorffiana* formed ephippia. This was independent of their density. However under day-and-night illumination, the number of ephippial females was density dependent (Stross and Kansas 1969).

The combined effect of density and photoperiod on transition to larval diapause was registered in desert rain pools with the copepod *Metacyclops minutus* (Fig 2.3).

Another population density phenomenon is the blockade of diapause induction for at least the first generation of cladoceran females hatching from ephippia. The

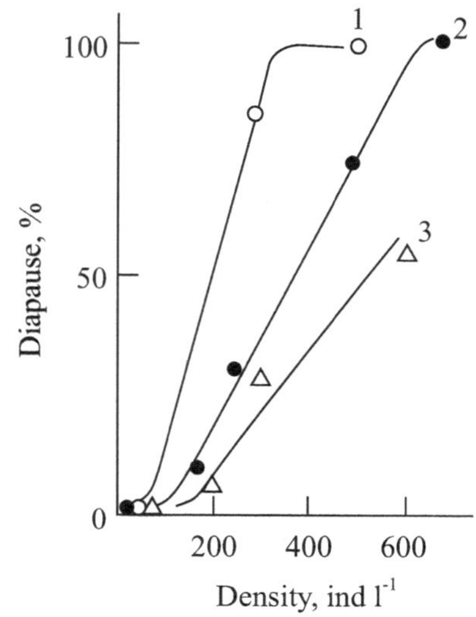

Fig. 2.3 Combined effect of population density and photoperiod on diapause induction in a cyclopoid *Metacyclops minutus*. 1. day length 12 h; 2. day length 16 h; 3. day length 20 h. (After Alekseev 1990)

theory of cladoceran immanent cyclic rhythms (Mocquard and Juchault 1985; Weismann 1880) was based on this phenomenon. Blocking the system responsible for changes in the type of reproduction is probably connected with the reserve materials in diapause eggs. It may even be transferred from the female who left ephippium to her offspring (Stross and Chisholm 1975). The reproductive system of the founder is protected (at least temporarily) from the consequences of sudden food cessation. Therefore, the real effect of trophic conditions is only felt later on. Such conservative reproductive strategies in crustaceans have definite adaptive significance and are directed to the fastest colonization of a basin at the very beginning of the vegetation period.

In nearly all cases where the combined effect of photoperiod and density on diapause induction has been studied, the density effect is limited by restriction of the optimal zone of PPR manifestation (see Stross (1987) for review). Thus, a species with a life strategy directed to rapid development in the presence of increasing food should also have a more rapid reaction to diapause induction. They must react to density changes faster than to photoperiod fluctuations. For example, *Moina macrocopa* actively develops under conditions of abundant organic matter. Its gamogenesis depends little on day length (Makrushin 1968).

Low population density also frequently suppresses the regulatory function of photoperiod. In arid zones, low density in the inhabitants of temporal basins (*Metacyclops minutus*, for instance) is a signal about not expiring of food resources in environment.

Organisms with only a short active period in their life cycle, however, cannot allow themselves such a "luxury" like low population density (Alekseev 1990). For example, most Cladocera fertilization proceeds to diapause, and therefore, it is necessary that opposite sexes meet (Smirnov 1971). Efficiency of such meeting also depends on density.

In more details, these effects of chemical signals are discussed in Chap. 10.

2.6 Temperature and Diapause Induction

Temperature is assumed to have a double function in relation to developing organisms. First, within a tolerance interval, it directly affects growth and metabolic rates by increasing or decreasing the intensity of these processes. Second, it can either decrease or fully interrupt these processes when temperature moves outside the tolerance limits (Ivleva 1981). Temperature also participates in induction, development, and termination of diapause (Danilevsky 1961).

Mechanisms by which temperature affects actively developing and diapausing organisms have been well studied, and many common features have been identified.

One example demonstrates the combined effect of both factors acting together and shows why they should not be considered separately. Photoperiod and temperature effects on gametogenesis were first studied in the cladoceran *Daphnia magna* (Fries 1964). Both the formation of ephippia by females and the appearance of males are regulated by different combinations of temperature and photoperiod. This happened in two combinations, either short days (4 h of light) at 18 °C or long days (20 h of light) at 11 °C (Fig. 2.4a). Photoperiod was influential on each response (appearance of males or ephippia) in a wide temperature range (8–30 °C), but gamogenesis in *D. magna* was most effective within a range of relatively low temperatures. Outside that range, light did not influence the process. Males did not appear at temperatures below 8 °C, while females could not lay ephippia at temperatures above 30 °C.

Temperature influences the formation of diapause eggs by the marine calanoid *Labidocera aestiva* (Marcus 1982). When temperature was reduced, the PPR threshold of this species within the range of tested temperatures shifted rather regularly to longer days. The gradient of this shift was about 2.5–3 h per every 5 °C.

Watson and Smallman demonstrated a temperature effect on the propensity to form diapause stages by the freshwater cyclopid *Diacyclops navus* (Watson and Smallman 1971). The temperature optimum for this response in *Diacyclops navus* was within the range of 10–25 °C (Fig. 2.4b). Above this range, transition to diapause was inhibited regardless of photoperiod. Temperature below this range induced diapause independently of day length.

Most investigations evaluating combined effects (temperature and photoperiod) on reproductive cycles deal with the termination of adult diapause (Bouchon et al. 1985; Branford 1978; March 1982). For example, the effect of photoperiod temperature conditions on the annual cycle of the crayfish *Astacus astacus* was studied by Westin and Gydemo (1986). By modeling annual temperature rhythms, the authors found two periods of copulation and breeding (March–April and August–September) instead of the one which occurs in natural conditions.

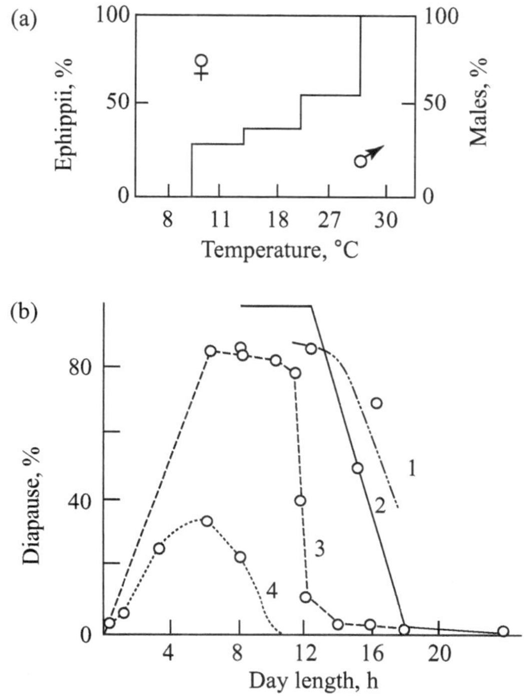

Fig. 2.4 Effect of temperature on male and ephippia production in *Daphnia magna* (**a**) (After Fries 1964) and transformation of photoperiodic response in *Diacyclops navus* under temperatures as following: 1. 10 °C; 2. 15 °C; 3. 20 °C; 4. 25 °C; (**b**) (After Watson and Smallman 1971)

Danilevsky asked what ecological value there was to PPR having a zone of optimal temperatures (Danilevsky 1961). He found that the transition to diapause tends to be accompanied by accumulation of reserves and reconstruction of some metabolic pathways. Other changes may occur under favorable conditions. Photoperiod signals the approach of seasonal environmental changes. The effectiveness of this signal increases when cold (for short-day responses) or warm (for long-day responses) conditions approach. However, the signal function of photoperiod loses its adaptive value at temperatures outside the optimal range.

Temperature changes in water have a greater biological significance than those in air. Water has a greater specific heat than air, and in any given day, its temperatures fluctuate less. This may help explain the two-fold difference in gradient of PPR-threshold change in crustaceans (0.6 h degree^{-1}) as compared with insects (0.28 h degree^{-1}) (Alekseev 1990). It seems possible that studies on the relationship of temperature and light on diapause induction of aquatic insects could help clarify the situation.

Although the temperature gradient of changes in PPR threshold and the existence of temperature optima suggest a similarity of temperature effects on both actively developing and diapausing organisms, there is one essential difference. The change of direction of the temperature transformation (growing or decreasing) is opposite on short- or long-day diapause types.

Combined effects, in summary, happen when there is an interplay of primary signal (e.g. photoperiod) and physiologically important factors (e.g. temperature). Temperature, for example, may either strengthen or lessen the influence of photoperiod. When this occurs, it is necessary to understand both factors well, even in cases where only one of these factors seems to be essential.

2.7 Reactivation Process in Organisms with Different Types of Diapause

Diapausing crustaceans in the marine environment often stay on the bottom of deep sea basins where light is almost absent, and temperature conditions stable (Uye et al. 1979). Diapausing stages of many freshwater crustaceans are primarily found in the deeper lightless parts of lakes, too (Behning 1941; Fryer and Smyly 1954; Sarvala 1979). What, then, triggers diapause reactivation? Literature reports are controversial. Meanwhile, photoperiod is a frequently mentioned candidate (Einsle 1967; Little 1968; Stross 1965). Consequently, research in which this factor was not considered or not controlled is now considered doubtful. Here, we review diapause reactivation mechanism for each type of diapause separately.

2.7.1 Embryonic Diapause

The first experimental reactivation of embryonic diapause with controlled photoperiod was done on *Daphnia pulex* ephippia (Pancella and Stross 1963). Summer ephippia were put in darkness immediately after they were laid. At particular intervals, subgroups of these ephippia were exposed to a light regime approximating the light spectrum of sunlight. Then the intensity of hatching was monitored. Time of hatching was found to depend on the previous period of darkness. When there had been 2–6 weeks of darkness, maximum reactivation occurred within the first day. When the period of darkness was less than 2 weeks, reactivation—or diapause termination—radically lengthened to 6 weeks. Embryos that were maintained in constant darkness (control group) did not hatch. The author speculated that light was also an obligatory condition of summer diapause termination for *D. pulex* (Pancella and Stross 1963).

Winter diapause in the embryos of this species was interrupted by a long exposure (5.5 months) of ephippia at a low temperature (3.5 °C). The photoperiod effect was considered poor, meaning that light was not an obligatory condition for winter diapause termination. Stross' work (1971), however, suggested that only one long-day light impulse (modeled by two short periods of illumination) might be enough to synchronize the end of embryonic winter diapause.

Uye et al. (1979) used marine calanoids (6 species) collected from the shelf-bottom. The strongest triggers of reactivation were temperature and oxygen concentration. Light conditions had no influence.

Using *Diaptomus oregonensis*, Cooley (1971) again found that temperature played a major role in triggering reactivation of diapausing eggs. Hatching rates corresponded to the amount of time embryos had spent under low temperatures. Maintenance of eggs under a constant high temperature prolonged the time before exiting from diapause, and reduced embryo survival. In both cases, however, embryos took 3 months to complete reactivation, even though in nature, the eggs would be maintained at a low temperature until the end of reactivation. The minimal mortality under the low-temperature regime suggests that this is the ecological norm of survival of diapausing embryos.

Nauplia of the brine shrimp *Artemia salina* (Phyllopoda) are economically important as a food in industrial fish breeding. Although Spectorova (1984) studied diapause termination in this species, her methods and experiments were developed from a pragmatic interest in improving the technological aspects of culture. Such essential data as light and gas regimes are absent in these publications, nor are dates of sampling or egg-laying typically mentioned. These omissions have led to some contradictions in the literature about *A. salina* diapause, and thus make additional experiments necessary. Despite the confusion, it is clear that oxygen is necessary for diapause termination in the embryos, and that light participates in synchronizing this process (Bogatova and Erofeeva 1985).

Early studies considered it necessary for eggs to be dried to accelerate reactivation of phyllopods. In 1964, however, Askerov and Sidorov reactivated eggs of *Apus (Triops) cancriformis* and *Leptasteria* sp. without drying the eggs (Askerov and Sidorov 1964). Instead, they used frequent water changes, which caused an increase in oxygen concentration.

Even though the factors that affect reactivation (temperature, photoperiod, oxygen concentration) are the same in winter and summer diapauses, their relative importance differs. For example, light is more important for summer diapause. Temperature sometimes strengthens, but sometimes weakens, the reactivation processes. The key factor in terminating winter embryo diapause is the duration of exposure to low temperatures. Light and oxygen then promote a more synchronized exit from diapause.

2.7.2 Larval Diapause

Termination of larval diapause was originally considered to be linked only with the seasonal dynamics of temperature (Sarojini and Gyananth 1985a). For example, Coker (1933) found that when copepodites of *Acanthocyclops vernalis* maintained in their mud habitat under low temperatures were changed to a high-temperature regime, diapause ended and the planktonic phase began.

More detailed data was obtained by Fryer and Smyly (1954). They collected diapausing cyclopoids (*Mesocyclops leuckarti*) in the pelagic zone in the normal winter diapause phase. However, regardless of whether they were kept under high or low temperatures, time of reactivation was the same. No record was kept of survival statistics.

Using the same species, Smyly (1962) established that reactivation of individuals collected mid-winter could be triggered either by a rapid increase of temperature or by continuous maintenance at 4 °C.

Einsle's studies (1967) on *C. vicinus* in Lake Constance showed that although oxygen depletion could delay reactivation, it could not interrupt it. Neither the presence of food nor light intensity had an effect on exit from diapause. Instead, day length had the greatest effect on accelerated reactivation. The critical value was 15 h of light.

Additional studies agreed that photoperiod was important in reactivation of the species (Spindler 1971). A combination of day-and-night illumination with a temperature of 20 °C was enough to cause *Diacyclops navus* to exit from larval diapause (Watson and Smallman 1971).

One area that has been poorly studied is the role of separate factors in the termination and reversibility of larval diapause. When studies were conducted to see how decreased temperatures interrupt early embryonic diapause, they did not control light conditions (Cooley 1971; Westin and Gydemo 1986). As these experiments were performed in summer, it is assumed that a long-day photoperiod facilitated this process.

Conclusions about larval diapause generally coincide with previous findings on embryonic diapause. The first stage of reactivation occurs under low temperatures. It is determined by the time of exposure to the temperatures. In the last stage, three factors may influence reactivation. Light (photoperiod), and/or a change of oxygen concentration, or temperature increase can accelerate reactivation.

2.7.3 Adult Diapause

Transition of the shrimp *Palaemonetes pugio* to diapause was reached by increasing the temperature from 10°C to 25°C (Little 1968). The efficiency of this process depended on the day length. Under a constant photoperiod (10.5 h), almost half (43%) of the experimental animals reactivated. Longevity of reactivation was determined by the date of sampling. Specimens collected mainly in November started to respond to photoperiod, and temperature increased no earlier than 2 months after being transferred to laboratory conditions.

Crisp and Patel (1969) found that light could affect acceleration or delay of maturation in the cirripedian genus *Balanus*. It was found that in three different species (*Balanus balanoides*, *B. balanus*, and *B. crenatus*), three different critical temperatures (10–12 °C, 10–14 °C, and 17 °C, respectively) were necessary for reactivation.

Tsukerzis and Shashtokas (1977) were able to induce early termination of diapause with the crayfish *Astacus astacus*. Females with eggs were collected at the end of December. Then they were kept at a constant temperature of 2–3 °C for 15 days. Gradually, the temperature was raised to 19–20 °C, and termination of diapause both in embryos and adults occurred in 45 days. That was almost 5 months earlier than observed in nature. Unfortunately, light conditions were not controlled or monitored in these experiments.

With the more southern species *Astacus leptodactylus cubanicus*, the period of temperature reduction needed to reactivate individuals was half as long (Tcherkashina and Karnaushenko 1982). When this species had been transferred from nature to a temperature of 14–15°C, its copulation began in January instead of April, as observed in nature. Again, the authors did not report on the photoperiodic conditions of this experiment.

Illumination and temperature were controlled in experiments by Westin and Gydemo (1986). Manipulations of water temperature allowed for a doubling of the number of reproductive periods of the crayfish *A. astacus*. The authors concluded that light conditions affected neither induction nor termination of diapause. However, their conclusion was apparently premature. Even though by that time others had demonstrated the efficiency of directional change (increasing or decreasing) in photoperiod (Juchault et al. 1980; Little 1968; Steele 1981), Westin and Gydemo did not consider directional change.

Such remarks may be applied also to reactivation of the shrimp *Macrobrachium australiense* (Lee and Fieder 1982). Symmetric values (0 and 36 h of illumination) of photoperiod did not affect the initiation of copulation. Copulation started only after temperatures increased to 25 °C. The time interval between first temperature increase and initiation of spawning depended on when the shrimp were collected. For those collected at the beginning of winter, the interval was about 50 days. When they were collected at the end of winter, it was 30 days. And, in the middle of spring, it was about 10 days.

Diapause termination has been studied in isopod *Armadillidium vulgare*. In different populations of this species, Juchault et al. (1980) were able to generalize the dependency relationship between reactivation time and latitude. Temperature had either a weak or nonexistent effect on reactivation time. Mocquard and Juchault (1985) demonstrated the significance of photoperiodic conditions for synchronization of reactivation processes in this species.

In gammarid *Gammarus lacustris*, March (1982) showed that during the last stage of diapause, light conditions had a significant effect on terminating the diapause. Therefore, as a trigger, conditions of light eclipse those of temperature.

Many histological studies have looked at ovarian cycles of Decapoda (Mocquard and Juchault 1985; Nilssen 1978; Sarojini and Gyananth 1985a). They find that temperature and photoperiod are important for maturation of these crustaceans.

In summary, reactivation of adult diapause is similar to other types of diapause. In nature, there is a rather long period under low temperatures. Transferring crustaceans to a high temperature during the refractory phase does not accelerate the process. Instead, it desynchronizes it and increases mortality (Mansingh 1971).

Using this property of diapause, artificial termination of diapause can be induced in crustaceans. Reversal of diapause may also be achieved. Time of diapause termination may be prolonged.

This review shows that both induction and termination of diapause are caused by the action of a similar complex of external factors (e.g., photoperiod, temperature), but the relative significance, mechanisms, thresholds of sensitivity, and limits of action are different for induction and reactivation.

2.8 Hormonal Control of Invertebrate Diapause

The published data on the involvement of hormones into the induction of diapause and termination in invertebrates are rather heterogeneous. On the one hand, about 100 papers describe the hormonal control of a delay in seasonal molting in immature individuals (larval diapause) and seasonal reproduction (induction and termination of imaginal diapause) in decapod crustaceans (Aiken 1981; Quackenbush 1986; Naya et al. 1989). On the other hand, the data on this problem for the diapause of lower crustaceans, including larval (Carlisle and Pitman 1961; Johnson 2003) and embryonic (Parker 1966; Van den Bosch de Aguilar 1969; MacRae 2005), should be regarded as clearly insufficient. Thus, there is no unified scheme on the role of hormones in the diapause induction and reactivation in crustaceans like that available for insects (Novak 1966; Denlinger 2002).

The very first studies into the neuroendocrine system of crustaceans and other hydrobionts demonstrated that it was similar to the insect system; however, they failed to indicate any particular similar and distinctive features or they only discussed individual elements of hormonal action in the regulation of the life cycle, in particular, reproduction (Carlisle 1957; Bliss 1966; Aiken 1981). The accumulation of comparable data allowed for a more precise discussion of the similarity and difference between the hormonal mechanisms involved in the control of diapause in a wide evolutionary range of organisms (Alekseev 1990; Oehlmann and Schulte Oehlmann 2003; Alekseev et al. 2007). The advance in this field opens an actual possibility of applying the schemes of diapause hormonal control available in entomology to crustaceans and other hydrobionts.

Novak (1966) comprehensively considered the hormonal regulation of seasonal insect development according to the specific features of various diapause types and defined three types of diapause similar in their neurohumoral regulatory mechanisms as follows:

Imaginal diapause is first and foremost associated with the cessation of the development of sex glands, which is determined by a deficiency in activation and juvenile (neotenin) hormones. The activation hormone is produced by cerebral neurosecretory cells, and the juvenile hormone is produced by corpora allata. In decapod crustaceans, maxillary organs produce methylfarnesoate, a hormone with a similar structure. The experimental application of juvenile hormone and its derivatives frequently results in the complete elimination of the symptoms of imaginal and embryonic diapauses (Olmstead and LeBlanc 2001).

Late embryonic, larval, and pupal diapauses appear as retardations in the development of the overall organism, which already has a sufficiently formed neurosecretory system. This type of diapause is the result of a deficiency in activation hormone and the molting hormone ecdysone, which in insects is secreted by the prothoracal glands.

Early embryonic diapause develops at the stage of the germ band and is accompanied by the cessation or delay of embryonic development. This is caused by the presence of the special diapause hormone, bombyxin, an insulin like substance produced by the neurosecretory cells of the maternal esophageal ganglion (Iwami 2000; Ichikawa 2003). Inducing factors (photoperiod) can activate these cells before sexual maturation of an individual.

Thus, a bihormonal model of diapause regulation is postulated for at least the first and second type of diapause; in this model, the neurohormone (activation hormone) controls the production of the hormone by the endocrine glands, that is, corpora allata or prothoracal glands.

Zaslavsky (1988) proposed a somewhat different two-level model based on a three-component system that controls diapause. This system comprises activating and inhibiting neurosecretory centers that interact and their common target, the endocrine gland. The fundamental issue here is that the result of action on the target is determined, not by the inhibition of a particular center, but rather by a certain ratio of the activity or titer of the corresponding hormones.

Both proposed schemes realize the principle of hierarchy in a hormonal system that controls diapause. Hierarchical subordination makes the control system more efficient and economical. First and foremost, this refers to spending activating and inhibiting hormones. Since these substances determine the function of the only target, their quantities can be very small. It is possible that this was among the factors that made it difficult to isolate and characterize neurohormones. Relative to the pervious step, endocrine (steroid terpenoid) hormones can be regarded as multifunctional active substances; they are required in considerably larger amounts, and it is not accidental that the juvenile hormone and ecdysone were not only the first to be characterized but also the first to be synthesized.

Many data on crustacean endocrinology match or approach the above-described schemes of insect hormone action. As for embryonic diapause (type 3), there are data on the increased neurosecretory activity of ventral cerebral cells of *Daphnia pulex* during the ephippium development (Van den Bosch de Aguilar 1969). Recollect that, in this species, the embryos in the mother's body display the highest sensitivity to the photoperiod (Stross and Hill 1968). The accumulated photoperiodic information was implemented after maturation. These data, as well as Parker's (1966) data on an increase in the content of neurosecretory hormones in the body of *Daphnia schodleri* females during the transition to gametogenesis, do not contradict Novak's (1966) scheme for early embryonic diapause. The involvement of steroid hormones in the molting cycle and the cyclic alternation of parthenogenesis and gametogenesis was also demonstrated for cladocerans using the example of *D. magna* (Baldwin and LeBlanc 1994).

A comprehensive study of ecdysteroids in the body of an ocean copepod, *Calanus pacificus*, demonstrated an increased ecdysone titer in the hemolymph before molting and its decrease between molts and during transition to larval diapause (Johnson 2003). These data fit well variant 2 in Novak's scheme (Novak 1966).

The role of hormones is known to the highest degree for the control of molting and reproduction (type 1 diapause) in decapods (for review, see Quackenbush 1986). Note that these data are only related to diapause if they were obtained when studying seasonal changes in reproductive or somatic growth. Unlike insects, higher crustaceans continue their linear and weight growth after maturation.

Many endocrinologists have demonstrated antagonism between reproductive and somatic growths in decapod crustaceans; this is among the major differences in the hormonal mechanisms of imaginal diapause in Crustacea and Insecta. Other differences are related to the topography of neuroendocrine centers in these arthropod groups. The most important center of the neurohormones involved in the control of seasonal molting in decapod crustaceans is the X organ/sinus gland complex, which is localized to the eyestalks (Passano 1951; Carlisle 1957). The ablation of the eyestalks accelerates molting, whereas the application of eyestalk extract, which contains molt inhibition hormone (MIH), delays this process (Zeleny 1905; Quackenbush 1986). It was recently found (Naya et al. 1989) that molting is slowed, not only by this hormone (determined as a derivative of xanthurenic acid), but also by the ratio of activities or titers of other hormones, as was postulated by Zaslavsky (1988).

The MIH target is the Y organ, a paired endocrine gland comparable in function to the prothoracal glands of insects (Carlisle 1957; Aiken 1969). The mold hormone secreted by the Y organ is an ecdysteroid (10 oxyecdysone); it is also very close to the insect ecdysone and acts in a similar manner on various tissues. Similar to insects, the system of hormonal control of seasonal molting in decapod crustaceans is a bihormonal two-level complex; however, no activating hormone produced by brain neurosecretory cells has been yet discovered in crustaceans.

Aiken (1969) proposed a somewhat different interpretation of the interrelations between molting and MIH by the example of seasonal molting of the immature river crayfish *Orconectes virilis*. Analyzing the reactivation from larval diapause (referred to by the author as resistance), the author concluded that there were two counterphase rhythms in the production (or activation) of molt inhibitor (MIH) and molt-stimulating hormone (MSH). According to Aiken, the production of MIH is determined by a decrease in the length of the day (its maximal content is reached by December); then, if *O. virilis* are kept in the darkness at a low temperature, the MIH titer steadily decreases. By the end of February, that is, 4–5 months after the induction of diapause, the content of the inhibitor reaches threshold levels. Concurrently, the MSH production grows and, starting from this period, it is possible to stimulate molt by increasing the day length and temperature. Any attempts to stimulate molt at earlier terms increase the crayfish mortality rate due to a hormonal imbalance, which is evidently manifested as incomplete ecdysis (uncompleted molting).

The inhibiting hormone has not been discovered in any decapod crustaceans, only in species whose development is interrupted due to heterogeneity of environment (Aiken 1981). Therefore, it can be regarded as the main factor of a hormonal nature that determines the seasonal pattern of crustacean growth and diapause. On the other hand, the acceleration in reactivation from larval diapause near the MIH threshold concentration (due to illumination and elevated temperature) observed in *O. virilis* favors the hypothesis on the existence of a neurohormonal center in crustaceans, which produce the hormone that activates the Y organ (Aiken 1969).

An illustrative and clear model, proposed by Aiken, nonetheless fails to explain many important events and facts in the lifecycle of decapod crustaceans. It is known that young individuals and males of river crayfish, for example, *Astacus astacus* L., in the temperate latitudes of Eurasia molt twice during one season (Alekseev 1989).

In warmer subtropical and tropical climates, the same or related species undergo three to four molts (Little 1968). The experiments with the ablation of the eyestalks, which are the hormonal centers that control the seasonal developmental pattern, performed with decapod populations in high and temperate latitudes demonstrated that the number of female molts also increased from one to two to three per year (Westin and Gydemo 1986). Finally, this model fails to explain the trimestrial rhythm of ecdysone (the molting hormone, which Aiken considered to be associated with ecdysis stimulation) production observed in crustaceans that inhabit caves deprived of light.

An analysis of the facts that contradict Aiken's model allowed us to assume that the antagonist hormones MIH and MSH differ, not only in the phase, but also in the rhythm frequency of their excretion (Alekseev 1989). Studying winter diapause in *A. astacus*, in particular respiration, feeding, and behavior, we proposed another model for seasonal hormone interactions in decapod crustaceans. Similar to Aiken, this model implies that the inhibiting hormone has an activity period of 1 year with its maximum in December and minimum in June, that, has a rhythm opposite to photoperiod. The activation hormone is secreted with a period of 3 months and displays maximal titers in May, November, and February.

Another hormone, an antagonist of ecdysis acting during reproduction, is involved in the control of molting and diapause in mature females (Carlisle 1957; Bliss 1966). This adaptation is connected with the fact that, in winter and, sometimes, in spring, females must breed developing eggs and younglings attached to the carapace. In our case, this effect of the hormone secreted by progeny was confirmed by depriving fertilized females of the eggs; this changed their physiological status to male and interrupted diapause in late February at the same time as the males. These females also displayed elevated oxygen consumption and male type feeding. The females with preserved eggs remained in diapause and, at this time, displayed a three-fold lower level of total metabolism in both the protein consumption and activity of Krebs cycle enzymes and succinate dehydrogenases (Alekseev 1989).

The interaction of three hormones—MIH, MSH, and reproductive hormone—explains practically all the events in the phenology of decapod crustaceans both in a moderate climate, where the activity of inhibiting hormone is the most pronounced, and in the lower latitudes, where the seasonal factor loses its importance.

First and foremost, this explains the two molts of males and immature females in May and August, which are determined by two peaks in activating hormone activity on a background of decreased titer of the inhibiting hormone (from March to September, when the day is longer than the dark period). Two other peaks of MSH activity in October and March (with an opposite ratio of the light and dark phases of photoperiod) are inhibited by a high MIH titer, with the resulting absence of molts in crayfish. Nonetheless, it is likely that their antagonism at the peaks of an activating hormone also contributes to other events in the life cycle of decapods. In particular, the copulation of crayfish (late fall) and the end of diapause for males and females deprived of eggs (early spring) take place during these periods.

Unlike males, the females with offspring have only one molting period in August and a longer diapause, continuing almost to June. As is discussed above, this is explainable by the effect of additional factor—the hormone excreted by eggs—which prevents untimely detachment of the offspring from the mother.

Our model also explains the diverse involvement of the calendar factor, that is, the photoperiod, in the control of the intricate seasonal cycle of decapods, whose lifespans amount to several decades and require a reliable external sensor of seasonal rhythm. As was already mentioned, the seasonal pattern of sexual maturation in decapod crustaceans is an imaginal diapause. Different authors have estimated the total number of hormones in various ways involved in implementing the reproductive function of crustaceans from six (Skinner 1985) to seven (Bliss 1966; Aiken and Waddy 1981) hormone species. The hormones involved in molt control are also included into this set due to the antagonism between growth and reproduction processes.

There are three additional sex-specific hormones that control male spermatogenesis and the development of female egg-producing structures (Quackenbush 1986). The remaining hormones are likely to determine the seasonal inhomogeneity of the reproductive function of crayfish, as was postulated by Otsu (1963) as long ago as the early 1960s.

The gonad-inhibiting hormone is produced by the X organ/sinus gland complex, which has been repeatedly confirmed experimentally by both eyestalk ablation and the application of eyestalk extracts (Sarojini and Gyananth 1985). Presumably, the gonad-stimulating hormone also has a neurosecretory nature. The application of extracts of brain and thoracic ganglia to the fresh water shrimp *Macrobrachium lammeri* accelerated the development of ovaries and testes (Sarojini and Gyananth 1985). Presumably, the interaction between inhibiting and stimulating hormones changed the activity of androgenic glands with the subsequent development or degradation of gonads, as confirmed by the experiments with eyestalk ablation (hormonal imbalance) in males of the crab *Carcinus maenas* (Demensy 1958).

These facts that demonstrate the involvement of two types of hormones in the formation of seasonal growth retardation and reproduction of decapod crayfish in general match the three-element, two-level model for the control of larval and imaginal diapauses in insects (Zaslavsky 1988). Currently, the lower level hormones in this model, that is, oxyecdysone, xanthurenic acid, and methyl farnesoate, have been characterized, and their roles in the development and termination of diapause in crustaceans and several other invertebrates have been determined (for review, see Chang 1984; Oehlmann and Schulte Oehlmann 2003).

The further study of this problem is first and foremost connected with the isolation and comprehensive characterization of signaling neurosecretory hormones; however, very important results can be also obtained in the experiments with the already discovered hormones of the lower level. An example here is the study of the effect of insect juvenile hormone and the pesticides that involve this hormone (methoprene and others) on the change in reproduction types and the induction of male production in cladocerans and other invertebrates (Laufer et al. 1993; Olmstead and LeBlanc 2000, 2001; Oehlmann and Schulte Oehlmann 2003).

2.9 Molecular-Genetic Mechanism of Diapause with Reference to Nematode *Caenorhabditis elegans*

As is mentioned above, diapause or other types of anabiosis that allow for survival during seasonal environmental instability are known for their very wide evolutionary range of organisms from bacteria to vertebrates (see Table 2.1). Interestingly, the photoperiod plays a leading role in the function of a seasonal biological clock in almost all groups of organisms studied from this standpoint; the information about the photoperiod can be transmitted in a series of generations through the maternal effect (Alekseev and Lampert 2001). Bunning (1936) was the first to propose the use of photoperiodic response to describe the biological clock in plants. Danilevsky (1961), the founder of the international school of photoperiodism in invertebrates, expanded this idea on the majority of the living organism and considered it to be one of the most intriguing problems in ecology and physiology. The central idea in photoperiodism is that organisms are able to perceive and adequately interpret changes in the daily periodicity of illumination to adjust themselves to seasonal environmental inhomogeneity. The study of the hormonal base of photoperiodism led to development of the modern concept of a two-level mechanism that invertebrates use to perceive and implement seasonal information (Otsu 1963; Novak 1966; Zaslavsky 1988).

Further study of photoperiodism is connected with the next control level in this intricate system providing for temporal orientation of living organisms, namely, molecular-genetic mechanisms coding for seasonal development in invertebrates. So far, little is known about the molecular processes involved in diapause, as well as about the genes that code for the induction and termination of diapause in hydrobionts (Tunnecliffe et al. 2005; Qiu and MacRae 2007). Terrestrial invertebrates, such as silk worms, blowflies, and, especially, the classical model of molecular studies, the soil nematode *Caenorhabditis elegans*, have been studied more comprehensively (Crag and Denlinger 2000; Gerisch and Antebi 2004). Taking into account the doubtless similarity between the mechanisms of diapause in a wide range of organisms, a review of the main molecular-genetic advances in decoding the gene determination of diapause in *C. elegans* is useful for those involved in studying this phenomenon in hydrobionts.

Despite the relative simplicity of the structural organization (the body of an adult nematode comprises about 1000 cells), *C. elegans* has a well-developed endocrine system, which determines the switch from a dauer larva to diapause or reproductive growth (Cassada and Russell 1975). A complex chain of physiological events commences with the perception of the signal from the environment, which triggers a cascade of genes associated with diapause. Environmental signals include pheromones, which indicate a high population density, the deterioration of feeding conditions, and high temperature (Golden and Riddle 1984). The metabolism of a larva transiting to diapause shortly before maturation reduces to 5–30% of the norm, and its feeding is halted, which allows for the survival of the larvae through adverse conditions with a maximal survival rate. After termination of diapause, which takes at least 2–3 months, larvae resume their development and commence reproduction. As is evident from comparison, the *C. elegans* diapause does not differ from that in insects and crustaceans in the majority of parameters and physiological manifestation.

Recent studies have demonstrated that the diapause in *C. elegans* is accompanied by changes in the expression of the specific gene group (*daf*) in the third instar. The choice between diapause and development is determined by the cascade of genes and their products, in particular, the gene *daf 9*, which encodes cytochrome P450, which is associated with steroidogenic hydroxylase, and the gene *daf 12*, which determines the production of lipophilic hormone, which in turn determines the physiological status of organism (Gerisch and Antebi 2004).

The following simple model describes this process. The *daf 9* gene produces a hormone that regulates expression of the *daf 12* gene. The products of the latter are responsible for a switch between diapause and reproductive development and for the lifespan. Presumably, this primary hormone is a sterol (Gerisch et al. 2001). The *daf 9* gene, which is expressed in a potentially endocrine tissue, is the central agent that controls the physiological changes in the overall organism and thus determines the life cycle of *C. elegans* (Gerisch and Antebi 2004).

The choice between development and diapause is also regulated by insulin, some specific peptides (transforming growth factor β, TGFβ), and serotoninergic signaling, which expand the selected program over the entire body, thereby changing the physiological status (Finch and Ruvkun 2001). Insulin and TGFβ are synthesized in response to the signals received by sensory neurons (Li et al. 2003). Note that the recently discovered insect bombyxin, a neurohormone of the silkworm larval diapause, is an insulin-like substance (Iwami 2000).

Presumably, under adverse conditions, the *daf* genes determine the transition to diapause, whereas TGF peptides inhibit *daf 9* and *daf 12*, which enables the continuation of reproductive development (Georgi et al. 1990; da Graca et al. 2003). According to our opinion, these results allow *daf* genes and their products to be associated with the MIH hormone, which induces the diapause in decapods, which, in this case, can appear similar to the insulin-like bombyxin of insects. Presumably, the effect of TGFβ peptides on *C. elegans* is similar to the action of MSH-activating hormone of crustaceans (Alekseev 1989, 1990; Aiken 1969).

It has been demonstrated for a wide range of organisms that insulin is not only involved in the regulation of diapause, but also regulates tissue growth and lifespan (Tatar et al. 2003). Other genes not mentioned above that belong to the same family of *daf* genes are also involved in this process. Insulin-like substances stimulate the *daf 2* gene, which in turn triggers the kinase cascade and inactivates the product involved in transcription of the *daf 16* gene (Morris et al. 1996; Pierce et al. 2001; Gerisch and Antebi 2004). This leads to the cytoplasmic immobilization of *daf 16* and, consequently, animals switch to completing development and reproduction, accompanied by a shortened lifespan. In an alternative scenario, a successful transcription of *daf 16* in the case of adverse conditions and synthesis of the corresponding products increase stress resistance, leading to diapause and the elongation of the lifespan (Henderson and Johnson 2001; Lee et al. 2001; Lin et al. 2001; Gerisch and Antebi 2004).

There are the data that demonstrate that the receptors of both insulin and TGFβ transduce the signals via the descending flows of the secondary hormonal excretion and that *daf 2* regulates diapause and lifespan via systemic signals (Apfeld and Kenyon 1989; Wolkow et al. 2000).

These results confirm the physiological observations on crustaceans and insects and match the above-discussed two-level mechanism of hormonal regulation of the arthropod diapause (Otsu 1963; Zaslavsky 1988). The gene *daf 9* produces cytochrome P450 and is involved in the production of steroids and enzymes, which enables xenobiotic detoxification and fatty acid oxidation. Presumably, this gene is the main factor of the primary level in the hormonal mechanism of diapause (Gerisch et al. 2001; Jia et al. 2002). The cascade of processes at this level produces the hormone that, in turn, activates the *daf 12* gene, which is associated with the production of vitamin D, a well-known participant in diapause induction in rotifers and other invertebrates (Gilbert and Thompson 1968; Antebi et al. 2000).

A great deal of molecular-genetic data were confirmed in experiments with *C. elegans* mutants who had lost their ability to activate cytochrome P450 (*daf 9*) or, on conversely, the nuclear receptor responsible for producing lipophilic hormone (*daf 12*). The null mutants at the *daf 9* gene developed into a diapausing larva under any conditions with a lifespan that exceeded a normal individual by ~25%. On the contrary, the null mutants at the *daf 12* gene lost their capacity for diapause and lived less than the normal individuals (Gerisch et al. 2001). These observations confirmed the association of these *daf* genes with diapause and lifespan.

Undoubtedly, recent basic discoveries in molecular genetics clarify which particular genes and environmental factors influence the basic metabolism, the choice between development and diapause alternatives, and many other seasonal adaptations, not only in *C. elegans* but also in other multicellular organisms. Numerous papers on the discovery and functional roles of *daf* genes in hydrobionts can be expected in the nearest future, similar to the genes recently discovered in bdelloid rotifers (Tunnecliffe et al. 2005). These studies will certainly lead to a deeper understanding of the fine mechanisms underlying photoperiodism and diapause in other hydrobionts. As a good example of such studies in insects, a large review on photoperiodism mechanism should be mentioned (Nelson et al. 2010). Some news in this field are summarized in following section.

2.10 The Modern Achievements in Cognition of the Diapause Regulation Mechanisms in Insects

The latest comprehensive review focused on mosquito diapause was published by Denlinger and Armbruster (2014). The authors discuss recent findings related to all the earlier-mentioned aspects of diapause in mosquitoes such as characteristics of the three types of diapause, the role of photoperiod (and in particular critical photoperiod), temperature and other environmental factors involved in diapause induction, geographical variation of diapause properties, maternal effects on diapause in progeny, and some other issues. Particularly interesting in the paper is the consideration of hormonal and genetic regulation of development and diapauses, and discussion of the latest advances in understanding of evolutionary dynamics and molecular physiology of diapause in mosquitos.

Based on the phylogenetic analysis, Denlinger and Armbruster (2014) suggest that variation among species within genera clearly indicates a degree of evolutionary flexibility of diapause stage, closely connected with life histories, that is, ecological conditions in different geographical regions. The diapause response in mosquitoes, especially diapause timing (critical photoperiod), has evolved rapidly during invasion and range expansion across a broad climatic gradient, that was confirmed by invasion and range expansion of *Ae. albopictus* in North America and by the latitudinal variation of critical photoperiod in *Wyeomiya smithii* during 30 years on the same continent (Bradshaw and Holzapfel 2001). These facts indicate that diapause is a crucial ecological adaptation to spatiotemporal climatic variation and to the current global climate change. Understanding of these responses is essential for predicting the spread of mosquito-borne diseases.

Hormonal control of development and diapause in mosquitoes includes prothoracicotropic hormone-stimulating synthesis of ecdysone by prothoracicotropic glands, juvenile hormone producing by corpora allata, acting together with ecdysteroids, and regulating molting, metamorphosis, and insulin signaling (a signal transduction pathway affecting growth, reproduction, and metabolism) (Denlinger et al. 2012). Hormonal control of adult diapause is studied well enough only in *Culex pipiens pipiens*. Shortly after eclosion in nondiapausing females, the release juvenile hormone triggers previtellogenic growth of ovarian follicles, and females are primed to seek a blood meal, after which eggs reach maturity. In contrast, females programmed for diapause do not release juvenile hormone after eclosion, their follicles remain small, and females do not seek a blood meal. Such pattern indicates that a lack of this hormone is one of the essential regulators of diapause in *Cx. pipiens*, as well as in adult diapause of other insects. Insulin signaling (activated by the presence of insulin or insulin-like peptide) is also implicated in regulatory scheme for diapause in these mosquitoes. Knocking down the insulin receptor in nondiapausing females with RNA interference stimulates diapause, but effect can be reversed with application of juvenile hormone (Sim & Denlinger 2008). Several lines of evidence suggest an interplay between insulin signaling pathway and juvenile hormone. At the same time, hormonal control of mosquito embryonic diapause is not studied. Transcriptome sequencing of mature oocytes and developing embryos also suggests a role of the ecdysteroid signaling in the control of this type of diapause in *Ae. albopictus*. Although the hormonal basis for larval diapause in mosquitoes has not been studied as well, probably lack of ecdysone is an important feature, as it was observed in studies on larval diapause in diverse insect species.

As to the genetic basis of diapause, the early crossing experiments are consistent with the hypothesis of polygenic control of adult diapause in *Cx. pipiens* (Wilton and Smith 1985). More recent results of line-crossing that simultaneously measure the critical photoperiod of parents, F1, F2, and backcross populations consistently indicate a complex genetic architecture differences in critical photoperiod between populations and variation of this parameter within populations of *Wy. smithii* (Lair et al. 1997).

The molecular basis of mosquito diapause is increasingly well characterized, and rapid advances in DNA sequencing technology will likely lead to exciting progress in the near future. Several studies have used quantitative trait locus mapping to identify specific regions of the genome (QTL) involved in diapause regulation in crossing between diapausing strain of *Cx. pipiens* and nondiapausing one *Cx. quinquefasciatus*. The results support four QTLs affecting diapause located in three chromosomes (Mori et al. 2007). The most extensive information on transcriptional changes associated with diapause is based on the recent high-throughput RNAseq studies, which examined the transcriptome at different stages of development in *Ae. albopictus* from prediapause to postdiapause. It revealed 10 upregulated transcripts, including genes involved in DNA replication and transcription, 2 genes connected with ecdysone signaling, a transcript previously associated with diapause in this species, and some others. A subsequent study of this species shows that number of differentially expressed genes increased from 71–78 h to 135–141 h postoviposition. The increased expression of lipid storage transcripts and the decrease in lipid mobilization and components of the citric acid cycle likely reflect the generally shutdown of metabolism connected with diapause (Poelchau et al. 2013).

It was also shown that knocking down of insulin receptor or insulin-like peptide1 in nondiapausing females of *Cx. pipiens* recapitulates the arrested development of ovaries characteristic of diapause. Conversely, knocking down the transcriptional factor FOXO (the downstream target of insulin signaling) in diapause-programmed females prevents accumulation of fat and reduces longevity and tolerance of oxidative stress, all of which are characteristics of the diapause syndrome. These results underscore the role of the insulin signaling/FOXO pathway as a key regulator of diapause (Sim & Denlinger 2008). Other genes connected to diapause likely include genes involved in regulating of huge accumulation of fat reserves associated with diapause. They switch to feeding on sugar rather than blood in diapause-programmed females, that is, upregulation of fatty acid synthase in the gene encoding the enzyme that converts sugars to fats, and the concurrent downregulation of genes encoding proteases needed to digest a blood meal, trypsin, and chymotrypsin-like protease (Robich and Denlinger 2005). In general, the transcriptional studies of *Ae. albopictus*, *Wy. Smithii*, and *Cx. pipiens*, despite differences in the diapausing stages, reveal many similarities. Genes encoding cuticular proteins have been implicated in diapause in *Ae. albopictus* and *Wy. smithii*, and differences in cuticular abundance have been revealed in *Cx. pipiens*, that may be connected to the enhanced desiccation resistance linked, in turn, to diapause. Transcriptional regulation of lipid metabolism in general, and beta-oxidation in particular, is common in the embryonic diapause of *Ae. albopictus* and adult diapause of *Cx. pipiens*. The fact that relatively few genes are differentially expressed during diapause suppose that the diapause phenotype can be attained by means of diverse molecular pathways (Denlinger and Armbruster 2014). It is necessary to conclude that progress in the discovery of molecular regulation of mosquito diapause shall likely generate desperately needed novel approaches to vector control based on chemical or genetic disruption of diapause. Clear sign similarity of hormonal and especially molecular-genetic mechanism in *C. elegans* and Arthropoda (insects) allows us to hope that these mechanisms are very close in a wide range of evolutionary distant organisms, such as rotifers, crustaceans, tardigrades, etc.

References

Aiken DE (1969) Photoperiod, endocrinology and the crustacean molt cycle. Science 164:149–155

Aiken DE (1981) Molting and growth. In: The biology and management of the Lobster, vol 1. Academic, New York, pp 136–163

Aiken DE, Waddy SL (1981) Reproductive biology. In: The biology and management of the Lobster, vol 1. Academic, New York, pp 215–276

Alekseev VR (1984) Effect of chlorine treatment on zooplankton in sturgeon nursery fish ponds. Trans State Lakes Rivers Res Inst 225:95–104. in Russian

Alekseev VR (1986) Role of diapause in acclimatization of crustaceans. Proc GosNIORKH 252:61–68. in Russian

Alekseev VR (1989) Effect of diapause on oxygen consumption in. Astacidae Tr GosNIORKh 300:80–90

Alekseev VR (1990) Diapauza rakoobraznykh: ekologo-fiziologicheskie aspekty (Diapause in Crustacea: ecological–physiological aspects). Nauka, Moscow, 144 pp (in Russian)

Alekseev VR (1998) Biochemical and physiological characteristics of Crustaceans in diapause: the internal mechanism of reactivation. Arch Hydrobiol 52:463–476

Alekseev VR (2004) Effects of dial vertical migration on ephippia production in Daphnia. J Limnol 63:1–6

Alekseev V, Lampert W (2001) Maternal control of resting egg production in Daphnia. Nature 414:899–901

Alekseev V, Lampert W (2004) Maternal effects of photoperiod and food level on life history characteristics of the cladoceran Daphnia pulicaria Forbes. Hydrobiologia 526:225–230

Alekseev VR, Starobogatov YI (1996) Types of diapause in Crustacea: definitions, distributions, evolution. Hydrobiologia 320:15–26

Alekseev VR, Sychev VN (2006) Effect of space station conditions on resting egg survivorship and parameters of life cycle in *D. magna*. Abst COSPAR Beijin July 2006

Alekseev VR, Pinel-Alloul B, Methot J (1999) Role of summer cyclopid diapause in lake meyobenthos forming in Quebec lakes (Canada). Abst. Annual scientific session of Zoological Institute, Academic Publishers, St. Petersburg 8–9 (in Russian)

Alekseev VR, Djenderedjan K, Fiks B (2001) Role of summer diapause in success of invasion of a new invertebrate predator into plankton ecosystem of a large mountain lake. In: Proceedings of 9th International Conference conservation and management of lakes, BIWAKO, Japan, pp 41–47

Alekseev V, Dumont H, Pensaert J, Baribwegure D, Vanfleteren JR (2006a) A redescription of *Eucyclops serrulatus* (Fischer, 1851) (Crustaceaa, Copepoda, Cyclopoida) and some related taxa, with a phylogeny of the *E. serrulatus*-group. Zooogica Scripta 35:123–158

Alekseev VR, Sychev VN, Novikova NI (2006b) Studying the phenomenon of dormancy: why it is important for space exploration. Abst. COSPAR Beijin July 2006

Alekseev VR, de Stasio BT, Gilbert JJ (eds) (2007) Diapause in aquatic invertebrates: theory and human use. Monographiae Biologicae 84. Springer, Dordrecht, 257 p.

Antebi A, Yeh WH, Tait D, Hedgecock EM, Riddle DL (2000) *daf-12* encodes a nuclear receptor that regulates the dauer diapause and developmental age in *C. elegans*. Genes Dev 14:1512–1527

Apfeld J, Kenyon C (1989) Cell Nonautonomy of *C. elegans daf 2* function in the regulation of diapause and lifespan. Cell 95:199–210

Arbaciauskas K (1998) Life history traits of Exephippial and par thenogenetically derived Daphnids: indicators of different life history strategies. Adv Limnol 52:339–358

Arbačiauskas K (2001) Life-history variation related to the first adult instar in daphnids derived from diapausing and subitaneous eggs. Hydrobiologia 442:157–164

Arbačiauskas K (2004) Seasonal phenotypes of Daphnia: post-diapause and directly developing offspring. J Limnol 63:7–15

Askerov MK, Sidorov PA (1964) Biology of phyllopods in sturgeon ponds and struggle against the crustaceans. Trans Azerbajdjan Res Inst Fish Ind 4:83–97. in Russian

Baldwin WS, LeBlanc GA (1994) Identification of multiple steroid hydroxylases in *Daphnia magna* and their modulation by xenobiotics. Environ Toxicol Chem 13:1013–1021

Banta AM, Brawn LA (1929) Control of sex in Cladocera. 1. Crowding the mothers as a means of controlling male production. Physiol Zool 2:80–92

Behning AL (1941) Cladocera of the Caucuses, Gruzmedizdat Tbilisi (in Russian)

Berg K (1934) Cyclic reproduction, sex determination and depression in Cladocera. Camb Biol Rev 9:1

Bliss DE (1966) Relation between reproduction and growth in decapod Crustaceans. Am Zool 6:231–233

Bogatova IB, Erofeeva GI (1985) Incubation of Artemia salina resting eggs without preliminary stimulation. Hydrobiol J 21:52–56 (in Russian)

Bouchon D, Remoissenent G, Mocquard JP (1985) Influence de la temperature sur l'entree en reproduction de Palaemonetes varians Leach (Crustace, Decapoda, Natantia). Bull Sci Zool Fr 110:439–447

Bradshaw WE, Holzapfel CM (2001) Genetic shift in photoperiodic response correlated with global warming. Proc Natl Acad Sci USA 98:14509–14511

Branford GK (1978) The influence of day-length, temperature and season on the hatching rhythm of Homarus gammarus. J Mar Biol Assoc UK 58:639–658

Bunning E (1936) Die Endonom Tagesrhythmik Als Grundlage der Photoperiodischen Reakton. Ber Deut Bot Ges 54:590–607

Burner HC, Halcrow K (1977) Experimental induction of the production of ephippia by Daphnia magna Straus (Cladocera). Crustaceana 32:77–86

Cáceres CE (1997) Temporal variation, dormancy, and coexistence: a field test of the storage effect. Proc Natl Acad Sci USA 94:9171–9175

Carlisle DB (1957) On the hormonal inhibition of moulting in decapod Crustacea. J Mar Biol Ass UK 36:291–307

Carlisle DB, Pitman WJ (1961) Diapause, neurosecretion and hormones in Copepoda. Nature 190:827–828

Carvalho GR, Wolf HG (1989) Resting eggs of lake-Daphnia I. Distribution, abundance and hatching of eggs collected from various depths in lake sediments. Freshw Biol 22:459–470

Cassada R, Russell R (1975) The Dauer Larva: a post embryonic developmental variant of the Nematode *C. elegans*. Dev Biol 46:326–342

Champeau A (1970) Etude de la vie latente chez des Calanoides (Copepodes) caracteristiques des eaux temperairres de Basse—Provence. Ann Fac Sci Marseille 44:155–189

Chang ES (1984) In: Engels W et al (eds) Ecdysteroids in Crustacea: role in reproduction, molting, and larval development advances in invertebrate reproduction, vol 3. Elsevier, Amsterdam, pp 223–249

Coker RE (1933) Arret du developpement chez les copepodes. Bull Biol 67:276–287

Cooley JM (1971) The effect of temperature on the development of resting eggs of Diaptomus oregonensis Lillj (Copepoda:Calanoida). Limnol Oceanogr 16:921–926

Crag TL, Denlinger DL (2000) Sequence and transcription patterns of 60S ribosomal protein P0 a diapause regulated AP endonuclease in the flesh fly, *Sarcophaga crassipalpis*. *Gene* 255:381–388

Crisp DJ, Patel B (1969) Environmental control of the breeding of three boreoarctic cirripedes. Mar Biol 2:283–295

da Graca LS, Zimmerman KK, Mitchell MC, Kozhan-Gorodetska M, Sekiewicz K, Morales Y, Patterson GI (2003) DAF-5 is a Ski oncoprotein homolog that functions in a neuronal TGFb pathway to regulate *C. elegans* dauer development. Development 131:435–446

Danilevsky AS (1961) *Fotoperiodizm i sezonnoe razvitie nasekomykh* (Photoperiodism and seasonal development of insects). Lening Gos University, Leningrad, 243 pp (In Russian)

De Stasio BT Jr (1990) The role of dormancy and emergence patterns in the dynamics of a freshwater zooplankton community. Limnol Oceanogr 35:1079–1090

Demensy N (1958) Recherches sur la Mue de Pubert du Decapoda *Brachyoure carcinus* Maeneas. Arch Zool Exp Gen 95:253 pp.

Denlinger DL (2002) Regulation of diapause. Annu Rev Entomol 47:93–122

Denlinger DL, Armbruster PA (2014) Mosquito diapause. Annu Rev Entomol 59:73–93. https://doi.org/10.1146/annurev-ento-011613-162023

Denlinger DL, Yocum GD, Rinehart JP (2012) Hormonal control of diapause. In: Gilbert LI (ed) Insect endocrinology. Academic, San Diego, pp 430–463

Einsle U (1967) Die ausseren Bedingungen der Diapause plankisch lebender Cyclops-Arten. Arch Hydrobiol 63:387–403

Finch CE, Ruvkun G (2001) The genetics of aging. Ann Rev Genom Hum Genet 2:435–462

Fries G (1964) Uber die Einwirkung der Tagesperiodik und der Temperatur auf den Generationswechsel, die Weibchengrosse und die Eir von Daphnia magna Straus. Ztschr Morphol und Okol Tierr 53:475–516

Fryer G (1996) Diapause a potent force in the evolution of freshwater crustaceans. Hydrobiologia 320:1–14

Fryer G, Smyly WIP (1954) Some remarks on the resting stagees of some fresh water cyclopoid and harpacticoid copepods. Ann Mag Nat Hist 7:65–72

Georgi LL, Albert PS, Riddle DL (1990) *daf-1*, a *C. elegans* gene controlling dauer larva development, encodes a novel receptor protein kinase. Cell 61:635–645

Gerisch B, Antebi A (2004) Hormonal signals produced by *daf* 9/cytochrome P450 regulate *C. elegans* Dauer Dia pause in response to environmental cues. Development 131:1765–1776

Gerisch B, Weitzel C, Kober-Eisermann C, Rottiers V, Antebi A (2001) A hormonal signaling pathway influencing *C. elegans* metabolism, reproductive development, and lifespan. *Dev Cell* 1:841–851

Gilbert JJ, Thompson GA Jr (1968) Alpha tocopherol control of sexuality and polymorphism in the rotifer *Asplanchna*. Science 159:734–736

Gliwicz ZM, Rowan MG (1984) Survival of *Cyclops Abyssorum tatricus* (Copepoda, Crustacea) in Alpine Lakes stocked with Planktivrous fish. Limnol Oceanogr 29:1290–1299
Golden JW, Riddle DL (1984) The *Caenorhabditis elegans* Dauer Larva: developmental effects of pheromone, food, and temperature. Dev Biol 102:368–378
Grosvener G, Smith G (1913) The life cycle of Moina rectirostris. Q J Microsc Soc 58:87–112
Hairston NG Jr, Cáceres C (1996) Distribution of crustacean diapause: Micro- and macroevolutionary pattern and process. Hydrobiologia 320:27–44
Hairston NG Jr, Olds EJ (1987) Population differences in the timing of diapause: A test of hypotheses. Oecologia 71:339–344
Henderson ST, Johnson TE (2001) *daf-16* integrates developmental and environmental inputs to mediate aging in the nematode *Caenorhabditis elegans*. Curr Biol 11:1975–1980
Herman AW, Sameoto DD, Longhurst AR (1981) Vertical and horizontal distribution patterns of copepods near the shelf .break front south of Nova Scotia, Canad. J Fish Aquatic Sci 38:1065–1076
Hirche HJ (1996) Diapause in the marine copepod, *Calanus finmarchicus* – a review. Ophelia 44:129–143
Ichikawa T (2003) Firing activities of neurosecretory cells producing diapause hormone and its related peptides in the female silkmoth, *Bombyx mori*. I. Labial cells. Zool Sci 20:971–978
Ivleva IV (1981) Temperature and metabolic rates in aquatic animals, Kiev. Naukova Dumka Publishers (in Russian)
Iwami M (2000) Bombyxin: an insect brain peptide that belongs to the insulin family. Zool Sci 17:1035–1044
Jassem W, Mocquard JP, Juchault P (1982) Determinisme de la reproduction saisonniere des femelles d'Armadillidium vulgare Latr. (Crustace, Isopode, Oniscoide) IV. Contribution a la connaissance de la perception du signal photopreriodique in duisant 1'entree en reproduction: mode de discrimination entre Ie Jour et la mit longueurs d'onde actives. Ann Sci Nat Zool et Biol Anim 4:85–90
Jia K, Albert PS, Riddle DL (2002) *daf* 9, a cytochrome P450 regulating *C. elegans* larval development and adult longevity. Development 129:221–231
Johnson CL (2003) Ecdysteroids in the oceanic copepod *Calanus pacificus*: variation during molt cycle and change associated with diapause. Mar Ecol 257:159–165
Juchault P, Pavese A, Mocquard JP (1980) Determinisme de la reproduction saisonniere des femelles d'Armadillidium vulgare Latr. (Crustacea, Isopode, Oniscoide) II. Etude phiques differentes. Ann Sci Nat Zool et Biol Anim 2:99–108
Kubersky ES (1977) Worldwide distribution and ecology of Alonopsis (Cladocera, Chydoridae) with a description of Alonopsis americana sp. nova, Intern Rev ges. Hydrobiology 62:649–685
Lair KP, Bradshaw WE, Holzapfel CM (1997) Evolutionary divergence of the genetic architecture underlying photoperiodism in the pitcher-plant mosquito, *Wyeomyia smithii*. Genetics 147:1873–1883
Lampert W (2003) Evolutionary ecology: natural selection in freshwater systems. In: Moya A, Font E (eds) Evolution from molecules to ecosystems. University Press, Oxford, pp 109–121
Laufer H, Ahl JSB, Sagi A (1993) The role of juvenile hormones in crustacean reproduction. Am Zool 33:365–374
Lee CL, Fieder DR (1982) Induced spawning in the freshwater prawn, Macrobrachium australiense Holthuis 1950 (Crustacea. Decapoda: Palaemonidae). Aquaculture 29:45–52
Lee RY, Hench J, Ruvkun G (2001) Regulation of *C. elegans daf* 16 and Its Human Ortholog FKHRL1 by the *daf* 2 Insulin Like Signaling Pathway. *Curr Biol* 11:1950–1957
Li W, Kennedy SG, Ruvku G (2003) *daf 28* encodes a *C. elegans* insulin superfamily member that is regulated by environmental cues and acts in the *daf* 2 signaling pathway. Genes Dev 17:844–858
Lin K, Hsin H, Libina N, Kenyon C (2001) Regulation of the *Caenorhabditis elegans* longevity protein *daf* 16 by insulin/IGF 1 and germline signaling. Nat Genet (2):139–145

Little G (1968) Induced Winter Breeding and Larval Development in the Shrimp *Palaemonetes pugio* (Holthius), *Crustaceana*. Suppl 2:19–26
MacRae TH (2005) Diapause: diverse states of developmental and metabolic arrest. J Biol Res 3:3–14
Makrushin AV (1968) Condition of ephippial female ovary in some Cladocera species. Trans State Lakes Rivers Res Inst 67: 365–369 (in Russian)
Makrushin AV, Stepanova IE (2003) Ob izberatenoj pronizaemosti obolochek latentnuh jaiz Moina macrocopa (Daphniiformes, Crustacea). (On the selective permeability of covering membrane in Moina macrocopa restin eggs). Zoologicheskij J 82:117–118
Mansingh A (1971) Physiological classification of dormencies in insects. Can Entomol 103:983–1009
March BGE (1982) Decreased day length and light intensity as factors inducing reproduction in Gammarus lacustris Sars. Can J Zool 60:2962–2965
Marcus NH (1982) Photoperiodic and temperature regulation of diapause in Labidocera aestiva (Copepoda: Calanoida). Hydrobiologia 162:45–52
Marcus NH (1986) Population dynamics of marine copepods: the importance of photoperiodism. Am Zool 26:469–477
Marcus NH (1996) Ecological and evolutionary significance of resting eggs in marine copepods: past, present and future studies. Hydrobiologia 320:141–152
Marcus NH, Lutz RV, Burnett W, Cable P (1994) Age, viability, and the vertical distribution of zooplankton resting eggs from an anoxic basin: evidence of an egg bank. Limnol Oceanogr 39:154–158
Mocquard JP, Juchault P (1985) Photoperiode et reproduction chez les femelles d'Armadillidium vulgare Latreille (Crustacea, Isopode, Oniscoide): variation en fonction de l'origine geographique des populations. Bull Soc Zool Fr 110:425–439
Monchenko VI (2003) Free living cyclopoid copepods of the Pontho-Caspian basin, Kiev. Naukova Dumka Academic Publishers (in Russian)
Mori A, Romero-Severson J, Severson DW (2007) Genetic basis for reproductive diapauses is correlated with life history traits within *Culex pipiens* complex. Insect Mol Biol 16:515–524
Morris JZ, Tissenbaum HA, Ruvkun G (1996) A phosphatidylinositol 3 OH Kinase family member regulating longevity and diapause in *C. elegans*. Nature 382:536–539
Mortimer CH (1936) Experimentelle und cytologische Untersuchungen liber den Generationswechsel der Cladoceren. Zool Jb Abt allg Zool und Physiol Tiere 56:323–388
Munuswamy N, Nazar AK, Dumont HJ (1992) Is pH(i) a factor for dormancy in freshwater fairy shrimps. Curr Sci 62:751–752
Naya Y, Mayumi O, Midori I, Wataru M (1989) What is molt inhibiting hormone? The role of an ecdysteroidogenesis inhibitor in the crustacean molting cycle. Proc Natl Acad Sci USA 86:6826–6829
Nelson RJ, Denlinger DL, Somers DE (eds) (2010) Photoperiodism: the biological calendar. Oxford University Press, New York, p 596
Nilssen JP (1978) On the evolution of life histories of limnetic cyclopoid copepods. Memorie dell'Istituto Italiano di Idrobiologia 36:193–214
Nilssen J, Elgmork K (1977) Cyclops abyssorum – life cycle dynamic and habitat selection. Memorie dell'Istituto Italiano di Idrobiologia 34:197–238
Novak VJ (1966) Insect hormones. Methuen, London, 478 pp.
Odum EP (1963) Ecology. New York/London
Oehlmann J, Schulte Oehlmann U (2003) Endocrine disruption in invertebrates. Pure Appl Chem 75:2207–2218
Olmstead A, LeBlanc G (2000) Effects of endocrine active chemicals on the development of sex characteristics of *Daphnia magna*. Environ Toxicol Chem 19:2107–2113
Olmstead A, LeBlanc GA (2001) Temporal and quantitative changes in sexual reproductive cycling of the Cladoceran *Daphnia magna* by a juvenile hormone analog. J Exp Zool 290:148–155

Olmstead AW, LeBlanc GA (2003) Insecticidal juvenile hormone analogs stimulate the production of male offspring in the crustacean Daphnia magna. Environ Health Perspect 111:919–924
Otsu T (1963) Bihonnonal control of the sexual cycle in the fresh water Crab *Potamon dehaani*. Embriologia 8:1–20
Owen RW (1981) Fronts and eddies in the sea: mechanism, interactions and biological effects. In: Longhurst AR (ed) Analysis of marine ecosystems. Academic Press, New York
Pancella JR, Stross RG (1963) Light induced hatching of Daphnia resting eggs. Chesap Sci 4:404–425
Panov VE, Krylov PI, Riccardi N (2004) Role of diapause in dispersal and invasion success by aquatic invertebrates. J Limnol 63(Suppl 1):56–69
Parker R (1966) The influence of photoperiod on reproduction and molting of *Daphnia schodleri* Sars. Physiol Zool 39:266–279
Passano LM (1951) The X Organ, a neurosecretory gland controlling molting in crab. Anat Rec 1: 559 pp.
Pierce SB, Costa M, Wisotzkey R, Devadhar S, Homburger SA, Buchman AR, Ferguson KC, Heller J, Platt DM, Pasquinelli AA (2001) Regulation of DAF-2 receptor signaling by human insulin and *ins-1*, a member of the unusually large and diverse *C. elegans* insulin gene family. Genes Dev 15:672–686
Pijanowska J (1997) Alarm signals in Daphnia. Oecologia 112:12–16
Poelchau MF, Reynolds JA, Elsik CG, Denlinger DL, Armbruster PA (2013) Deep sequencing reveals complex mechanism of diapauses preparation in the invasive mosquito, Aedes albopictus. Proc R Soc B 280:20130143
Pourriot R, Clement P (1973) Photoperiodisme et cycle heterogonique chez Notommata copeus (Rotifere, Monogonorte) II. In fluence de gualite de la lumiere Spectres d'action. Arch Zool Exp et Gen 114:277–300
Qiu Z, MacRae TH (2007) Developmentally Regulated Synthesis of P8, a Stress Associated Transcription Cofactor, in Diapause Destined Embryos of *Artemia franciscana*. Cell Stress Chaperon 12:255–264
Quackenbush LS (1986) Crustacean endocrinology: a review. Can J Fish Aquat Sci 43:2271–2282
Robich RM, Denlinger DL (2005) Proc Natl Acad Sci USA 102:1512–1517
Sarojini R, Gyananth G (1985) Hormonal Control of Reproduction in the Freshwater prawn *Macrobrachium lamerri*. J Curr Biosci 2:111–116
Sarvala J (1979) Bentic resting periods of pelagic cyclopods in an oligotrophic lake. Holarct Ecol 2:88–100
Scharfenberg V (1914) Weitere Untersuchungen an Cladoceren liber die experimentel Lebeeinflussung des Geschlechts und der Dauereibildung. Intern Rev Ges Hydrobiol Biol Suppl 6:1–34
Shan RK (1974) Reproduction in laboratory stocks of Pleuroxus (Chydoridae, Cladocera) under influence of photoperiod and light intensity. Intern Rev Ges Hydrobiol 59:643–666
Shan RK, Frey DG (1968) Induced interbreeding between two stocks of a Chydorid Cladoceran. Bioscience 18:203–205
Shull AF (1943) Origin of diverse strains of an aphid species within a limited area. Pap Mich Acad Sci Arts Lett Pt II Zool 28:425–431
Sim C, Denlinger DL (2008) Insulin signaling and FOXO regulate the overwintering diapauses of the mosquito *Culex pipiens*. Proc Natl Acad Sci USA 105:6777–6781
Skinner DM (1985) Interacting factors in the control of the Crustacean Molt cycle. Am Zool 25:275–284
Slusarczyk M (1995) Predator induced diapause in *Daphnia*. Ecology 76:1008–1013
Smirnov NN (1971) Chydoridae of the World. Nauka Academic Publishers, Leningrad (in Russian)
Smyly WJP (1962) Laboratory experiments with stage V copepodids of the freshwater copepod, Cyclops leuckarti Claus, from Windemere and Easthwaite water. Crustaceana 4:273–280
Spectrova LV (1984) Recommendations for Artemia culturing and using in aquaculture. VINITI Center, Moscow, p N629px. [in Russian]

Spindler KD (1971) Untersuchungen fiber den Einfiup auperer Faktoren auf die Darner der Embryonalentwicklung und der Hantungsrhythuns von Cyclops vicinus. Oecologia 7:342–355
Steele VJ (1981) The effect of photoperiod on the reproductive cycle of Gammarus lawrencianus Bousfield. J Exp Mar Biol Ecol 53:1–7
Stross RG (1965) Termination of summer and winter diapause in Daphnia. Am Zool 15:701
Stross RG (1966) Light and temperature requirement for diapause development and release in Daphnia. Ecology 47:368–374
Stross RG (1969) Photoperiod control of diapause 142 in Daphnia. II. Induction of winter diapause in the arctic. Biol Bull 136:264–273
Stross RG (1971) Photoperiodism and diapause in Daphnia: a strategy for all seasons. Trans Am Microsc Soc 90:110–112
Stross RG (1987) Photoperiodism and phased growth in Daphnia populations: coactions in perspective. In: Peters RH, de Bernardi R (eds) Daphnia. Memorie dell'Istituto Italiano di Idrobiologia, vol. 45, pp 413–437
Stross RG, Chisholm SW (1975) Density stabilization in arctic populations of Daphnia. Verhandlungen Int Vereiningung Limnol 19:2879–2884
Stross RG, Hill JC (1968) Photoperiod control of winter diapause in the fresh water Crustacean, *Daphnia*. Biol Bull 134:176–198
Stross RG, Kansas DA (1969) The reproductive cycle of Daphnia in an arctic pool. Ecology 50:457–460
Stuart C, Banta A (1931) Available bacteria and the sex ratio in Moina. Physiol Zool 4:654–696
Tatar M, Bartke A, Antebi A (2003) The endocrine regulation of aging by insulin like signals. Science 299:1346–1351
Tcherkashina NJA, Karnaushenko IV (1982) Before-embryonic diapause in cray-fish (Astacus leptodactilis cubanicus Bir. et Win.). J Obshej Biologii 43: 687–689 (in Russian)
Thiriot A (1978) Zooplankton communities in the West African upwelling area. In: Boje R, Tomezak M (eds) Upwelling ecosystems. Springer, New York
Tsukerzis JAM Shashtokas IA (1977) Embryonic diapause in the noble cray-fish (Astacus astacus L.). J Obshej Biologii 38: 929–933 (in Russian)
Tunnecliffe A, Lapinski J, McGee B (2005) A Putative LEA protein, but no Trehalosa is present in anhydrobiotic bdelloid rotifers. Hydrobiologia 542:315–321
Tyshchenko VP (1977) Fiziologiya fotoperiodizma u nasekomykh (Physiology of Photoperiodism in Insects). Nauka, Leningrad, 156 pp (In Russian)
Ulomsky SN (1953) News in ecology of some Mesocyclops. Doklady Acad Sci USSR 90:295–297 (in Russian)
Uye S, Kasahara S, Onbe T (1979) Calanoid copepod eggs in sea-bottom muds. IV. Effects of some environmental factors on the hatching of resting eggs. Mar Biol 51:151–156
Van den Bosch de Aguilar P (1969) Nouvelles Donnees Morphologiques et Hypothises sur le r61e du Systeme Neuro secreteue chez *Daphnia pulex* (Crustacea: Cladocera). Ann Soc R Zool Belgique 99:27–44
Bertalanfy, L Von (1969) Study on common theory of systems. Progress, Moscow (in Russian)
Watson NHF, Smallman BN (1971) The role of photoperiod and temperature in the induction and termination of an arrested development in two species of freshwater cyclopid copepods. Can J Zool 49:855–862
Weismann A (1880) Beitrage zur Naturgeschichteder Daphnoiden. Ztschr wiss Zool 33:55–270
Westin L, Gydemo R (1986) Influence of light and temperature on reproduction and moulting frequency of the Crayfish *Astacus astacus* L. Aquaculture 52:43–50
Williams JA (1980) The light-response rhythm and seasonal entrainment of the endogenous circadian locomotor rhythm of Talitrus saltator (Crustacea, Amphipoda). J Mar Biol Assoc UK 60:773–785
Wilton DP, Smith GC (1985) Ovarian diapause in three geographic strains of *Culex pipiens* (Diptera, Culicidae). J Med Entomol 22:524–528
Winberg GG (1936) Cyclic breeding in Cladocera. Usp Sovrem Biol 5:201–202 (in Russian)

Wolkow CA, Kimura KD, Lee MS, Ruvkun G (2000) Regulation of *C. elegans* life span by insulin like signaling in the nervous system. Science 290:147–150
Woltereck R (1911) Ober Veranderung der Sexualitat Bei Daphniden. Int Rev Hydrobiol 4:91–128
Zaffagnini F (1987) Reproduction in Daphnia. Memorie dell'Istituto Italiano di Idrobiologia 45:245–284
Zaslavsky VA (1988) Insect development. photoperiodic and temperature control. Springer, Berlin, p 187
Zeleny N (1905) The relation of the degree of injury to the rate of regeneration. J Exp Zool 2:347–369

Chapter 3
Dormancy in Freshwater Tardigrades

Roberto Bertolani, Roberto Guidetti, Tiziana Altiero, Diane R. Nelson, and Lorena Rebecchi

Abstract For more than two centuries, tardigrades have been well known for their ability to undergo dormancy. However, this capability has been well studied mainly in the so-called limnoterrestrial species, i.e., in the species colonizing moist terrestrial habitats, such as mosses, lichens, and leaf litter. In these kinds of substrates, tardigrades are active only when a film of water is available around their body so in this condition they behave like aquatic animals. When the substrate dries or freezes, tardigrades achieve dormancy (quiescence) by entering cryptobiosis, specifically anhydrobiosis or cryobiosis, respectively. In freshwater habitats, both forms of cryptobiosis have been verified only in species able to live both in freshwater and terrestrial habitats. In the truly freshwater (or limnic) species, anhydrobiosis has not been verified, while cryobiosis has been confirmed in a few species. Another dormancy phenomenon bound to diapause is frequent in freshwater species: encystment (sometimes found even in limnoterrestrial species). The cyst state, which involves deep structural and physiological modifications, has been known from the beginning of the past century, but only recently has its morphology and inducing factors been studied in depth. Although data on molecular mechanisms allowing cryptobiosis are available, this information does not exist for encystment.

Keywords Tardigrades · Diapause · Encystment · Cryobiosis · Anhydrobiosis

---

R. Bertolani (✉)
Department of Education and Humanities, University of Modena and Reggio Emilia, Modena, Italy
e-mail: roberto.bertolani@unimore.it

R. Guidetti · L. Rebecchi
Department of Life Sciences, University of Modena and Reggio Emilia, Modena, Italy

T. Altiero
Department of Education and Humanities, University of Modena and Reggio Emilia, Modena, Italy

D. R. Nelson
Department of Biological Sciences, East Tennessee State University, Johnson City, TN, USA

V. R. Alekseev, B. Pinel-Alloul (eds.), *Dormancy in Aquatic Organisms. Theory, Human Use and Modeling*, Monographiae Biologicae 92,
https://doi.org/10.1007/978-3-030-21213-1_3

3.1 Introduction

Together with rotifers and nematodes, tardigrades have been well known for more than two centuries for their capability to undergo dormancy at any stage of their life history, from the egg up to the adult. In tardigrades, dormancy has been mainly studied in the so-called limnoterrestrial species, i.e., those colonizing terrestrial habitats, such as bryophytes, lichens, grassland, and leaf litter, in which they are active only when surrounded by at least a film of water. Tardigrades, rotifers, and nematodes do not have specific structural adaptations for living as terrestrial organisms. When the substrate dries or freezes, these animals become dormant (quiescent) by entering cryptobiosis: anhydrobiosis or cryobiosis, respectively (Rebecchi et al. 2007; Guidetti et al. 2011b). In freshwater habitats, cryptobiosis has been verified only in a few tardigrade species, e.g., *Hypsibius dujardini* (Doyère 1840), which can also colonize terrestrial habitats (Boothby et al. 2017). In the truly limnic tardigrade species, anhydrobiosis has never been verified, while cryobiosis has been identified in a few species (Guidetti et al. 2011a) and for the marine heterotardigrade *Echiniscoides sigismundi* (Schultze 1865) (see Clausen et al. 2014). Another dormancy phenomenon, encystment, can be present in freshwater species, bound to diapause. To date, this type of dormancy has been verified only in a limited number of limnic species, but it is also known for some limnoterrestrial and marine species (see Guidetti and Møbjerg 2018). Moreover, in tardigrades, another phenomenon bound to diapause has been identified: resting eggs, which were described in the species *Paramacrobiotus fairbanski* (Schill et al. 2010) that mainly colonize leaf litter, but sometimes are also found in freshwater (Altiero et al. 2010).

3.2 Diapause in Tardigrades

The first case of encystment in tardigrades was described at the beginning of the past century (Lauterborn 1906) and confirmed a few years later (Murray 1907a, b, c; Heinis 1910; von Wenck 1914). The species cited by Lauterborn (1906), *Macrobiotus macronyx* Dujardin 1851 (currently *Dactylobiotus macronyx*), is difficult to identify, leading some to question its existence. Subsequent studies involved more easily identifiable species. One year after Lauterborn's paper, Murray (1907a, c) described encystment in *Macrobiotus dispar* (Murray 1907a), a species still accepted and currently named *Dactylobiotus dispar* (Murray 1907a). Wanda von Wenck (1914) described in detail the encystment in *Macrobiotus lacustris* Dujardin, a species no longer accepted, but from the drawings of the claws and the placoids, this species is easily attributable to *Pseudobiotus megalonyx* (Thulin 1928). As reported by Murray (1907c) and later by many other authors (for review, see Guidetti and Møbjerg 2018), cyst formation is reported for other species, both eutardigrades and heterotardigrades. Encystment is very rare among moss-dwelling tardigrades (except for the species in the genus *Bertolanius*) but more frequent in species dwelling in substrates such as leaf litter, grassland, and especially freshwater sediments, in which heterotardigrades are extremely rare (such as the true limnic genus *Carphania* Binda 1978). Encystment is probably occasional in species from mosses and lichens surrounding freshwater basins. The limnic species that frequently enter encystment belong to different genera and families: *Dactylobiotus* (Murrayidae), *Pseudobiotus* and *Isohypsibius* (Isohypsibiidae), *Hypsibius* (Hypsibiidae), and *Bertolanius* (Eohypsibiidae).

Rahm (1925) noted the difference between the "tun" (anhydrobiotic state) and the cyst, verifying that the tun had a single cuticle and was absolutely unable to move, while the cyst had at least two cuticles (one thickened) within which the animal could also move. Contrary to what was reported by Murray (1907c), von Wenck (1914) concluded that histolysis does not occur during encystment. The absence of histolysis was confirmed by Węglarska (1957) and Szymańska (1995), both working on reared specimens of the freshwater eutardigrade *D. dispar*, and by Guidetti et al. (2006), working with the freshwater *Dactylobiotus parthenogeneticus* Bertolani 1982a and the limnoterrestrial *Bertolanius volubilis* (Durante Pasa and Maucci 1975).

Węglarska (1957) was the first author to compare cyst formation with molting. In particular, she noted that both phenomena begin by expelling the sclerified parts of the buccal-pharyngeal apparatus. In comparison, during molting, the old cuticle is subsequently abandoned, while during encystment, the old cuticle is retained. Węglarska (1957) and Szymańska (1995) reported information on the histochemistry and ultrastructure, respectively, of the encysted tardigrades. Other authors examined the steps involved in encystment. Westh and Kristensen (1992) and Hansen and Katholm (2002) analyzed the steps of cyst formation, revealing the presence of two types of cysts in two species of *Bertolanius* (Eohypsibiidae) collected in freshwater (submersed or bank mosses in Greenland springs): *Bertolanius nebulosus* (Dastych 1983) and *Bertolanius weglarskae* (Dastych 1972). A complete picture of the various steps of encystment was provided by Guidetti et al. (2006), who considered two species belonging to different superfamilies and living in different habitats: *D. parthenogeneticus* (Macrobiotoidea: Murrayidae) (Fig. 3.1), an exclusively limnic species, and *B. volubilis* (Eohypsibioidea: Eohypsibiidae) (Fig. 3.2), a moss-dwelling species. Cysts are also known in a related limnic species, *Bertolanius smreczynskii* (Węglarska 1970), in addition to the two species found both in freshwater and in terrestrial habitats: *B. weglarskae* and *B. nebulosus* (Bertolani 1982b; Dastych 1983; Westh and Kristensen 1992; Bertolani and Rebecchi 1996). *Dactylobiotus* has only one type of cyst (Guidetti et al. 2006, 2008), whereas *Bertolanius* has two types of cysts that Westh and Kristensen (1992) called "white" and "red" and correlated with seasonal environmental changes. In both genera, the cyst is the result of a series of successive morphological changes that, as in molting, begins by discharging the sclerified parts of the buccal-pharyngeal apparatus (but not the body cuticle, as during molting), reducing the animal size by contracting the body longitudinally, and decreasing the body movements to a complete standstill (Guidetti et al. 2006, 2008). At this stage, any openings of the animal (i.e. mouth and cloaca) are closed. Including the steps before and after encystment, Guidetti and Møbjerg (2018) recognized nine

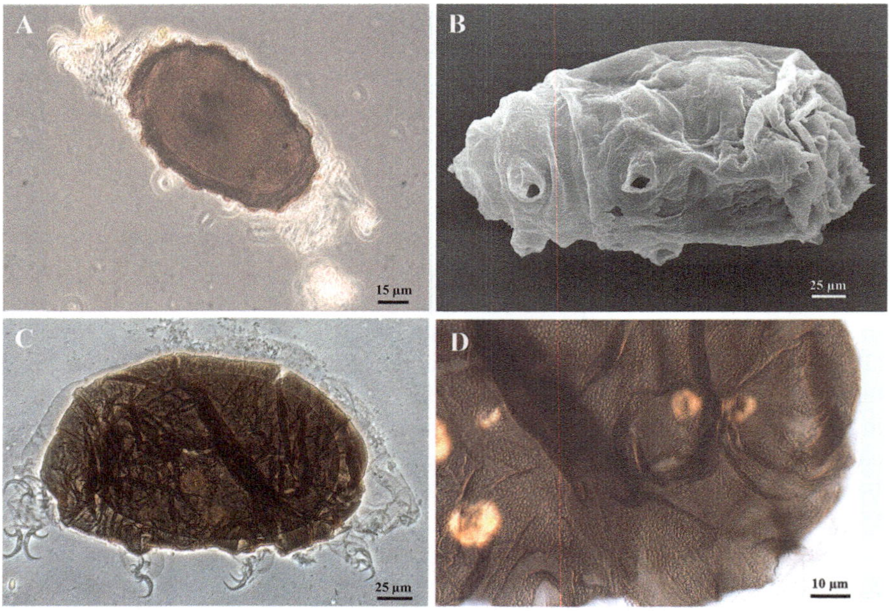

Fig. 3.1 *Dactylobiotus parthenogeneticus*. (**a**) cyst with the old cuticle; eye spots are still visible (phase contrast); (**b**) cyst (SEM); (**c**) cyst containing the mummy cuticle (phase contrast): D: detail of the cyst cuticle (differential interference contrast)

steps of the encystment process in *D. parthenogeneticus*, seven in the "type 2" cysts of *B. volubilis*, and five in the "type 1" cysts of the same species. After the discharge of the sclerified parts of the buccal-pharyngeal apparatus and the body contraction, the synthesis of cuticular material begins with the production of the so-called sarcophagus cuticle with a folded surface, common to all the three types of cysts and characterized by the absence of claws (Guidetti et al. 2006). In *Dactylobiotus*, the sarcophagus cuticle becomes harder and darker. Its formation is followed by the synthesis of a new, highly modified and nonfunctional buccal-pharyngeal apparatus. Successively, the animal synthesizes another wrinkled cuticle, the mummy cuticle (name proposed by Westh and Kristensen 1992, for a step in the formation of *Bertolanius* cysts), without both legs and claws. The subsequent step in the *Dactylobiotus* cyst is the discharge of the sclerified structures of the buccal-pharyngeal apparatus, followed by a new synthesis of them, but always nonfunctional. Further subsequent steps include yet another discharge of the buccal-pharyngeal sclerified structures, followed by the synthesis of a new body cuticle with legs and claws and the cuticular parts of a functional buccal-pharyngeal apparatus, like those of the non-encysted animals. At this time, the cyst is ovoid and the external involucra are dark-brown, almost black; the animal is ready to leave the cyst by breaking all the cuticular coats (onion or matryoshka-like stage; Guidetti et al. 2006).

In the moss-dwelling *B. volubilis*, the "type 2" cyst (Fig. 3.2a), corresponding to the "red" cyst of the limnic species *B. nebulosus* and *B. weglarskae*, after the forma-

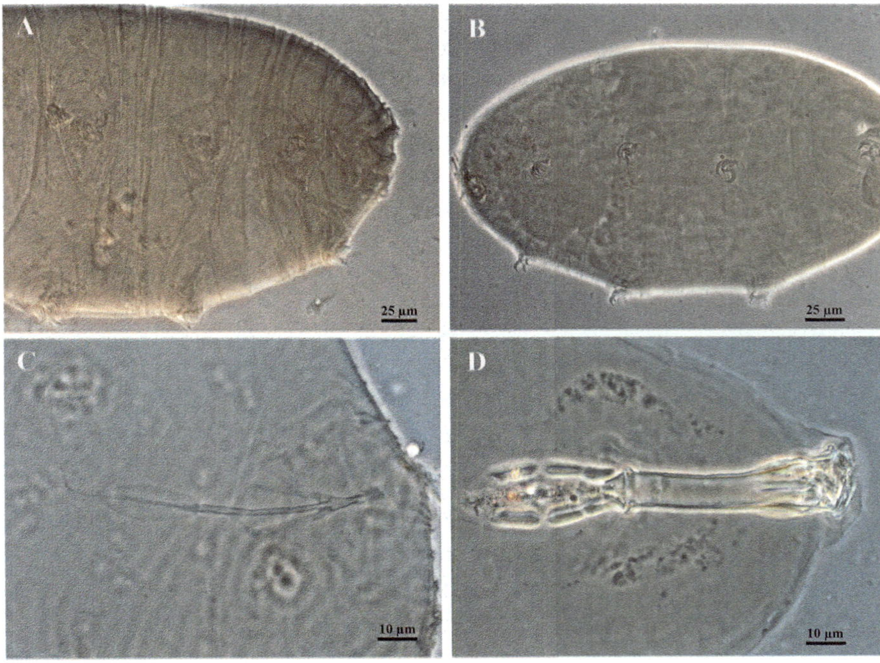

Fig. 3.2 *Bertolanius volubilis*. (**a**): red or type 2 cyst (phase contrast); (**b**): white or type 2 cyst (phase contrast); (**c**): modified buccal-pharyngeal apparatus (phase contrast); (**d**): functional buccal-pharyngeal apparatus (phase contrast)

tion of the sarcophagus cuticle (however absent in *B. nebulosus* and *B. weglarskae*; see Westh and Kristensen 1992; Hansen and Katholm 2002), a mummy cuticle is formed together with modified and nonfunctional circular parts of the buccal-pharyngeal apparatus (Fig. 3.2c). These parts are subsequently discharged, and a fourth cuticle and a functional buccal-pharyngeal apparatus (Fig. 3.2d) are produced (matryoshka-like stage). This step is followed by the exit of a complete and non-encysted animal. After the sarcophagus cuticle formation, the "type 1" cyst of *B. volubilis* (Fig. 3.2b) produces a third cuticle (mummy cuticle) and a complete and functional buccal-pharyngeal apparatus. This step is followed by the exit of a complete and non-encysted animal. For a more detailed description of the encystment steps, see Guidetti et al. (2006, 2008) and Guidetti and Møbjerg (2018).

In laboratory conditions, encystment was induced in limnic species by Węglarska (1957) and Szymańska (1995) (both using *D. dispar*) and by Guidetti et al. (2006) (in *D. parthenogeneticus*). Guidetti et al. (2006, 2008) also induced encystment in the moss-dwelling species *B. volubilis* (which has related limnic species). Węglarska (1957) and Szymańska (1995) reared their animals with decomposed vegetable materials. Węglarska's observations led her to believe that encystment was caused by gradually deteriorating environmental conditions, whereas Szymańska concluded that a decrease in environmental pH was an important factor leading to encystment. Nelson et al. (2016) stated that the factors inducing cyst formation were unknown, even though stressful environmental conditions (e.g., reserve or oxygen depletion, pH alteration, temperature variation, and pollution) have been suggested. Węglarska (1957) noted that *D. dispar* needed approximately 12 days to produce a cyst, but once formed, it could stay for more than 9 months in this state. She also observed that the placement of an encysting *D. dispar* specimen in clean water induced an active normal state rather than the completion of encystment, whereas when left in water with decaying leaves, it completed encystment. Węglarska (1957) also noted that cysts placed in water containing a tangle of algae did not encyst, but when they were subsequently placed in tap water, the animals left the cyst within 6–48 h.

To obtain cysts, Guidetti et al. (2006) kept *D. parthenogeneticus* in water with algae at 14 °C for at least 90 days, whereas they kept *B. volubilis* in water at 14 °C for at least 20 days. Further laboratory experiments were conducted by Guidetti et al. (2008) using *B. volubilis*. The study of cysts and animals collected in nature (Italy) and kept at 6, 14, or 20 °C and at 12 h/12 h light/dark demonstrated that active tardigrades collected in April produced mainly "type 2" cysts, whereas animals collected in November produced mainly "type 1" cysts, indicating that the different responses are functions of the physiological state at the time they were collected. Photoperiod did not seem to affect the type of cysts formed by *B. volubilis*. The dynamics of the two types of cysts in nature show opposite seasonal trends: "type 2" cysts are present only during the warm season, and "type 1" cysts are present during the cold season (Guidetti et al. 2008). These data confirmed what was observed in nature for the congeneric limnic species *B. nebulosus* from Greenland (Westh and Kristensen 1992; Hansen and Katholm 2002). Temperature was the environmental factor involved in induction, maintenance, and termination of the cyst. Guidetti et al. (2008) hypothesized that endogenous factors, due to exogenous factors, can be involved in tardigrade encystment. Moreover, animals and cysts of *B. volubilis* survived freezing (18 of 20 animals and 48 of 53 cysts) after the following protocol: frozen at −9 °C for 24 h, then placed at −80 °C, kept frozen at −80 °C for 61 days, and then directly transferred to 20 °C for thawing (Guidetti et al. 2008). Importantly, *B. nebulosus* overwinters as frozen cysts or, in a few cases, as eggs or adults (Westh and Kristensen 1992).

We have unpublished data on cysts of tardigrades living in submerged mosses (sample labeled as C2573) from a spring located in Val de la Mare, Pian Venezia (Italian Alps, Stelvio National Park, Trentino, NE Italy, at 2270 m a.s.l.). The samples collected in November 2004 were frozen at −20 °C and maintained in this state up to October 2008. The species present in the sample (all eutardigrades) were *Borealibius zetlandicus* (Murray 1907b), *Mixibius saracenus* (Pilato 1973), *Murrayon pullari* (Murray 1907b), and *Isohypsibius granulifer* (Thulin 1928). After thawing, all the specimens of the first three species were extended and did not show any movement. Only two specimens, recognizable as *I. granulifer* by the typically granulated cuticle, placoids, and claws, were present as live cysts (Fig. 3.3a). After putting them in a

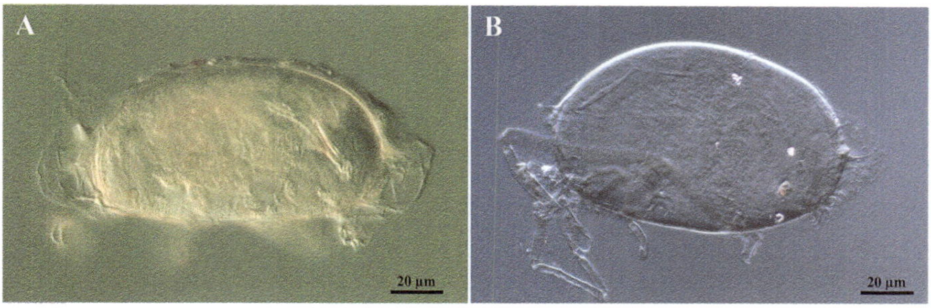

Fig. 3.3 (**a**): cyst of *Isohypsibius granulifer* from Pian Venezia that survived freezing (differential interference contrast): (**b**): cyst of freshwater *Thulinius ruffoi* from type locality (phase contrast)

drop of water and observing them under a light microscope, we were able to recognize slow and continuous movement in each animal within its cyst involucra. Therefore, only one species (*I. granulifer*) survived being frozen and maintained in the cyst state for 4 years. The Greenlander freshwater specimens of *B. nebulosus* seem to survive mainly as cysts, but also as non-encysted animals and eggs (Westh and Kristensen 1992). *Isohypsibius granulifer* is a limnic species and to date encystment is the only known way for it to survive freezing. Schill et al. (2007) reported that *Thulinius ruffoi* (Bertolani 1982a, b), a species normally colonizing freshwater, was found exclusively as cysts in 46 moss samples during winter (when temperature can easily drop below 0 °C at night) on the Sinai Mountains (Egypt, 1750 m a.s.l). Cysts of this species were already found in freshwater (Fig. 3.3b). Based on these data, we conclude that the formation of a cyst is certainly a way to survive freezing, even though we cannot be sure that other ways do not exist, despite the presence of all the dead extended animals. Encystment could also be involved in overcoming other seasons without freezing temperatures, since limnic tardigrade species often disappear during the summer, even when the water is still present in lakes and rivers where they live (Bertolani 1982b).

Although we now have a better understanding of cryptobiosis (anhydrobiosis, cryobiosis) in tardigrades, the molecular mediators of desiccation tolerance are not fully known. Boothby et al. (2017) demonstrated that tardigrade-specific intrinsically disordered proteins (TDPs) are essential for desiccation tolerance in *Hypsibius dujardini*, which colonizes both freshwater and terrestrial habitats. In contrast, our knowledge of the molecular mechanisms involved in encystment (diapause) are unknown, although the morphological stages of cyst formation have been described thoroughly (Guidetti et al. 2006, 2008; Guidetti and Møbjerg 2018). Encystment is thought to be induced by exogenous factors (temperature, oxygen) and possibly endogenous stimuli, but further studies are essential to confirm the ability of various tardigrade species in different habitats to form cysts and to determine the molecular mechanisms involved.

References

Altiero T, Bertolani R, Rebecchi L (2010) Hatching phenology and resting eggs in tardigrades. J Zool 280:290–296

Bertolani R (1982a) A new genus and five new species of Italian fresh-water tardigrades. Bollettino del Museo Civico di Storia Naturale di Verona 8:249–254

Bertolani R (1982b) *Tardigradi (Tardigrada)*. Guide per il riconoscimento delle specie animali delle acque interne italiane. Quaderni CNR, AQ/l/168 15:1–104.

Bertolani R, Rebecchi L (1996) The tardigrades of Emilia (Italy). II. Monte Rondinaio. A multi-habitat study on a high altitude valley of the northern Apennines. Zool J Linnean Soc 116:3–12

Binda MG (1978) Risistemazione di alcuni Tardigradi con l'istituzione di un nuovo genere di Oreellidae e della nuova famiglia Archechiniscidae. Animalia 5:307–314

Boothby TC, Tapia H, Brozena AH, Piszkiewicz S, Smith AE, Giovannini I, Rebecchi L, Pielak GJ, Koshland D, Goldstein B (2017) Tardigrades use intrinsically disordered proteins to survive desiccation. Mol Cell 65:975–984

Clausen LKB, Andersen KN, Hygum TL, Jørgensen A, Møbjerg N (2014) First record of cysts in the tidal tardigrade *Echiniscoides sigismundi*. Helgol Mar Res 68:531–537

Dastych H (1972) *Isohypsibius weglarskae* sp. n. a new species of Tardigrada from Poland. Bull Acad Pol Sci Ser Sci Biol 11:761–764

Dastych H (1983) Two new Eutardigrada species from West Spitsbergen and the Tatra Mts. Bull Soc Amis Sci Lett Pozn 23:195–200

Doyère M (1840) Mémoire sur les Tardigrades. Annales Des Sciences Naturelles 14:269–362

Dujardin F (1851) Sur les Tardigrades et sur une espèce a longs pieds dans l'eau de mer. Annales Des Sciences Naturelles 15:160–166

Durante Pasa MV, Maucci W (1975) Descrizione di tre nuove specie di Tardigradi della Scandinavia. Atti Soc Ital Sci Nat Mus Civ Stor Nat Milano 116:244–250

Guidetti R, Møbjerg N (2018) Environmental adaptations: encystment and cyclomorphosis. In: Schill RO (ed) Water bears: the biology of tardigrades, zoological monographs, vol 2, pp 249–271

Guidetti R, Boschini D, Rebecchi L, Bertolani R (2006) Encystment processes and the "Matryoshka-like stage" in a moss-dwelling and in a limnic species of eutardigrades (Tardigrada). Hydrobiologia 558:9–21

Guidetti R, Boschini D, Altiero T, Bertolani R, Rebecchi L (2008) Diapause in tardigrades: a study of factors involved in encystment. J Exp Biol 211:2296–2302

Guidetti R, Altiero T, Bertolani R, Grazioso P, Rebecchi L (2011a) Survival of freezing by hydrated tardigrades inhabiting terrestrial and freshwater habitats. Zoology 114(2):123–128

Guidetti R, Altiero T, Rebecchi L (2011b) On dormancy strategies in tardigrades. J Insect Physiol 57:567–576

Hansen JG, Katholm AK (2002) A study of the genus *Amphibolus* from Disko Island with special attention on the life cycle of *Amphibolus nebulosus* (Eutardigrada: Eohypsibiidae). In: Hansen JG (ed) *Arctic biology field course Quqertarsuaq 2002*. Zoological Museum University of Copenhagen, Copenhagen, pp 129–163

Heinis F (1910) Systematik und Biologie der Moosebewohnen den Rhizopoden, Rotatorien und Tardigraden usw. Archiv für Hydrobiologie und Planktonkund 5:1–115

Lauterborn R (1906) Demostrationen aus der Fauna ders Oberreheins und seiner Umgebung. Verh Dtsch Zool Ges:265–268

Murray J (1907a) The encystment of *Macrobiotus*. The Zoologist 11:4–11

Murray J (1907b) Scottish Tardigrada, collected by the lake survey. Trans R Soc Edinb 45: 641–668 (with 3 plates)

Murray J (1907c) Encystment of Tardigrada. Trans R Soc Edinburg 45: 837–854 (with 2 plates)

Nelson DR, Guidetti R, Rebecchi L (2016) Phylum Tardigrada. In: Thorp, Covich (eds) Key to Nearctic Fauna, vol. 15, pp 277–290

Pilato G (1973) Tardigradi delle acque dolci siciliane. Nota seconda. Boll Accad Gioenia Sci Nat Catania 12:177–186

Rahm G (1925) Die Cystenbildung bei den wasserbewohnenden Tardigraden. In: Verh. intern. Vereinig. theoret. angew. Limnologie, Bd. 3, Moskau, pp 364–371

Rebecchi L, Altiero T, Guidetti R (2007) Anhydrobiosis: the extreme limit of desiccation tolerance. Invertebr Surviv J 4:65–81

Schill RO, Huhn F, Köhler HR (2007) The first record of tardigrades (Tardigrada) from the Sinai Peninsula, Egypt. Zool Middle East 42:83–88

Schill RO, Förster F, Dandekar T, Wolf M (2010) Using compensatory base change analysis of internal transcribed spacer 2 secondary structures to identify three new species in *Paramacrobiotus* (Tardigrada). Org Divers Evol 10:286.296

Schultze M (1865) *Echiniscus sigismundi*, ein Arctiscoide der Nordsee. Arch Mikr Anat 1:428–436

Szymańska B (1995) Encystment in the tardigrade *Dactylobiotus dispar* (Murray, 1907) (Tardigrada: Eutardigrada). Part 1: Observation of living animals and structure of cyst. Zoologica Poloniae 40:91–102

Thulin G (1928) Über die Phylogenie und das System der Tardigraden. Hereditas 11:207–266

von Wenck W (1914) Entwicklungsgeschichtliche Untersuchungen an Tardigraden (*Macrobiotus lacustris* Duj.). Zool Jahrb Abt Anat Ontog Tiere 37:465–514

Węglarska B (1957) On the encystation in Tardigrada. Zoologica Poloniae 8:315–325

Węglarska B (1970) *Hypsibius* (*Isohypsibius*) *smreczynskii* spec. nov., a new species of freshwater tardigrade. Zeszyty Naukowe Uniwersytetu Jagiellońskiego Prace zooliczne 16:107–114

Westh P, Kristensen RM (1992) Ice formation in the freeze-tolerant eutardigrades *Adorybiotus coronifer* and *Amphibolos nebulosus* studied by differential scanning calorimetry. Polar Biol 12:693–699

Chapter 4
Fish: Diapause, Dormancy, Aestivation, and Delay in Gonad Development

Dmitry L. Lajus and Victor R. Alekseev

Abstract The review addresses different types of dormancy on fish. It can be subdivided into two groups. The first is adult aestivation. This allows fish to survive droughts in low latitude freshwater environments (several species from various families), or ice coverage periods causing anoxic conditions in high latitudes (cyprinid *Carassius spp.*). It is a facultative phenomenon induced by various environmental cues (but not photoperiods), accompanied by specific biochemical adaptations that slow down the metabolism of aestivating individuals and allow their bodies to tolerate pollution from metabolic waste products. The second, more variable group is embryonic diapauses. These obligatory processes are well-integrated into the life cycle of fishes. It occurs in more than 30 killifish species (order Cyprinodontiformes) as a key adaptation that significantly expands their environmental range to include ephemeral pools. Diapause in several elasmobranch species occurs in the mother's body. In bitterling *Acheilognathus*, diapause is a key adaptation to effectively use their limiting resource—the large bivalves where bitterlings deposit their eggs. Cessation of female gonad development after completing vitellogenesis is common in many species of high latitudes. This phenomenon has not been considered in the context of fish dormancy before, but biologically it fulfills the same functions as embryonic diapause and thus should be considered within the same framework. The effectiveness of life strategies containing diapause depends very much on both the sensitivity of dormant organisms to environmental stresses and the accuracy of their responses to environmental cues—primary changes in photoperiod or temperature to induce or terminate dormancy. The optimal strategy should balance phenotypic plasticity and bet-hedging as tools to maximize the fitness of diapausing organisms in partially predicted environments.

Keywords Fish · Diapause · Aestivation · Delay of gonad development

---

D. L. Lajus
Saint Petersburg State University, St. Petersburg, Russia

V. R. Alekseev (✉)
Zoological Institute of Russian Academy of Sciences, St. Petersburg, Russia
e-mail: alekseev@zin.ru

V. R. Alekseev, B. Pinel-Alloul (eds.), *Dormancy in Aquatic Organisms. Theory, Human Use and Modeling*, Monographiae Biologicae 92,
https://doi.org/10.1007/978-3-030-21213-1_4

4.1 Introduction

Fishes aim to better adapt to their environment as do other organisms. Operationally, this means situating their most active development stages, in terms of growth and reproduction, in the most optimal environmental conditions. As environments are not always optimal, an organism's task is to reduce the effect of unfavorable environmental conditions on its most sensitive (and usually most important) stages of development. When adverse conditions are impossible to avoid by changing location, fish must tolerate them in situ, often by delaying development in order to match less sensitive life stages with unfavorable environmental conditions. Delayed development strategies have different names—diapause, dormancy, and aestivation—even though it is not always possible to draw clear borders between them. Nevertheless, their evolutionary significance in all cases is to adjust the most important life stages to optimal environmental conditions.

This chapter is subdivided in the following sections:

1. Embryonic diapause, as it occurs in three species groups. First, adult killifish have adapted to ephemeral environments where water is available for a short time during the year, sometimes only for two months; their embryos survive for the rest of the year buried in sediments. Second, free embryos of autumn-spawning bitterlings diapause just after hatching. Third, several species of viviparous elasmobranchs arrest embryo development inside a female's body just after embryogenesis has begun.
2. Aestivation of adult fish from various taxa, which survive extremely cold or dry periods by burying in the mud.
3. Seasonal delays in female gonad development in species living in moderate to high latitudes that use warm, food-rich periods for adult growth and adjust juvenile appearance to the period most favorable for their feeding.

4.2 Embryonic Diapause

4.2.1 Killifish

By far, embryo diapause is most commonly studied in killifish. Killifish are small, sexually dimorphic fish that belong to several families (Aplocheilidae, Cyprinodontidae, Fundulidae, Nothobranchiidae, Profundulidae, Rivulidae, and Valenciidae) of the order Cyprinodontiformes. This order contains 1264 species in several families; except for killifish, some families include live-bearers. Killifish typically inhabit shallow waters or marginal aquatic habitats (Scheel 1990; Huber 2000; Wildekamp 2004). They are distributed in Africa (family Nothobranchiidae) and South America (family Rivulidae) and have adapted to life in seasonally ephemeral water bodies that dry out completely on an annual basis (Wourms 1972c; Murphy and Collier 1999; Hrbek and Larson 1999). Small size makes killifish very popular in aquaria and in scientific experiments, hence the enormous progress in research on their diapause.

Embryonic diapause is the most prominent biological feature of killifish, a unique phenomenon among animals in general that has attracted the attention of researchers for several decades (Peters 1963; Wourms 1972a, b, c). This adaptation is so unique that Furness (2016), in reviewing killifish diapause adaptions, chose terminology usually applied to the biology of ephemeral plants adapted to arid or semiarid regions: seed dormancy, bet-hedging, and plasticity in the timing of germination, rapid growth, and development. This unique adaptation allows killifish to inhabit areas that are dry most of the year, thus considerably expanding the range of environmental conditions where fish as a taxon are able to live. In diapause, killifish can maintain viability over many months, even years. Eggs of *Nothobranchius furzeri* have survived at least 3 years in storage (Cellerino et al. 2016).

Long survival is only possible when fish diapause at a particular stage of ontogenesis, after the formation of the body axis, when embryos have about 38 pairs of somites. This is called Diapause II. Diapause is possible at two other stages of embryogenesis, called Diapause I and Diapause III by Wourms (1972c). Diapause I occurs just before the formation of the embryonic axis and is also called the dispersed cell phase. Diapause III, or delayed hatching, occurs as the fully developed embryos become ready to hatch. However, these types of diapause are usually shorter and not characterized by such high resistance to environmental stresses.

An important question is whether or not diapause in different groups of killifish originated independently. As it is not characteristic of all killifish species, some authors consider that diapause was present in the ancestral form but disappeared in some daughter forms (Murphy and Collier 1999; Hrbek and Larson 1999). An alternative hypothesis, based on thorough genetic analysis of numerous species of killifish, argues that diapause adaptation appeared independently three times in Rivulinidae and three times in Nothobranchiidae (Furness et al. 2015). This hypothesis is supported by extensive genetic and morphological data and, thus, seems better justified.

Not all killifish species undergo diapause in their life cycle. These species are called non-annual killifish, as opposed to the annual ones for which diapause is obligatory. Non-annual killifish live in permanent waterbodies and thus have no need for diapause. Yet, they still experience diapause-like events during development, such as delayed hatching occurring either spontaneously or because conditions are substandard for several days. This represents a predisposition toward diapause in killifish as a group and indicates that it is an ancient adaptation.

Why does diapause in different species occurs only in specific developmental stages? This behavioral similarity is used to support the loss of adaptation hypothesis pertaining to some killifish (Murphy and Collier 1999). At the same time, there are reasons to think that some developmental stages have particular characteristics that make them more suitable for diapause than the others.

Diapause I occurs after the blastula has formed and then dispersed and before cellular reaggregation occurs to form the primary embryonic body axis. It is unique in annual killifish. Possibly due to its dispersed non-differentiated cells, an embryo at this stage is rather resistant to damage: other cells can relatively easily compensate for any damaged cells (Wourms 1972a).

Diapause II takes place just after the foundations of the central nervous system, optic cups, olfactory and lens placodes, and otic vesicles and a functional tubular heart begin to form (Podrabsky et al. 2010b). The principal organ systems appear at this stage but are still small and not yet costly to maintain. Therefore, this developmental period is less sensitive and presents an ideal time for arresting development (Wourms 1972c). Diapause II is associated with important adaptation. Compared with non-annual killifish (not possessing diapause), annual killifish (possessing diapause) display significantly reduced development of the cranial region and circulatory system, which require a lot of energy to maintain (Podrabsky et al. 2010a; Furness et al. 2015). This reduced development may reduce energy expenditure by shunting resources away from energetically costly organs (the brain and heart) and, thus, increase survival chances during diapause. Furness et al. (2015) tested this hypothesis on *N. furzeri* embryos and found a correspondence between entering Diapause II, associated with metabolic depression (beneficial for survival), and morphological characteristics. Perhaps the most remarkable features of this adaptation are that, on the one hand, it occurs during the very early stages of embryogenesis, which is supposedly highly conserved among vertebrates (Raff 1996), and on the other, the process is similar for several annual killifish species that originated independently. The evolutionary cost of this adaptation may have been quite high, but this has not yet been studied.

Diapause III occurs just before hatching when the embryos are fully developed. This is a logical stage for possible diapause because newly hatched larvae may be the most vulnerable stage in the entire life cycle of fish. Therefore, Diapause III is a way to optimally mitigate the risks associated with hatching in unsuitable environments.

Different types of diapause are not mutually exclusive. Often, annual killifish can produce eggs that follow different developmental trajectories, ranging from direct development (proceeding directly to Diapause III and hatching, as do non-annual killifish) to entering each successive diapause stage (I, II, and III) in turn (Wourms 1972c; Podrabsky et al. 2010a). One interesting example of non-annual killifish entering only Diapause III is *Kryptolebias marmoratus*, the only known selfing hermaphroditic vertebrate. With almost amphibious behavior, it is extremely tolerant to a range of environmental factors (Taylor 2012). Fully developed eggs of this fish hatch right after their pool is flooded.

Initiation of Diapause I and II is sensitive to a variety of environmental factors, including oxygen levels, temperature, photoperiod, and the presence of adult fish (reviewed in Podrabsky et al. 2010a). Inducing different types of diapause depends on how environmental factors affect different species. For instance, *N. korthausae* embryos easily enter Diapause II and III when exposed to low temperatures, short photoperiods, or constant darkness, and diapause is terminated by longer photoperiods and higher temperatures (Levels and Denucé 1988). It is more difficult to induce Diapause I in this species: Diapause I only becomes possible with very low temperatures (18–20° C). The authors did not find that maternal hormones affected these responses. Diapause terminated by a lengthened photoperiod, accompanied with a 200-fold increase in metabolic rate, was observed for *Austrofundulus limnaeus* (Hand and Podrabsky 2000).

Maternal hormones may contribute to the induction of diapause also. Levels of steroid hormones in the fertilized eggs of *A. limnaeus* were not correlated with entry or escape from Diapause II, although levels of androstenedione tend to be higher in embryos not entering diapause (Pri-Tal et al. 2011). These embryos were found to exhibit faster hormone metabolism and earlier hormone synthesis than embryos entering Diapause II. Incubation of embryos in exogenous 17-β-estradiol was associated with a sevenfold increase in the number of nondiapausing embryos and significantly elevated androstenedione levels. These data suggest that maternal steroid hormones are the critical factors involved in determining developmental pathways in embryos of killifish. At the same time, more research is needed in this area: for instance, a recent review of rhythms in the endocrine systems of fish (Cowan et al. 2017) does not even mention diapause.

A special study was devoted to determining whether or not killifish enter diapause trying to avoid predators, because it is known that some animals respond to the presence of predators in this way (see Polačik and Janáč (2017) and references herein). The authors studied several populations of *Nothobranchius* to see if the initiation of Diapause III depended on the presence of chemical cues from predators (kairomones), but no evidence for such effects was found. They concluded that, in the killifish's unpredictable environment, antipredator defense strategies are too costly to be used.

The sensitivity of diapausing embryos to environmental factors was considered in several papers. Matias and Markofsky (1978) studied the effect of temperatures ranging from 8 to 40°C on *N. guentheri*. They found that diapausing embryos are more resistant to temperature effects than nondiapausing embryos and Diapause II conferred much more resistance than Diapause I. Matias (1984) found that chorion sensitivity of diapausing eggs to proteinase in *N. guentheri* differs at different stages of development and is highest in developmental stages coinciding with Diapause I, II, and III. Podrabsky et al. (2007) found extremely high tolerance in *A. limnaeus* to oxygen deprivation, which peaked in Diapause II. Early embryos were able to tolerate anoxic conditions for one day, whereas Diapause II embryos tolerated such conditions for 62 days. During that time, metabolism is supported mainly by the production of lactate. The authors concluded that this was the first report of such extreme anoxia tolerance in a vertebrate embryo. High anoxia tolerance was not only because undifferentiated cells better survive anoxia. In addition, these embryos are specially adapted to such events and likely contain functional neurons and cardiomyocytes that are incredibly resistant to anoxia.

One important observation over diapausing embryos is that the time when diapause terminates is highly variable. Embryos from the same clutch, incubated under common garden conditions at cue levels that are uninformative with regard to seasonal timing, exhibit heterogeneity in terms of entering diapause; the length of diapause, if entered; and when hatching occurs (Podrabsky et al. 2010a; Furness et al. 2015). This means that the variance cannot be explained by phenotypic plasticity alone, which would cause similar changes in individuals under common garden conditions, and works perfectly as an adaptation strategy only when cues perfectly predict subsequent environmental conditions (Wong and Ackerly 2005). To explain this high level of variance, the concept of bet-hedging has been applied, a risk-spreading strategy borrowed from plant biology (equivalent to not placing all your eggs in one basket). Originally bet-hedging was used to explain the seed bank phenomenon, a reservoir of ungerminated seeds in the soil, which allows plants to mitigate risks of unfavorable conditions. If a drought kills the germinated plants, those with seeds remaining in the seed bank will have a fitness advantage in comparison with plants whose seeds germinate immediately. Simons and Johntson (1997) linked bet-hedging to developmental instability, representing the inability of a developmental organism to perfectly follow genetically predetermined developmental trajectories (Waddington 1957; Zakharov 1992). Developmental instability is expressed as stochastic variation, which represents a third component of phenotypic variation standing on an equal footing with genotypic and environmental variation (Lajus et al. 2003). This type of phenotypic variation is ubiquitous and explains differences between clonal individuals growing in the same environment, or between left and right bilaterally symmetrical structures, called fluctuating asymmetry. It also explains most of the phenotypic variance of *Daphnia* when forming diapausing eggs in response to changing photoperiods (Lajus and Alekseev 2004).

An important pattern of developmental instability, and thus stochastic variation, is its increase under more stressful conditions (Parsons 1990; Graham et al. 1993). This pattern, to the best of our knowledge, was not discussed in relation to bet-hedging, but is highly relevant because organisms may benefit, as variance in diapause timing may increase under more stressful conditions, exactly when it is most important for the organism. This may be an example of when the phenomenon has little adaptive value by itself, but may have evolutionary significance. In studying the contribution of phenotypic plasticity and bet-hedging in killifish diapause, Furness and coauthors (2015) found that both strategies are involved.

4.2.2 Bitterling Free-Embryo Diapause

Besides killifish, only one other teleost group possesses diapause in early developmental stages—autumn-spawning bitterling of the genus *Acheilognathus* (Kawamura et al. 2014). Bitterlings are small cyprinids that deposit their fertilized eggs in the mantle cavity of freshwater bivalves. In *A. rhombeus*, developmental arrest occurs soon after hatching at a stage characterized by a visible auditory vesicle appearing behind the optic vesicle and a blood vascular system, including the heart, appearing in the dorsal region of the yolk (Kawamura and Uehara 2005). Very slow development takes place during diapausing. When the formation of the eyes progresses enough and minute scute-like projections on the body surface gradually degenerate, it terminates. Autumn-spawning bitterlings benefit from diapause because it allows free embryos to survive during cold winters, just as killifish diapause allows embryos to survive during dry periods.

Of special interest is the developmental stage at which diapause in bitterling occurs. Killifish research has clearly identified the developmental stages at which the diapause occurs in these fishes. But the studies reviewed above have advantages rooted in processes occurring during the ontogenesis of fish and vertebrates in general. The developmental stage of diapausing bitterling does not conform to the same logic. It would be more logical for bitterling to diapause just before hatching, i.e., at the equivalent of Diapause III in killifish. In reality, bitterling diapause occurs several days after hatching. We assume that this unusual pattern of development can be explained by very unusual biology of these fish. Bitterling spawn in bivalves. Postponing diapause to later stages of development may help to adjust the time when bitterling larvae exit the mantle cavity to optimize conditions for their growth. Diapause allows autumn-spawning bitterlings to share a limited number of mollusk hosts with their sympatric spring-spawning relatives. For bitterlings, the strong limitation of their spawning habitat is confirmed by findings that larvae survival is strongly density dependent (Smith et al. 2004; Kitamura 2005).

Bitterling diapause persists about 6 months, irrespective of the time the treatment was initiated. This period appears to be genetically predetermined, although exposure to low temperatures is necessary to terminate it. The genetic underpinnings of bitterling diapause have been confirmed by hybridization experiments (Kawamura and Uehara 2005). Hybrids of *A. rhombeus* and spring-spawning bitterling, representing most of species of this group of fish, do not possess diapause. Thus, the ability to diapause may be a recessive phenotypic character. Korean *A. rhombeus*, unlike the Japanese fish described above, diapause at 22°C and terminate diapause about two months after hatching, with no treatment (Suzuki and Jeon 1991).

In addition to *A. rhombeus*, similar diapause patterns were also observed in two other *Acheilognathus* species—*A. typus* and *A. longipinnis* (Nakamura 1969, cite by Kawamura and Uehara 2005). Notably, these three species were shown to be closely related on a molecular tree of mitochondrial DNA (Okazaki et al. 2001). Diapause is likely the key adaptation that permits bitterling of genus *Acheilognathus* to spawn in autumn. Since all other species in this genus spawn in spring, *Acheilognathus* are unique among the huge Cyprinidae family.

4.2.3 Elasmobranch Embryonic Diapause

Embryonic diapause has been observed in or suggested for many elasmobranch species (Wyffels 2009; Waltrick et al. 2012). Lessa (1982) first hypothesized this process for *Rhinobatos horkelli*, and 10 years later Simpfendorfer (1992) first demonstrated its occurrence in *Rhizoprionodon taylori*. In all known cases, diapause has been observed in the early stages of embryo development. Embryonic diapause has only been confirmed so far in two elasmobranch species, but evidence suggests its occurrence in as many as fourteen other species: twelve batoids from two orders and five families and two selachians from two orders and two families (Waltrick et al. 2012, Table 1). In all these cases, diapause is obligate, i.e., occurring regularly at a specific stage of development and for a particular period in every generation of a given species, irrespective of environmental conditions. No facultative diapause, which occurs from case to case, has been found in elasmobranches. In both species studied in detail (Simpfendorfer 1992; Morris 1999, cit. by Waltrick et al. 2012), the development was arrested at the blastodisc stage (similar to the stage to Diapause I in killifish) and lasted for 7–9 months. In shovelnose guitarfish *Pseudobatos productus*, embryonal diapause lasts for about 8 months (Romo-Curiel et al. 2017) and may range from 4 to 10 months in other species. Diapause was observed in both lecithotrophic (when the embryo uses only its own resources) and matrotrophic (when the embryo partly uses its mother's resources) species.

Little is known about mechanisms regulating diapause in elasmobranchs, but Waltrick et al. (2012) hypothesized an important role of progesterone in arresting embryonic development. This is highly probably given that diapause is an obligatory part of their life cycle. It is also likely environmental cues such as temperature changes may be involved in regulating diapause, as they may trigger hormone activity.

Elasmobranch diapause may occur in different seasons, but, according to authors who have studied it, in every case it evolved as a mechanism which guarantees that parturition will occur when newborns are more likely to make it through their early life stages and reach maturity. It is possible that a period of embryonic diapause benefits a species by allowing reproductive events to be synchronized (parturition and mating) or prolonged (fertilization and parturition). However, there has been no rigorous testing of these hypotheses so far (Waltrick et al. 2012).

4.3 Aestivation of Adult Teleosts

The best known case of aestivation in adult fishes is in lungfishes, an archaic group of sarcopterygian fishes characterized by possession of a lung opening off the ventral side of the esophagus. They attract the attention of researches because they possess adaptations which are important in the ancient transition from water to land: air-breathing, urea synthesis, heart partitioning, and redirection of blood flow. Such adaptations likely resulted in the appearance of tetrapods. The ability of lungfishes to survive in dry sediments for long periods of time is one of their most prominent adaptations.

African lungfish (*Protopterus*) coexist with annual killifish across most of their range. Thus, this species faces some of the same environmental challenges but meets them in a different way. At the beginning of dry season, lungfish bury into the soil, secreting a slime coat that hardens into a cocoon, and aestivate over a period of months or even years until the rainy season returns (Smith 1931). Aestivation (or estivation) is defined a state of corporal torpor adopted by some animals to survive for an extended period in arid conditions at high temperature without food and water intake. Three phases of aestivation are distinguished (Hiong et al. 2015). During the induction phase, the fish detects environmental cues and respond to them via changes at behavioral, structural, physiological, and biochemical levels. In particular, it secretes a substantial amount of mucus which turns into a dry cocoon over

6–8 days. The induction phase ends when the fish is completely encased in its dried mucus cocoon and feeding and locomotion have been terminated. During the maintenance phase, the fish must preserve its biological structure and maintain a slow rate of waste production to avoid polluting its internal environment. Interestingly, aestivating fish are still able to respond to environmental changes. For instance, Fishmen and coauthors (1986), working on *P. aethiopicus*, noted that heartbeats slowed from 15 to 6 per minute as aestivating fish responded to laboratory noise. The arousal phase is initiated by the addition of water. After arousal, the fish excretes accumulated waste products and resumes normal activity, including feeding, approximately 7–10 days after arousal (Hiong et al. 2015).

Triggers for aestivation have not been identified. It has been suggested that environmental stresses such as starvation and dehydration are the most probable factors (Fishman et al. 1986), but some researchers believe that both initiating mechanisms and perpetuating factors are poorly understood. As lungfish aestivation is not an obligatory part of their seasonal cycle, but is episodic instead, it is not part of a chronobiological rhythm and, thus, does not depend on a photoperiod.

Researchers have studied changes in the physiological parameters of aestivating lungfish. DeLaney et al. (1977), working with *P. aethiopicus*, found substantial decrease in oxygen consumption (about fivefold), heart rate (some twofold), and mean blood pressure (less than twofold) in aestivating individuals. At the same time, Fishman and coauthors (1986) reported that the level of oxygen consumption in the same species is similar during aestivation and starvation. Page and coauthors (2010) found that, in African *P. dolloi* aestivating for 2 months, most of the intracellular antioxidant enzyme levels and activities were significantly upregulated in the brain and, to a lesser extent, in heart tissue. This enhances oxidative stress resistance in the brain and heart of aestivating fish. Importantly, these changes were not due to food deprivation, as they did not occur in starving fish maintained in water for the same period of time. In general, observed physiological patterns of lungfish aestivation are in line with the aestivation patterns of other vertebrates.

A series of research studies was conducted over the last few years on the molecular analysis of various structures and functions involved in *P. annectens* aestivation (Chng et al. 2014, 2016, 2017a, b; Loong et al. 2012; Ong et al. 2017). In general, this molecular research described the complex nature of aestivation, which is based on changes in the expression of multiple genes, and argues that aestivation cannot be regarded as solely the result of general metabolism depression. Instead, it involves the complex interplay between upregulation and downregulation of diverse cellular activities.

Most of the research on fish aestivation has been done on lungfish, in particularly on *Protopterus*, where this phenomenon is best known. Yet, several other fish species are able to aestivate by burying themselves in sediments: black-stripe Australian minnow *Galaxiella nigrostriata* (Gill and Morgan 2003; Galeotti et al. 2010); *Monopterus cuchia* from Pakistan (Rahman et al. 2005); swamp eel *Synbranchus marmoratus* from Brazil (Moraes et al. 2005); South American lungfish *Lepidosiren paradoxa*, relative to African lungfish also occurring in Brazil (Mesquita-Saad et al. 2002); Australian salamanderfish *Lepidogalaxias salamandroides* (Pusey 1990; Collin 1998; Morgan et al. 2000); New Zealand mudfish *Neochanna burrowsius* (Eldon 1979); and the endemic Uruguayan Gymnotiformes *Brachyhypopomus bombilla*, for which aestivation is regulated by temperature and photoperiod cues—during 5 months of the lowest temperatures and shortest photoperiod, these fish bury in mud (Eldon 1979).

Aestivation is also well-known in the genus *Carassius*, which are common in the Eurasian north. The anaerobic metabolism of the goldfish *C. auratus* has been an object of intensive physiological research (i.e., Shoubridge and Hochachka 1980), but its close relative, the crucian carp *C. carassius*, is even more hardy, tolerating months of total anoxia in winter (Blažka 1958, 1960; Piironen and Holopainen 1986). During winter and dry periods, crucian carp burrow in the mud (Holopainen and Hyvärinen 1984; Allardi and Keith 1991). In experiments, these fish have been maintained under anoxic conditions for 140 days (Allardi and Keith 1991; Holopainen and Hyvärinen 1984). According to Hyvärinen and coauthors (1985), when wintering in anaerobic conditions in Finnish lakes, crucian carp used glycogen for metabolism. Exceptionally high glycogen concentrations, together with increased size of the main storage organ, the liver, make crucian carp a unique vertebrate. During the summer, the fish may also use this metabolic expedient to survive anaerobic conditions, though for a more limited time period (Shoubridge and Hochachka 1980). Ethanol is the major end product of this anaerobic metabolism, which is unusual among vertebrates. Alcohol production occurs mainly in the muscle tissues, but also in the liver, where the process is thought to have originated.

Present findings (Tikkanen et al. 2017) show that a seasonal drop in temperature activates a gene expression program that preconditions the crucian carp's heart for the maintenance of regulated activity in the winter. When anoxia develops, the homeostatic response to oxygen shortage involves a strong bradycardic reflex and a lengthening of the potential duration of ventricular action. These remarkable adaptations of the heart are likely necessary for the exceptional anoxia tolerance of the crucian carp, driven by selection pressure in shallow lakes which may freeze down to the bottom during long northern winters.

Despite the wide distribution of aestivation among lungfish, special adaptations to severe stress-producing circumstances are notably absent in some of them. For instance, Alaska blackfish *Dallia pectoralis* inhabits the Arctic, but is as much an air-breathing fish as the other species (Lefevre et al. 2014). There are unanswered questions regarding the physiological acclimation and adaptation of Alaska blackfish to hypoxia and cold, particularly mechanisms involved in hypoxic survival during prolonged winter submergence, but so far no evidence suggests adaptations similar to those of *Carassius*.

4.4 Delays of Oocyte Development

Diapause in early stages of fish development allows them to adjust juvenile appearance to optimal conditions. In all the cases discussed above, diapause took place in embryos. In many fish species occurring in moderate and high latitudes, this task is approached by arresting oocyte development during the late stages of oogenesis, which has biological consequences similar to arresting embryo development.

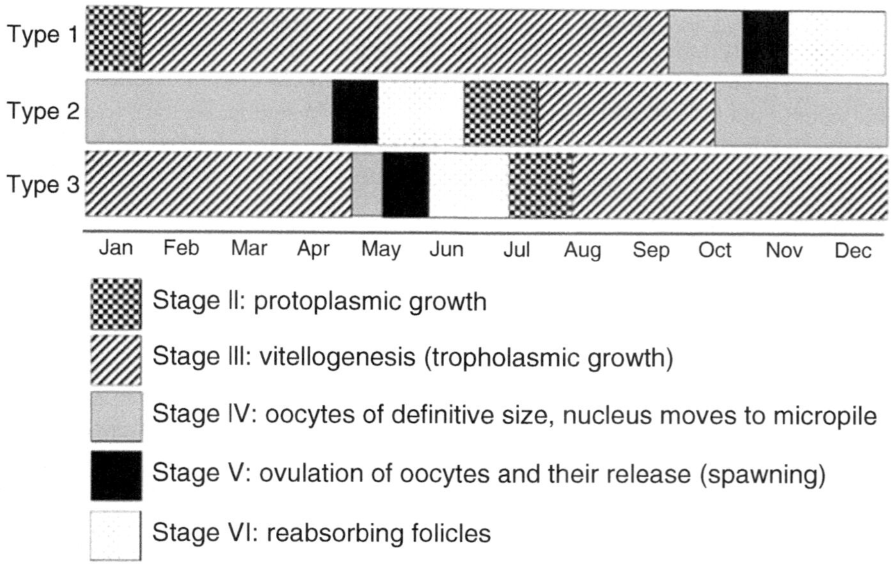

Fig. 4.1 Duration of different stages of female gonad development in teleosts, separated into three types according to seasonal cycle (see text for more explanations). (From Petlina 1987, with modifications)

However, this similarity has not attracted much research attention and has never been addressed as dormancy or diapause.

According to Sakun and Butskaya (1963), the development of female fish gonads can be separated into six stages: (I) beginning of the protoplasmic growth of the oocytes; (II) end of most protoplasmic growth; (III) period of vitellogenesis (trophoplasmic growth) or yolk deposition (oocyte size quickly increases due to accumulation of lipids and granules of yolk); (IV) end of vitellogenesis and approaching of eggs to their final size, with oocytes maturing (nucleus moves to micropile); (V) ovulation of oocytes and their release during spawning; and (VI) reabsorbing broken follicles in the ovary and beginning of a new sexual cycle. These stages are typical of fractional spawners that release all their eggs during several hours or days. For batch spawners, which are also numerous and release their eggs over much longer periods, oocyte development is not as synchronous, but this changes little in the general picture. Active processes occur in all stages except IV.

Three types of sexual cycles in fish differ in the duration of the above stages of female gonad development (Fig. 4.1). The first type of cycle is typical for autumn-spawning salmonids and coregonides. After a very short stage VI, the ovary move to stage III. This very long stage starts in winter and ends the next autumn, with the most active accumulation of lipids and yolk in late summer. The second type is typical of the majority of freshwater and some marine fishes that spawn in spring and summer: pikeperch *Stizostedion lucioperca*, perch *Perca fluviatilis*, roach *Rutilus rutilus*, pike *Esox lucius*, sturgeons Acipenseridae, herring *Clupea* sp., and many others. In these species, stages VI, II, and III are relatively short, and stage IV is very long, starting in autumn, continuing through the winter, and finishing shortly before spawning in the spring. The third type is typical of some freshwater fish (gudgeon, stickleback) and many spring- or summer-spawning marine teleosts. This sexual cycle differs from type two in that the fish have a very long stage III, which finishes in spring shortly before spawning, and, thus, a very short stage IV.

Fish possessing the second type of yearly sexual cycle experience a long stage IV with no notable changes. It lasts (approximately) from the second half of autumn, when feeding conditions for adults decline, up to the spawning period, which is adjusted to the best conditions for hatching larvae. No considerable changes occur in gonads during this stage. For a long time, it has been known that hormones can induce the spawning of common perch *P. fluviatilis* in autumn, about 6 months before their natural spawning time (Skadovsky and Parfenova 1937). At the same time, long delays in gonad maturation can be achieved using low temperatures (Kazansky 1956; Podushka 1976; Feklov 1997). Mechanisms regulating oogenesis have been reviewed in several papers (Tyler and Sumpter 1966; Nagahama and Yamashita 2008). Researchers are investigating the possibility using various agents to delay or speed up oocyte maturation and thus adjust spawning to the most convenient time (i.e., Zarski et al. 2013 on pikeperch, Schaefer et al. 2016 on burbot *Lota lota*). These approaches become possible due to the natural delay of female gonad development, which is, by its nature, very close to the phenomenon of dormancy.

4.5 Conclusions

Fish dormancy can be subdivided into two groups. The first is adult aestivation. This allows fish to survive droughts in low-latitude freshwater environments, or ice coverage periods causing anoxic conditions in high-latitude lakes and pounds. It is a facultative phenomenon induced by various environmental cues (but not photoperiods), accompanied by specific biochemical adaptations that slow down the metabolism of aestivating individuals and allow their bodies to tolerate pollution from metabolic waste products.

The second, more variable group is embryonic diapauses. These obligatory processes are well-integrated into the life cycle of fishes. Embryonic diapause occurs in more than 30 killifish species as a key adaptation that significantly expands their environmental range to include ephemeral pools filled with water only 2–3 months per year. Such adaptations have resulted in morphological changes in early developmental stages, which are extremely conservative in vertebrates. Diapause only occurs in particular developmental stages: (I) the dispersal cell phase, (II) after formation of the body axis, and (III) just before hatching. Diapause in several elasmobranch species occurs in the mother's body during the developmental stage corresponding to Diapause I in killifish. In bitterling, diapause is a key adaptation that optimizes the use of a resource that may limit the species' abundance—the large bivalves where bitterlings deposit their eggs to protect the embryos and larvae. Unlike nondiapausing bitterling that spawns in spring, diapausing bitterling spawns in autumn, and this behavior facilitates using the bivalves as nurseries for longer period. This diapause occurs after the hatching, probably because of the very specific environment where the embryos develop.

Many freshwater and marine fish from high latitudes experience a cessation of female gonad development after completing vitellogenesis that may continue during the entire cold season. Spawning can be artificially initiated at any time during this period. To the best of our knowledge, this phenomenon has not been considered in the context of fish dormancy before, but biologically it fulfills the same functions as embryonic diapause and thus should be considered within the same framework. In comparison with other types of diapause in fish, the delay of gonad development is the most common, least specialized, and occurs during earlier stages of oocyte development, even before fertilization.

The effectiveness of life strategies containing diapause depends very much on both the sensitivity of dormant organisms to environmental stresses and the accuracy of their responses to environmental cues—primary, changes in photoperiod or temperature—to induce or terminate dormancy. The main problem with that response is unpredictability of environmental conditions due to stochasticity. This limits the importance of phenotypic plasticity in the evolution of diapause-containing life strategies and requires other mechanisms for risk management, such as bet-hedging. The optimal strategy should balance phenotypic plasticity and bet-hedging as tools to maximize the fitness of diapausing organisms in partially predicted environments. So far, this seems to occur in fish species studied to address this question.

Acknowledgments The authors thank Karen Aleksander for improving the language of this manuscript. D.L. was supported by the Russian Science Foundation grant 19-14-00092, and V.A. got support from the Russian Foundation for Basic research grant 17-04-00027.

References

Allardi J, Keith P (1991) Atlas préliminaire des poissons d'eau douce de France, Coll. Patrimoines Naturels, vol. 4. Secrétariat Faune Flore, Muséum national d'Histoire naturelle, Paris, 234p

Blažka B (1958) The anaerobic metabolism of fish. Physiol Zool 31:117–128

Blažka B (1960) On the biology of the crucian carp (*Carassius carassius* L. morpha humilis Heckel). Zool Zhurnal 9:1384–1389. (In Russian)

Cellerino A, Valenzano DR, Reichard M (2016) From the bush to the bench: the annual Nothobranchius fishes as a new model system in biology. Biol Rev 91(2):511–533

Chng YR, Ong JL, Ching B, Chen XL, Wong WP, Chew SF et al (2014) Molecular characterization of argininosuccinate synthase and argininosuccinate lyase from the liver of the African lungfish Protopterus annectens, and their mRNA expression levels in the liver, kidney, brain and skeletal muscle during aestivation. J Comp Physiol B 184(7):835–853. https://doi.org/10.1007/s00360-014-0842-z. PMID: 25034132

Chng YR, Ong JL, Ching B, Chen XL, Hiong KC, Wong WP et al (2016) Molecular characterization of aquaporin 1 and aquaporin 3 from the gills of the African lungfish, Protopterus annectens, and changes in their branchial mRNA expression levels and protein abundance during three phases of aestivation. Front Physiol 7:532. https://doi.org/10.3389/fphys.2016.00532. PMID: 27891097

Chng YR, Ong JL, Ching B, Chen XL, Hiong KC, Wong WP et al (2017a) Aestivation induces changes in the mRNA expression levels and protein abundance of two isoforms of urea transporters in the gills of the African lungfish, Protopterus annectens. Front Physiol 8:71. https://doi.org/10.3389/fphys.2017.00071. PMID: 28261105

Chng YR, Ong JLY, Ching B, Chen XL, Hiong KC, Wong WP et al (2017b) Molecular characterization of three Rhesus glycoproteins from the gills of the African lungfish, Protopterus annectens, and effects of aestivation on their mRNA expression levels and protein abundance. PLoS ONE 12(10):e0185814. https://doi.org/10.1371/journal.pone.0185814

Collin SP (1998) Retinal and lenticular ultrastructure in the aestivating salamanderfish, Lepidogalaxias salamandroides (Galaxiidae, Teleostei) with special reference to a new type of photoreceptor mosaic. Histol Histopathol 13(4):1037–1048

Cowan M, Azpeleta C, López-Olmeda JF (2017) Rhythms in the endocrine system of fish: a review. J Comp Physiol B Biochem Syst Environ Physiol 187(8):1057–1089

DeLaney RG, Lahiri S, Hamilton B, Fishman AP (1977) Acid-base balance and plasma composition in the aestivating lungfish (Protopterus). Am J Phys 232:RlO–R17

Eldon GA (1979) Breeding, growth, and aestivation of the canterbury mudfish, Neochanna burrowsius (salmoniformes: Galaxiidae). N Z J Mar Freshw Res 13(3):331–346

Feklov YA (1997) Eco-hystophysiological analysis of oogenesis of peled. In: Problemy nadezhnosti funktsionirovania reproduktivnoi sistemy u ryb Trudy BINII, vol 44, pp 39–49

Fishman AP, Pack AI, Delaney RG, Galante RJ (1986) Estivation in Protopterus. J Morphol 190(Suppl. 1):237–248

Furness AI (2016) The evolution of an annual life cycle in killifish: adaptation to ephemeral aquatic environments through embryonic diapause. Biol Rev Camb Philos Soc 91(3):796–812

Furness AI, Reznick DN, Springer MS, Meredith RW (2015) Convergent evolution of alternative developmental trajectories associated with diapause in African and South American killifish. Proc R Soc Lond Ser B Biol Sci 282:1–9

Galeotti DM, McCullough CD, Lund MA (2010) Black-stripe minnow Galaxiella nigrostriata (Shipway 1953) (Pisces: Galaxiidae), a review and discussion. J R Soc West Aust 93(1):13–20

Gill HS, Morgan DL (2003) Ontogenetic changes in the diet of the black-stripe minnow Galaxiella nigrostriata (Shipway, 1953) (Galaxiidae) and the salamanderfish Lepidogalaxias salamandroides (Mees, 1961) (Lepidogalaxiidae). Ecol Freshw Fish 12(2):151–158

Graham JH, Freeman DC, Emlen JM (1993) Developmental stability: a sensitive indicator of populations under stress. In: Landis WG, Hughes JS, Lewis MA (eds) Environmental toxicology and risk assessment. American Society for Testing and Materials, Philadelphia, pp 136–158

Hand SC, Podrabsky JE (2000) Bioenergetics of diapause and quiescence in aquatic animals. Thermochim Acta 349(1-2):31–42

Hiong KC, Ip YK, Wong WP, Chew SF (2015) Differential gene expression in the liver of the African Lungfish, Protopterus annectens, after 6 months of aestivation in air or 1 day of arousal from 6 months of aestivation. PLoS ONE 10(3):e0121224. https://doi.org/10.1371/journal.pone.0121224

Holopainen IJ, Hyvärinen H (1984) Ecology and physiology of Crucian carp (Carassius carassius L.) in small Finnish ponds with anoxic conditions in winter. Verhandlungen Internationale Vereiningung fur theoretische und angewandte Limnology 22(4):2566–2570

Hrbek T, Larson A (1999) The evolution of diapause in the killifish family Rivulidae (Atherinomorpha, Cyprinodontiformes): a molecular phylogenetic and biogeographic perspective. Evolution 53:1200–1216

Huber JH (2000) Killi-Data 2000. Updated checklist of taxonomic names, collecting localities and bibliographic references of oviparous Cyprinodont fishes (Cyprinodontiformes). Société Française d'Ichtyologie, Paris

Hyvärinen H, Holopainen H, Piironen J (1985) Anaerobic wintering of crucian carp (Carassius carassius L.)-I. Annual dynamics of glycogen reserves in nature. Comp Biochem Physiol A Physiol 82(4):797–803

Kawamura K, Uehara K (2005) Effects of temperature on free-embryonic diapause in the autumn-spawning bitterling Acheilognathus rhombeus (Teleostei: Cyprinidae). J Fish Biol 67:684–695
Kawamura K, Ueda T, Arai R, Smith C (2014) Phylogenetic relationships of bitterling fishes (Teleostei: Cypriniformes: Acheilognathinae), inferred from mitochondrial cytochrome b sequences. Zool Sci 31(5):321–329
Kazansky BN (1956) Ovogenez and adaptatsii, sviazannye s razmnozheniem u ryb. Doctoral dissertation thesis, Leningrad State University, Leningrad
Kitamura J (2005) Factors affecting seasonal mortality of rosy bitterling (Rhodeus ocellatus kurumeus) embryos on the gills of their host mussel. Popul Ecol 47(1):41–51. https://doi.org/10.1007/s10144-004-0201-0
Lajus DL, Alekseev VR (2004) Phenotypic variation and developmental instability of life-history traits: a theory and a case study on within-population variation of resting eggs formation in Daphnia. J Limnol 63(Suppl. 1):37–44
Lajus DL, Graham JH, Kozhara AV (2003) Developmental instability and the stochastic component of total phenotypic variance. In: Polak M (ed) Developmental instability: causes and consequences. Oxford University Press, Oxford, pp 343–363
Lefevre SL, Damsgaard C, Pascale DR, Nilsson GE, Stecyk JAW (2014) Air breathing in the Arctic: influence of temperature, hypoxia, activity and restricted air access on respiratory physiology of the Alaska blackfish Dallia pectoralis. J Exp Biol 217:4387–4398. https://doi.org/10.1242/jeb.105023
Lessa RPT (1982) Biologie et dynamique des populations de Rhinobatos horkelii du plateau continental du Rio Grande do Sul (Bre´sil). PhD thesis, Universite de Bretagne Occidentale, Brest
Levels PJ, Denucé JM (1988) Intrinsic variability in the frequency of embryonic diapauses of the annual fish Nothobranchius korthausae, regulated by light: dark cycle and temperature. Environ Biol Fish 22(3):211–224
Loong AM, Chng YR, Chew SF, Wong WP, Ip YK (2012) Molecular characterization and mRNA expression of carbamoyl phosphate synthetase III in the liver of the African lungfish, Protopterus annectens, during aestivation or exposure to ammonia. J Comp Physiol B 182:367–379. https://doi.org/10.1007/s00360-011-0626-7. PMID: 22038021
Matias JR (1984) The stage-dependent resistance of the chorion to external chemical damage and its relationship to embryonic diapause in the annual fish, Nothobranchius guentheri. Experientia 40:753–754
Matias JR, Markofsky J (1978) The survival of embryos of the annual fish Nothobranchius guentheri exposed to temperature extremes and the subsequent effects on embryonic Diapause. J Exp Zool 204(2):219–227
Mesquita-Saad LSB, Leitão MAB, Paula-Silva MN, Chippari-Gomes AR, Almeida-Val VMF (2002) Specialized metabolism and biochemical suppression during aestivation of the extant South American lungfish – Lepidosiren paradoxa. Braz J Biol 62(3):495–501
Moraes G, Altran AE, Avilez IM, Barbosa CC, Bidinotto PM (2005) Metabolic adjustments during semi-aestivation of the marble swamp eel (Synbranchus marmoratus, Bloch 1795) – a facultative air breathing fish. Brazilian J Biol 65(2):305–312
Morgan DL, Gill HS, Potter IC (2000) Age composition, growth and reproductive biology of the salamanderfish Lepidogalaxias salamandroides: A re-examination. Environ Biol Fish 57(2):191–204
Morris JA (1999) Aspects of the reproductive biology of the Bluntnose stingray, Dasyatis say, in the Indian River lagoon system. University of Central Florida, Orlando
Murphy WJ, Collier GE (1999) Phylogenetic relationships of African killifishes in the genera Aphyosemion and Fundulopanchax inferred from mitochondrial DNA sequences. Mol Phylogenet Evol 11:351–360
Nagahama Y, Yamashita M (2008) Regulation of oocyte maturation in fish. Dev Growth Differ 50(Suppl):S195–S219
Okazaki M, Naruse K, Shima A, Arai R (2001) Phylogenetic relationships of bitterlings based on mitochondrial 12S ribosomal DNA sequences. J Fish Biol 58:89–106. https://doi.org/10.1006/jfbi.2000.1430

Ong JL, Chng YR, Ching B, Chen XL, Hiong KC, Wong WP et al (2017) Molecular characterization of myostatin from the skeletal muscle of the African lungfish, Protopterus annectens, and changes in its mRNA and protein expression levels during three phases of aestivation. J Comp Physiol B 187(4):575–589. https://doi.org/10.1007/s00360-017-1057-x. PMID: 28184997

Page MM, Salway KD, Ip YK, Chew SF, Warren SA, Ballantyne JS, Stuart JA (2010) Upregulation of intracellular antioxidant enzymes in brain and heart during estivation in the African lungfish Protopterus dolloi. J Comp Physiol B Biochem Syst Environ Physiol 180(3):361–369

Parsons PA (1990) Fluctuating asymmetry: an epigenetic measure of stress. Biol Rev 65:131–145

Peters N (1963) Embryonic adaptations of oviparous cyprinodonts from waters that are periodically desiccated. Int Rev Gesamten Hydrobiol 48:257–313

Petlina AP (1987) Opredelenie plodovitosti I stadia zrelosti ryb: uchebnoe posobie. Izdatelstvo Tomskogo universiteta, Tomsk

Piironen J, Holopainen IJ (1986) A note on seasonality in anoxia tolerance of crucian carp (*Carassius carassius* L.) in the laboratory. Ann Zool Fenn 23:335–338

Podrabsky JE, Lopez JP, Fan TWM, Higashi R, Somero GN (2007) Extreme anoxia tolerance in embryos of the annual killifish Austrofundulus limnaeus: Insights from a metabolomics analysis. J Exp Biol 210(13):2253–2266

Podrabsky JE, Garrett IDF, Kohl ZF (2010a) Alternative developmental pathways associated with diapause regulated by temperature and maternal influences in embryos of the annual killifish Austrofundulus limnaeus. J Exp Biol 213:3280–3288

Podrabsky JE, Tingaud-Sequeira A, Cerda J (2010b) Metabolic dormancy and responses to environmental desiccation in fish embryos. In: Lubzens E, Cerda J, Clark M (eds) Dormancy and resistance in harsh environments. Topics in current genetics. Springer, Berlin, pp 203–226

Podushka SB (1976) Eksperimentalnyi analiz sezonnosta razmnozhenia plotvy I trekhigloi koliushki v sviazi s ekologo-fiziologicheskoi periodizatsiei oogeneza kostistykh ryb. PhD thesis, Leningrad State university, Leningrad

Polačik M, Janáč M (2017) Costly defense in a fluctuating environment—sensitivity of annual Nothobranchius fishes to predator kairomones. Ecol Evol 7(12):4289–4298

Pri-Tal BM, Blue S, Pau FK-Y, Podrabsky JE (2011) Hormonal components of altered developmental pathways in the annual killifish, Austrofundulus limnaeus. Gen Comp Endocrinol 174(2):166–174

Pusey BJ (1990) Seasonality, aestivation and the life history of the salamanderfish Lepidogalaxias salamandroides (Pisces: Lepidogalaxiidae). Environ Biol Fish 29(1):15–26

Raff RA (1996) The shape of life: genes, development, and the evolution of animal form. The University of Chicago Press, Chicago, p 520

Rahman MM, Narejo NT, Ahmed GU, Bashar MR, Rahmatullah SM (2005) Effect of temperature on food, growth and survival rate of freshwater mud eel, Monopterus cuchia (Hamilton) during aestivation period. Pak J Zool 37(3):181–185

Romo-Curiel AE, Sosa-Nishizaki O, Pérez-Jiménez JC, Rodríguez-Medrano MC (2017) Reproductive cycle and maternal–embryonic nutritional relationship of shovelnose guitarfish Pseudobatos productus in the Gulf of California. J Fish Biol 90(3):889–905

Sakun OF, Butskaya NA (1963) Opredelenie stadii zrelosti i izuchenie polovykh tsiklov ryb [Determination of gonad maturity and study of sexual cycles of fish]. Nauka, Moscow

Schaefer FJ, Hermelink B, Husmann P, Meeus W, Adriaen J, Wuertz S (2016) Induction of gonadal maturation at different temperatures in burbot Lota lota. J Fish Biol 89(5):2268–2281

Scheel JJ (1990) Atlas of killifishes of the old world. T.F.H. Publications, Inc., Neptune City

Shoubridge EA, Hochachka PW (1980) Ethanol end product of vertebrate anaerobic metabolism. Science 209:307–308

Simons AM, Johnston MO (1997) Developmental instability as a bet-hedging strategy. Oikos 80:401–406

Simpfendorfer CA (1992) Reproductive strategy of the Australian sharpnose shark, Rhizoprionodon taylori (Elasmobranchii: Carcharhinidae), from Cleveland Bay, Northern Queensland. Aust J Mar Freshwat Res 43:67–75. https://doi.org/10.1071/MF9920067

Skadovsky SN, Parfenova OI (1937) Effect of gravidan on ovary maturation and spawning of perch. Uchenye zapiski Moskovskogo universitita 9:139

Smith HW (1931) Observations on the African lung-Wsh, Protopterus æthiopicus, and on evolution from water to land environments. Ecology 12:164–181

Smith C, Reichard M, Jurajda P, Przybylski M (2004) The reproductive ecology of the European bitterling (Rhodeus sericeus). J Zool 262:107–124

Suzuki N, Jeon SR (1991) Development of bitterling, Acheilognathus rhombeus (Cyprinidae), from Korea. J Basic Sci 5:53–62

Taylor DS (2012) Twenty-four years in the mud: What have we learned about the natural history and ecology of the mangrove rivulus, Kryptolebias marmoratus? Integr Comp Biol 52(6):724–736

Tikkanen E, Haverinen J, Egginton S, Hassinen M, Vornanen M (2017) Effects of prolonged anoxia on electrical activity of the heart in Crucian carp (Carassius carassius). J Exp Biol 220(3):445–454. https://doi.org/10.1242/jeb.145177

Tyler CR, Sumpter JP (1966) Oocytes growth and development in teleosts. Rev Fish Biol Fisher 6:287–318

Waddington CH (1957) The strategy of the genes. George Allen and Unwin, London

Waltrick D, Awruch C, Simpfendorfer C (2012) Embryonic diapause in the elasmobranchs. Rev Fish Biol Fish 22:849–859. https://doi.org/10.1007/s11160-012-9267-5

Wildekamp RH (2004) A world of killies. Atlas of the Oviparous Cyprinodontiform fishes of the world. American Killifish Association, Mishawaka

WongTG ADD (2005) Optimal reproductive allocation in annuals and an informational constraint on plasticity. New Phytol 166:159–172

Wourms JP (1972a) Developmental biology of annual fishes. I. Stages in the normal development of Austrofundulus myersi Dahl. J Exp Zool 182:143–167

Wourms JP (1972b) The developmental biology of annual fishes. II. Naturally occurring dispersion and reaggregation of blastomeres during the development of annual fish eggs. J Exp Zool 182:169–200

Wourms JP (1972c) The developmental biology of annual fishes. III. Pre-embryonic and embryonic diapause of variable duration in the eggs of annual fishes. J Exp Zool 182:389–414

Wyffels JT (2009) Embryonic development of chondrichthyan fishes—a review. In: Kunz YW, Luer CA, Kapoor BG (eds) Development of non-teleost fishes. Science Publishers, Enfield, p 301

Zakharov VM (1992) Population phenogenetics: analysis of developmental stability in natural populations. Acta Zool Fenn 191:7–30

Zarski D, Targońska K, Kaszubowski R, Kestemont P, Fontaine P, Krejszeff S, Kupren K, Kucharczyk D (2013) Effect of different commercial spawning agents and thermal regime on the effectiveness of pikeperch, Sander lucioperca (L.), reproduction under controlled conditions. Aquac Int 21(4):819–828

Chapter 5
Cysts and Resting Eggs from Marine Zooplankton: Dimension of the Phenomenon, Physiology of Rest, and Ecological and Biogeographic Implications

Genuario Belmonte and Fernando Rubino

Abstract Encystment is a common strategy adopted by coastal marine plankton species to avoid adverse conditions. Spiny cysts have been identified in marine sediments of more than 600 millions years ago, thus suggesting that this has been an original trait of metazoan life cycles. Protista, Chromista, and Metazoa share this trait which is not the result of a convergent evolution, but a plesiomorphy, probably dictated by low oxygen concentrations. During the evolution of life on the Earth, the morphology has been modified, and today some taxa show typical and recognizable patterns. The affirmation of parental cares (brood protection) is suspected to have provoked the disappearing of spiny covering of eggs and/or cysts. Cysts produced in the pelagos sink to the bottom sediments where their dormancy may extend for decades. In polar seas, cysts are suspected to be entrapped within sea ice and released with ice melting to refuel a new plankton community in the complex frame of the so-called resurrection ecology. Confined marine coastal areas accumulate cysts in sediments due to many drivers as (i) reduced water movement/hydrodynamics, (ii) concentration of cyst-producing species with high population densities, and (iii) absence and/or scarcity of possible cyst consumers in the benthos. The pelagic-benthic nexus, which affects both the sediments and the water column (and possibly sea ice), is still poorly understood. The presence of cysts in the life cycle is likely to have considerable consequences for the ecology of coastal plankton as well as the evolution and biogeography of species.

G. Belmonte (✉)
Department of Biological and Environmental Sciences and Technologies,
University of the Salento, Lecce, Italy
e-mail: genuario.belmonte@unisalento.it

F. Rubino
Water Research Institute, Unit Talassografico "A. Cerruti", National Research Council,
Taranto, Italy
e-mail: rubino@irsa.cnr.it

V. R. Alekseev, B. Pinel-Alloul (eds.), *Dormancy in Aquatic Organisms. Theory, Human Use and Modeling*, Monographiae Biologicae 92,
https://doi.org/10.1007/978-3-030-21213-1_5

Keywords Life cycles · Dormancy · Biogeography · Potential biodiversity · Resurrection ecology · Confined marine environments · Cyst banks · Zooplankton

5.1 Introduction

The number of species in plankton communities is so surprisingly high which led Hutchinson (1961) to write a scientific article where he urged biologists to think about the reasons for the seemingly illogical coexistence of so many species in the apparently isotropic water habitat.

Certain species are perennial and are always present albeit with variable abundances. Other species, referred to as "seasonal," show a presence interrupted by absences that are sometimes longer than the periods when they are present.

The existence of such a strategy for many planktonic organisms is already well known in both freshwater and marine organisms but has been little studied. In detail, if the disappearance of freshwater organisms could be linked to excessively worst condition (seasonal) up to the complete disappearance from the water habitat, this has never been considered as a possibility for the sea. Marine plankton, indeed, is composed by species with alternate presences. The recognition that species potentially have resting stages in their life cycles is an important contributor to understanding the periodical disappearance of species from the plankton.

The production of encysted resting stages is typical of unicellular organisms, while it is not the only way to rest for planktonic Metazoa which, apart from resting eggs, can enter a lethargic phase also as larvae, juveniles, or adults (see Williams-Howze (1997) and Baumgartner and Tarrant (2017), for marine copepods).

Marine dormant stages of hundreds of species of Protista and Metazoa, however, are unified in a few typical morphologies (mostly spherical and spiny) and characteristically different from the freshwater counterparts being generally free and not sheltered by the producer organism (the mother, in the case of Metazoa). As a consequence, they are equally affected by the dynamic processes of sea bottom sediments, making it possible to adopt methodologies of investigations coming from techniques adopted for the study of meiobenthos.

The ecological role of cysts is still underestimated in marine biology textbooks, up to the point that terms like "cyst," "resting," or "diapause" are absent from the list of arguments at the end of each volume (see, e.g., Valiela (1995), Barnes and Hughes (1999), Levington (2001), and Kaiser et al. (2005)). Nevertheless, resting stages are considered responsible for many of the intermittent occurrences of species and for their abundance/rarity cycles in the marine coastal environment (Boero 1994; Giangrande et al. 1994; Boero et al. 1996).

The germination of active stages from bottom sediments was named "resurrection ecology" (Kerfoot and Weider 2004), and, in contrast to supply side ecology (Gaines and Roughgarden 1985; Lewin 1986), where propagules for new populations arrive from adjacent areas, it pivots on species which subtract themselves from plankton functioning, remaining (although inactive) as cysts in the same area of active forms.

During recent years, another element was added to the plankton ecology framework, i.e. the finding of resting stages that are contained within ice formed in periodically frozen seas (Horner et al. 1992; Brierley and Thomas 2002). When the ice melts, the resting stages return in the water and germinate to seed new plankton populations (Rozanska et al. 2008). The rest condition, contrarily to the ecological origin of resting stages, is the common feature which probably allows to consider together phytoplankton and zooplankton in a shared study approach. This notwithstanding, the present chapter, will be focused only on the etherotrophic part of the plankton (the zooplankton) with the following aims:

(a) To summarize the state of art of studies on the presence of encysted stages in the life cycles of marine zooplankton, about 50 years from their recognition (Sazhina 1968)
(b) To point out, on the basis of the rich literature available, the ecological role, evolutionary importance, and biogeographic implications of the presence of resting stages in marine zooplankton organisms
(c) To model the dynamics of the bottom sediment cyst bank, considering the problems of recruitment (encystment), survival/mortality, and export (germination) of resting stages, with a focus on the sediment community (although dormant) and not on that of the water column

5.2 Resting Stages in Marine Zooplankton

Currently, the production of resting stages is known to be a common life cycle trait for hundreds of coastal plankton organisms belonging to various realms (see Belmonte and Rossi (1998) and Lennon and Jones (2011)). Among marine Protista, the most studied are Bacillariophyceae and Dinophyceae, typically considered as phytoplankton. As a consequence, these groups comprise the majority of cyst-producing taxa known so far, and many zooplankton cysts could be compared with them. Dozens of Ciliophora (Reid and John 1983) belong to the category of cyst-producing organisms. Marine Metazoa that produce cysts include Rotifera (Gilbert 2004), Cladocera (Onbé 1974), and Copepoda Calanoida (Mauchline 1998). An analysis of published literature on the topic allowed us to report a total of 500 marine species among the so-called holoplankters, with 88 Ciliophora and 73 Metazoa (Belmonte and Rubino 2019).

5.3 Morphological Aspects

Encystment is considered as a regular trait of the life cycle in Ciliophora, but of 141 described species of aloricate Oligotrichaea of marine and brackish waters, the cyst-active stage correspondence has been fully described in only three species until now: *Strombidium oculatum*, *Helicostomella subulata*, and *Cyrtostrombidium boreale*

(Kim et al. 2002). Tintinnina cysts are typically contained within the lorica of the preceding active stage. They are spherical or elliptic in shape. Hypotrichida cysts are spherical, while the cysts of Heterotrichida and Oligotrichida show an apical bubble-like cap from which the active stages emerge at the end of their dormancy.

These kinds of cysts were named *papulifera* by Meunier (1910), due to the presence of this "bubble" (*papula*, in Latin) on their apex. Meunier grouped these cysts according to their shape under the names Fusopsis (spindle-like), Piropsis (pear-like), and Sphaeropsis (spherical or ovoid) (Reid and John 1978). The papula is considered an archaic feature because it is very common among cyst-producing ciliates. Although *papulifera* were described from marine plankton samples as a well-differentiated group (50 different morphotypes in Meunier 1910) and their identification as ciliate cysts is given for sure, the correspondence cyst-active stage has been ascertained only for three species, and all the others still retain the name attributed to them by Meunier on the basis of their morphology. This probably also remarks the difficulty to obtain active stages by cysts or vice versa with marine ciliate species. Also recently found cysts, typically with a papula, are attributed to Ciliophora, although without any successful germination (Rubino et al. 2016; Rubino and Belmonte 2019).

Among Rotifera, having an ecologically important role only in coastal to brackish waters, some species (e.g., those of the genera *Brachionus*, *Synchaeta*, and *Trichocerca*) produce cysts (resting eggs) (see, for a review, Ricci (2001)) generally associated with the sexual phase of their life cycle, often after many parthenogenetic generations. According to Gilbert (2007), the parthenogenetic multiplication is evolutionarily affirmed with the aim to favor the maximum production of resting eggs. Commonly, the resting eggs of marine rotifers are ovoid (80–150 μm in length) and easily recognizable in the cyst assemblages of bottom sediments (Marcus 1990; Viitasalo and Katajisto 1994; Belmonte et al. 1995).

Unlike subitaneous eggs of Rotifera, the surface of resting eggs is often sculptured and/or rough. The surface features never assume the shape of long spines however. The cyst is spherical and smooth in species found in continental (fresh water) habitats, even those hyperhaline (Ricci 2001).

The internal embryo is often seen to be partially distanced from one side of the cyst wall, so forming an empty space that accounts for up to 25% of the total cyst volume.

In *Synchaeta* and *Brachionus*, the germination of active stages occurs through a narrow hole which opens like a door at one of the extremities of the cyst wall (Piscia et al. 2016).

Each female produces only one/two resting eggs, and consequently, the abundance of resting eggs in sediment cyst assemblages is numerically inferior to that of unicellular organisms and comparable with that of other Metazoa (see, e.g., Rubino et al. (2013)).

Among marine Cladocera, similarly to Rotifera, reproduction is parthenogenetic during most of the year, and each female produces a relatively high number of parthenogenetic eggs (generally incubated in the ephippium, a dorsal pouch). The appearance of males in the population follows the same pattern as Rotifera and seems to be directly induced by the numerical growth of specimens just before the arrival of the unfavorable season (Alekseev 2007). The fertilized eggs are resting (diapausal) and appear in small numbers (one to two per female). The appearance of males in the population seems directly connected with the strategy of producing fertilized/diapausal eggs, and Miyashita et al. (2011) demonstrated that local populations can completely lose their males for years.

Marine Cladocera (eight species in total) lay their eggs in the water column together with the exuvia of the molted females or the bodies of the dead ones.

The low number of diapause eggs produced by each female practically corresponds to the egg production rate of Rotifera. However, egg abundance in the sediment assemblages is dependent on the site and the period, with Cladocera being particularly abundant only for short periods or in large bays. This leads to the overall low numerical importance of Cladocera eggs in the sediments of confined environments, although their biomass may be important due to the relatively larger size (average diameter = 150–300 μm).

The diapause eggs of Cladocera are generally larger than all the others (Onbé 1985, 1991). They can be spherical, as in *Evadne* and *Podon*, or ovoid and typically concave at one side, as in *Penilia*. Indeed, such eggs are the largest to be found in sediment cyst assemblages (Marcus 1990; Viitasalo and Katajisto 1994; Belmonte et al. 1995; Madhupratap et al. 1996).

They can also exhibit calcification of the external layer, which results in higher sinking rates (Braiko 1966). At the same time, this layer enables the eggs to better resist to pressure and scraping.

Most of these cysts exhibit an evident stratification of the envelope, which is typically thick. The robust external covering is probably a defense against predators. Indeed, the *Mysis* shrimps of the northern Baltic Sea feed selectively on the eggs of *Bosmina longispina maritima* (contained in an ephippium), ignoring the eggs of *Podon polyphemoides* (not ephippiated) which have a thicker external covering (Viitasalo and Viitasalo 2004; Viitasalo 2007).

Copepoda (mostly Calanoida) are the most diverse and abundant Metazoa taxon in the marine meso-zooplankton.

In order to overcome adverse periods, species of the Centropagoidea superfamily commonly produce resting eggs (Mauchline 1998).

Sazhina (1968) was the first to report the existence of a resting egg (morphologically distinguishable) in the life cycle of two marine Calanoida (*Pontella mediterranea* and *Centropages ponticus*) from the Black Sea.

Since then, the list of marine Calanoida species able to produce resting eggs has been repeatedly updated and today includes more than 54 species (Belmonte and Rubino 2019). However, the list contains only coastal species, and they are all phylogenetically cognate (belonging to the same superfamily, Centropagoidea). All freshwater Calanoida belong to this superfamily, and the presence of a resting egg in their life cycle has been proposed as being responsible for the adaptation of this marine taxon to freshwater habitats (Lindley 1992; Belmonte 1998a).

The diapause eggs of Calanoida are always spherical (diameter, 60–200 μm) and commonly show ornamentations on the surface.

Marine Centropagoidea lay eggs directly in the water column (they have no ovisacs) (Sazhina 1987), and eggs inevitably accumulate on the seabed. At all temperate latitudes, the coastal sediments of shallow areas contain more copepod eggs than pelagic and/or deep seabeds. In addition, egg abundance in the sediments is often higher than that of active stages in the water column (Uye 1985; Belmonte et al. 1995; Dahms et al. 2006).

Calanoida can be very abundant in coastal areas, but it should be noted that species are not parthenogenetic and only a small portion of the population (adult females) is involved in egg production. Nevertheless, production can be high and continuous over many days. Indeed, unlike Rotifera and Cladocera, the number of Calanoida resting eggs in the sediment is higher, conditioned by the fact that fertilized females lay eggs throughout their life span (25–45 days), at a rate which can reach 100 eggs per day in *Centropages typicus* (Ianora 1998). More commonly, the egg production rate of species in coastal-confined habitats ranges from 0 to 78 eggs per female per day (Belmonte and Pati 2007; Drillet et al. 2008). This explains why the Calanoida eggs (about 10^4–10^6 m^{-2} of bottom area) are more abundant than those of Rotifera (about 10^4 m^{-2}) and Cladocera (about 10^3–10^5 m^{-2}) in sediment assemblages (Hairston 1996).

Among some marine Calanoida, diapause eggs are sometimes undistinguishable from subitaneous ones, as is commonly the case among freshwater species. Grice and Marcus (1981) suggested that, in such cases, resting eggs can only be identified by their typical delayed hatching, showed even if exposed to suitable conditions for active stages. However, many marine Centropagoidea produce diapause eggs that are morphologically distinct from the subitaneous ones (e.g., Sazhina 1968; Santella and Ianora 1990; Ianora and Santella 1991; Belmonte 1992, 1997; Belmonte and Puce 1994; Onoué et al. 2004). In these more common cases, diapause eggs show chorionic spines and processes. Many species also show chorionic processes in some subitaneous eggs of their clutches (Hansen et al. 2010), but generally the diapause eggs are never smooth. In one case, smooth diapause eggs of *Paracartia latisetosa* have been reported among the more common spiny ones by Posi and Belmonte (2011), but the authors demonstrated that this was due to a tendency of some diapause eggs, laid as spiny, to loss their external spiny covering, thus becoming successively smooth. The morphology of the processes on the chorion of subitaneous eggs can vary from small tubercles to long, thick, and branching spines, but the general rule is that there are more abundant and robust spines on the chorion of diapause eggs than on subitaneous ones (see Belmonte (1997)).

Subitaneous and diapause eggs do not differ in terms of morphology alone. Wang et al. (2005) noted that the dry weight of subitaneous eggs in *Centropages tenuiremis* was substantially lower than that of diapause eggs, as was the lipid, protein, and carbohydrate content. The levels of aspartic acid, glutamic acid, glycine, methionine, isoleucine, leucine, norleucine, lysine, ammonium chloride, and arginine were significantly higher in diapause eggs than in subitaneous ones.

Belmonte et al. (1997) noted that spines are typical of marine species (being absent in freshwater ones) and suggested that this trait could be an adaptation to marine coastal habitats. Most marine Centropagoidea have no ovisacs and lay their eggs directly in the water column. Spines on the chorion probably delay sinking in the deep, often hypoxic, water layers commonly found in confined coastal environments. This delay is indispensable to early embryonic development, which needs oxygen, before entering the insensible lethargic state (Lutz et al. 1992). In addition, the spines of resting eggs can protect the egg from scraping by mineral particles in the sediment, enlarge the communication surface between the inside and outside of the egg, and probably avoid predation by chewer species (Belmonte et al. 1997).

As already reported for Ciliophora (Foissner et al. 2007) and some Dinophyceae (Kokinos and Anderson 1995; Mertens et al. 2011), the number and length of spines are not constant even within the same species. Among Copepoda, there is a continuum of measurements between the "short" and the "long" extremes, which generally corresponds to different development destinies (see Belmonte (1992) and Hansen et al. (2010)).

Due to the spines, Calanoida eggs in recent marine sediments were initially interpreted by micropalaeontologists as fossil cysts of Dinophyceae, and for this reason they were reported as "hystricosphaeres." The birth of nauplii from these cysts (McMinn et al. 1992), however, has definitively resolved the puzzle.

On hatching, Calanoida egg chorion fracturates along an equatorial line, producing two hemispherical parts. The fracture of the outer envelope happens as a consequence of enhanced internal hydrostatic pressure and is not due to mechanical action by the organism within, because the newly born nauplius is still enveloped in a thin y membrane when exiting from the egg envelope.

Due to intraspecific morphological variability, it is still difficult to identify what species resting eggs belong to. In fact, some spiny patterns are not so typical as proposed (e.g., Belmonte 1998a, b; Belmonte et al. 1997,), but rather variable in each species (or even in each clutch) (Mura 1986; Foissner et al. 2007), appearing as adaptive and even inducible by environment (Dumont et al. 2002).

The covering of Calanoida resting eggs is, in addition, thick, but never calcified, and spines appear rapidly, just after the egg laying (Santella and Ianora 1992).

5.4 Physiology of the Rest

The rest can be distinguished in two different periods: insensitive and competent (Mansingh 1971; Grice and Marcus 1981). The former has a species-specific duration, varying from hours to months. During this time span, the organism does not perceive external stimuli and rests even if external conditions are appropriate for active life. During the second period, the resting organism perceives external stimuli, and the egg can hatch if conditions are suitable for the active life (Rathaille and Raine 2011). In these terms, each species can undergo a diapause period only in a precise moment of its life cycle (cysts and spores among Protista and embryonic early stages among the most common planktonic Metazoa) (Alekseev 1990).

Many cues, depending on the season, can induce dormancy, but the only one common to all cyst-producing groups seems to be day length (i.e., photoperiod) (see

Alekseev (Alekseev 1990, 2007) for Crustacea and Kamiyama et al. (1995) for Ciliophora).

For dinoflagellates Kremp et al. (2009) proposed the existence of signals (the proximate causes) as triggers of the advent of the diapause. This topic is directly related to the concept of an endogenous clock, i.e., a genetically controlled mechanism capable of activating diapause (Boero 1994; Anderson 1998).

Among Rotifera Monogononta, overcrowding seems to be an important signal (Gilbert 1974, 1992, 2007). This rule is also valid for the copepod *Eurytemora affinis*, whose females produce resting eggs in overcrowded conditions (Ban 1994).

Among Crustacea, photoperiod seems to have a similarly important role (Alekseev 1990).

In many species of the copepod family Acartiidae, each single female can produce subitaneous and diapause eggs contemporaneously (Onoué et al. 2004, Posi and Belmonte 2011), and females of the same population show oscillations in the relative abundances of the two types of eggs over a relatively long period (e.g., some months).

These observations suggest that the genetic character of the phenomenon is subjected to high variability, in order to allow the populations to better adjust physiologically to unpredictable environmental variations.

The hostile period can be recognized in different moments of the year (Marcus 1979; Uye 1985). Some species are seasonal only at certain latitudes and perennial at others.

Given the presence of a period of obligatory rest (the refractory or insensitive phase), resting eggs are able to "perceive" the moment of their reactivation only at the end of this phase, generally stimulated by environmental (external) factors (Kasahara et al. 1974; Uye 1985; Marcus 1987; Alekseev 1990).

Concerning the duration of the resting state, two types can be distinguished (see Dahms 1995 for Copepoda or Gyllström and Hansson 2004 for freshwater zooplankton): a long-term diapause, dependent from astronomical date and not from climate, controlled by an endogenous clock, and a short-term rest (not properly diapause), acting as a rapid-response strategy to deal with sudden changes in the environment, allowing the species to quickly shift between pelagic and benthic habitats. The former could be used by species living in deep waters where cysts must face more constant temperatures and light conditions (Anderson 1998), while the latter may be more advantageous for neritic strains that need to be able to perceive stimuli in a more competitive way.

Given the passivity of diapause eggs, the duration of rest is likely to be determined by a pre-existing organism (the grandmother among Rotifera and Cladocera, the mother among Copepoda, the active stage among Protista). Among Metazoa, the maternal role in reactivation automatically derives from the programming of the refractory phase duration. In such cases, external stimuli play a merely interfering role (e.g., as reactivation inhibitors), affecting but not determining an already programmed reactivation time (Marcus 1987; Ban 1992; Belmonte and Pati 2007). Marcus (1987) found different diapause durations in eggs produced by cospecific females in different seasons. Ban (1992) clearly spoke of a "mother effect" in

Eurytemora affinis, where ovogenesis of adult females was affected by temperature and photoperiod when they were still at the nauplius stage. In *Paracartia latisetosa*, Belmonte and Pati (2007) demonstrated that the mother set the clock for the duration of diapause in each egg. Diapause eggs obtained in the laboratory and reared under constant conditions until hatching (e.g., without temperature or photoperiod variations) hatched on average in March, independently from the fact that they were produced in September or in December. This suggested that diapause interruption and/or synchronous germination were under environmental conditions perceived by the mother during the ovogenesis.

Regardless of the season (mainly photoperiod and temperature) or overcrowding that determines the start and duration of diapause, resting eggs hatch at the end of their refractory phase in the presence of adequate oxygen concentration. Calanoida diapause eggs do not hatch in hypoxic environments even if the refractory period has terminated. Furthermore, oxygen seems to be absolutely essential for the complete development of embryos (Lutz et al. 1992; Engel 2005).

Cysts and resting eggs may accumulate on the bottom of water basins where the photoperiod could not be perceived and where temperature variations may be insignificant. Such seabeds, especially in deep and confined coastal sites, may be periodically hypoxic or anoxic, due to the absence of water mixing and photosynthesis (oxygen producing) at least in some periods of the year. This, as a consequence, can inhibit the resumption of the oxidative metabolism of embryos that are awaking from diapause but have not hatched yet. In this framework, the spiny surface of resting eggs has been considered as useful for the entrance into diapause. Indeed, by delaying egg sinking, the "parachute effect" determined by the presence of spines could allow each egg to reach the bottom only after the embryo's initial development, which needs high oxygen levels. This hypothesis is suggested by a study on Calanoida, where Lutz et al. (1992) demonstrated that low oxygen concentrations (typical of bottom mud in confined areas) during initial embryonic development impede the correct entrance into diapause, causing abortion of eggs.

In addition to low oxygen concentration, egg hatching can be inhibited by temperature (Komazawa and Endo 2002). However, synchronous hatching can be triggered by external stimuli, as in the case of Cladocera eggs in the Guanabara Bay (Mugrabe et al. 2007), but it can be delayed or staggered (non-synchronous) also for endogenous reasons (Marcus et al. 1994). This might even be the result of a deliberate strategy by the species, which engineers non-synchronous hatching of the cysts so as to distribute the risk of unsuccessful hatches over time.

During the rest period, Rotifera profit of the relative inactivity of the metabolism, to repair damaged portions of DNA. This gives hatchlings with a greater fitness than that of the generation they derive from (Ricci and Fontaneto 2009).

Among Crustacea, the metabolism of cysts in the refractory phase is at undetectable levels. This can be obtained by deep dehydration of the embryo and/or substitution of water molecules with trealose. In *Artemia* endocellular chaperon proteins vitrify the entire molecular apparatus with reversible alterations (Clegg 1997). Tunnacliffe et al. (2005) found that dehydration/vitrifying of freshwater Rotifera is obtained by producing small hydrophilic proteins. This type of metabolic vitrifying, however, has not been ascertained in the diapause of marine species, which never leave the water. In any case, the molecules involved are simple, they are not exclusive of particular taxa, and chaperon proteins are a widespread and available molecular tool in living organisms.

Generally, the cyst wall is thicker and more resistant than the covering of the corresponding active stage. This is considered sufficient to protect the diapause subject, although Stabili et al. (1999) found lysozime-like and trypsin-like activity on the cyst wall of *Artemia franciscana*, highlighting the existence of a passive immunity, as well as mechanical defense, in that particular resting stage. In addition to the physical barrier and the possible immunological defense, the environment contributes to the preservation of resting stages. A chemically reducing and anoxic environment, such as the muddy bottoms of confined coastal areas, is hostile to the majority of living organisms, thus preventing other species from damaging or preying on cyst assemblages (Persson 2000). On the other hand, the chemical aggression of such an environment does not in itself damage cysts which are particularly insensitive to external toxicity. The most common disinfectants (such as those typically used in aquaculture) have only a partial effect on cyst germination in certain species of Dinophyceae, Ciliophora, Rotifera, and Crustacea (Pati and Belmonte 2003; Belmonte and Pati 2007).

This particular and generalized resistance allows cysts to pass throughout the gut of Metazoa with no consequences for their viability. Cysts and resting eggs have been hatched in the laboratory after ingestion by birds (Figuerola and Green 2002), polychaetes (Marcus 1984; Kremp et al. 2003), fishes (Redden and Daborn 1991), and small crustaceans (Montresor et al. 2003; Kuwata and Tsuda 2005).

This feature is particularly important for the persistence of species, although inactive, in the environment, and their eventual re-proposition in the pelagic domain after hostile periods.

5.5 Evolutionary and Adaptive Aspects

Resting eggs and cysts are a winning strategy for short living organisms which have to face with variable recurrent harsh conditions. Being such organisms impeded to subtract themselves to bad season by migrating elsewhere, the transversal solution adopted is the subtraction by active presence and the entering of a resting period and waiting for the return of suitable conditions.

Marine spiny cysts are known also in the fossil record, from very ancient times, and represent a plesiomorphy for different realms of plankton living beings: Protozoa, Chromista, and Metazoa. The spiny surface is a multitask device against sinking, scraping, and predation and contemporaneously in favor of dispersal. What is evident in the comparison between species even in the same family or genus, is the lacking of spines with the adoption of egg sacs and other types of containers. Cysts and resting eggs which are someway protected by the mother organism are not spiny (as the rule for Rotifera, Anostraca, Cladocera, Cyclopoida, and Harpacticoida, whose females bring eggs with themselves in few numbers or in egg sacs). The same can be observed for unicellular organisms which encyst inside a pre-existing lorica (e.g., the Ciliophora Tintinnina). The spiny surface, consequently, appears as a defense strategy in the absence of parental cares (Belmonte 2018; Belmonte and Rubino 2019), or, to be precise, parental cares should be evolved successively in the invention of spiny surface and were responsible of the spine loosing. The morphological differences detectable among the surface features of modern resting eggs in species with no egg cares may represent the result of the evolutionary divergence of taxa.

This interpretation of spines evolution is supported by Cohen et al. (2009) who suggested that the organic walled spiny surface of early marine Metazoa (the large ornamented Ediacaran microfossils, LOEM, interpreted as resting stages) of about 600 million years ago is truly primitive and would have been indispensable for survival in the shallow hypoxic seas of the Precambrian. The fact that many phyla show resting stages with spiny surfaces (Belmonte et al. 1997) may thus be the result not of evolutionary convergence but of plesiomorphy.

The spiny surface is realized immediately after the laying in marine Copepoda Calanoida which have not ovisacs (Santella and Ianora 1992). As already stressed, the spiny envelope of such eggs gives them some advantages (defense and floating) and offers to the species the possibility to invest more energy in egg production because each female is not limited by the filling in of one egg sac in producing eggs.

The cyst banks in sediments contain cysts deriving from overlapping generations of free-swimming organisms. For this reason, at any one time, cyst banks host a wider genetic diversity than that realised in water column communities (Gyllström and Hansson 2004). Cyst banks actually produce a generation overlap which together with environmental fluctuations can maintain species coexistence enriching the interpretation of the so-called storage effect of Chesson and Warner (1981).

Rich cyst banks, together with the existence of an extra long diapause (ELD), increase the resilience of each population and enhance the biodiversity of an area. The survival of cysts in bottom sediments has been evaluated as decades in Copepoda (Marcus et al. 1994; Jiang et al. 2004; Dahms et al. 2006) and as more than 300 years in freshwater Calanoida (Hairston et al. 1995).

Càceres and Tessier (2003), studying freshwater species, reported that only a portion (from 6 to 50% in *Daphnia*) of cysts produced every year germinate at the first suitable opportunity. The remaining cysts prosecute dormancy, transferring the population into future scenarios. Even smaller percentages of germination (3.5%) were found by Glippa et al. (2014) in the egg bank of three copepod species in the Seine estuary (France). This variable timing of hatching pays off in habitats where occasionally conditions are so bad that recruitment from dormant pools totally fails (Cohen 1966; Seger and Brockmann 1987; Philippi and Sen 1989). Tarazona et al. (2017) empirically demonstrated the existence of the bet-hedging strategy in resting egg production and hatching of the rotifer *Brachionus plicatilis*. In this recent study, populations experiencing unpredictable environmental variations started earlier producing resting eggs and a smaller fraction hatched, in comparison to populations reared under predictable variations.

In this framework, diapause duration may simply represent the average value of a distribution (wide or narrow) of many different time values, with some of them being potentially very far apart.

ELD opens up genetic flows between different generations. For this reason, it has been described as a delaying factor in the species' evolutionary process (Hairston and De Stasio 1988). When cysts germinate, they introduce genetic variability comparable to immigration from an external source (the "migration from the past" of Templeton and Levin 1979). Bohonack et al. (2006) suggested that this could result in higher genetic uniformity across different populations in each region.

An ELD of phytoplankton cysts can be a powerful strategy for the permanent residence of a species' population in a region. Estimated as more than 100 years by Ribeiro et al. (2011), an ELD has been proposed as a possible way for the recovery of photosynthetic marine communities after the asteroid impact at Chicxulub, Yucatan, and the resulting period of prolonged darkness that occurred ~65.5 million years ago at the Cretaceous-Tertiary transition.

An ELD is the consequence of both environmental conditions and genetic programming. In basins with variable water volumes (mainly freshwater), Ellner et al. (1999) found that the timing of diapause and success of germination from the cyst bank affects the plankton population more strongly than any selective factor acting directly on the active stages.

In any case, the existence of cyst banks and possible successive injections of propagules of the same population is responsible for the adaptive success of ELD (Hairston 1996). The most adapted genotypes (the ones that produce the most cysts) are repeatedly proposed, even in the absence of suitable conditions, as a means of ensuring the survival of active stages, and this results in the coexistence of many, potentially competitive, species (Hairston and Fox 2010).

Any understanding of the mechanisms of adaptation to variable environments such as confined coastal waters, as well as the correct interpretation of plankton community dynamics, must take into account all the biological aspects of benthic-pelagic coupling. For this to be achieved, the traditional compartments separating scientists studying plankton from those studying benthos have to be overcome (Boero et al. 1996, 2004).

5.6 Ecological Implications: The Resurrection Ecology

Bottom sediments seem to be the obvious site for accumulation of cysts from the overlying water column, but in cold seas cysts could also be trapped in the ice. Both the benthic and the ice compartments are studied due to the existence of complex autochthonous active communities (meiobenthos within sediments, sympagic biota within ice). The non-active biotic pools, however, should instead represent an essential link between different activity seasons and are responsible for biomass and energy transfer between different ecological compartments.

Such abundant, yet disregarded, organism pool allowed Marcus and Boero (1998) to propose the term "supply-vertical ecology" inspired by the supply-side ecology of Lewin (1986) (see also Underwood and Fairweather (1989)). That term was successively substituted with the more impactful "resurrection ecology" proposed by Kerfoot and Weider (2004).

Large numbers of resting stages are captured and accumulated mainly in sediments of confined coastal areas (Belmonte et al. 1995). Cyst production and accumulation are particularly evident during the abundance peaks of plankton populations.

In contrast, open neritic habitats, where cyst-producing species are present, tend to have coarser sediments (i.e., the area is subjected to high hydrodynamism) and consequently have numerically poorer cyst assemblages. This links the cyst abundance to the size of sediment particles. According to Cohen et al. (2009), the abundance of cysts in confined environments could be also enhanced by the absence of potential predators, which are poorly adapted to hypoxic sediments.

Modig and Òlafsson (2001) demonstrated that the amphipod *Monoporeia affinis* can rework sulfide sediments and thus reoxygenate them down to a depth of 10 mm, thereby playing an indirect role (through oxygenation and bio-turbulence) in the germination of cysts. *M. affinis* has been considered by Albertsson and Leonardsson (2000) as possibly affecting the recruitment of active stages to the water column. More recently, Viitasalo and Viitasalo (2004) and Viitasalo (2007) demonstrated that Mysida feed selectively only on certain cyst types, thus possibly affecting plankton composition by differentiated subtraction of propagules.

In many cases, however, cysts pass through the gut of deposit feeders without consequences for their viability. Redden and Daborn (1991) even proposed that being preyed upon (with a consequent passage through the gut) could be a strategy for egg-carrying Calanoida to enhance their reproductive success at the population level because embryos contained in the egg sac of *Eurytemora herdmanni* are not digested by the predator fish *Menidia menidia*. In addition, the selective predation of egg-carrying females enhances the likelihood that non-mature females will survive and produce their own clutch of eggs.

Limivore Polychaeta rework sediment, with substantial implications for the overlying water column, similar to the impact of soil reworkers on seed banks in terrestrial forests (Meysman et al. 2006). For example, Giangrande et al. (2002) found that limicole Polychaeta rework sediment particles together with Dinophyceae cysts down to 7 cm below the sediment surface. This reworking leads to rapid burial of newly arrived cysts, but at the same time allows the uplift of deeply buried cysts to the surface.

This reworking of the sediment can also be produced by human activities. Although this field has not been extensively studied, some data (Giannakourou et al. 2005; Siokou-Frangou et al. 2005) suggest that trawling causes the resuspension (and consequently germination) of cysts in the Gulf of Thermaikos. This aspect of the interaction between fishery and cyst banks not only involves resuspension and possible germination of cysts but also their damage and the consequent reduction of their germination rate, as demonstrated by Drillet et al. (2014).

Whatever the destiny of cysts in the superficial sediment layer, the mechanisms of their subtraction (e.g. through germination) are responsible for the characteristic vertical profile of their abundance in the sediment column. Indeed, abundance peaks are typically found just below the most superficial layer (Hairston and Fox 2010; Sichlau et al. 2011). Some findings of abundance peaks at the sediment surface, on the contrary (Dahms et al. 2006), can easily be attributed to a proxy period/episode of massive production from the plankton. In any case, the top layer of sediments is that mostly interested by encystment/germination dynamics, showing strong fluctuations of cyst abundance according to the season (Rubino et al. 2016).

Resurrection ecology is at the basis of the functioning of confined environments (sensu Guelorget and Perthuisot (1992)). Although such environments are known to be the most productive on the planet (Odum 1971), for a long time they were believed to depend on marine influence ("vivification" sensu Sacchi (1985)) for the restoration of biological communities seasonally suppressed by environmental stress. The marine vivification hypothesis assumed the colonization of a highly productive, although stressed, area from a less productive one.

Today, resurrection ecology may explain the reappearance of species in accordance with the seasons; for many species this reappearance does not depend on neighboring environments (supply-side ecology) but mainly on the bottom sediments of a confined area where they rest during the adverse season. Resurrection ecology in confined areas is a typical example of ecological resilience. The plankton of confined environments appear to be autochthonous, and the presence of similar species in neighboring neritic areas is now believed to be the consequence of the export of specimens from the assemblages of a confined area, the exact opposite of what was thought some times ago.

In the last 20 years, scientists suggested that ice habitats also provide a biological reservoir for the plankton community in cold seas (Gradinger 1999; Brierley and Thomas 2002). The trapped plankton organisms, both as a planktonic stage and cysts, may remain dormant in the ice during the adverse season and repopulate the water column when the ice melts. Although scientists admit the data need confirmation, they consider the biodiversity found as sufficient to hypothesize that cysts provide an injection to initiate plankton blooms at the start of new ice-melting season. Gradinger (1999) found also that "In some cores (e.g., Station 204 of ARK 10/1) cysts of unknown origin, which were not attributed to either diatoms or flagellates, contributed a high proportion of the algal abundance."

The overwintering of plankton species within ice is an admitted possibility (see Horner (1985), for a review of sea ice communities), and Heikkila et al. (2016) propose the rich amount of cysts collected by their sediment traps as coming (at least partly) also from the melt-mediated release of sea ice-dweller cysts to the water column in Hudson Bay.

The research needs to clarify from where the rich assemblages of sympagic benthic species originate, when the sympagic assemblages appear in the ice formation, and where the organisms go when the ice melts.

Taxa potentially able to produce cysts are well represented in the sea ice biota, and together with taxa found in that habitat, some papers (Buck et al. 1992; Ikavalko and Thomsen 1997, Gradinger 1999, Ikavalko 2001) list unspecified "cysts." Other studies (Garrison and Buck 1989; Schnack-Schiel et al. 2001; Meiers et al. 2002; Ikavalko et al. 2004) have recognized planktonic taxa in the sea ice biota, and Friedrich and De Smet (2000) have reported the production of cysts but, before the melting period, by species commonly active in the ice during the winter. The extreme conditions present within the ice during the long-lasting polar winters (low temperature, low light or complete darkness, high salt concentration in liquid water segregated in ice crevices) suggest the suitability of a resting strategy for at least generalist species (Thomas and Dieckmann 2002; Werner et al. 2007).

The sea ice biota is a peculiar compartment where planktonic species, trapped during the autumn in the forming ice, may have evolved the possibility of producing cysts to survive the extreme winter conditions. Conversely, it could be hypothesized that specialized ice species survive in the sympagic biota and produce cysts after the ice melt to wait for the formation of a new ice habitat (Ratkova and Wassmann 2005). In any case, the presence of cysts could allow the immediate restoration of planktonic communities when favorable conditions return (Belmonte and Rubino 2019).

As with terrestrial forest seed banks, in order to supply a decisive contribution to our understanding of plankton functioning in coastal areas, the ecology of marine cyst banks must evaluate input (fecundity/fertility, import), storage (short and long term), and subtraction (death, predation, recruitment, export) for each of the species involved. Persson and Rosenberg (2003) demonstrated that the relative abundance of a cyst type is affected by the composition of the predator community which feeds on cysts in sediment.

Viitasalo (2007) proposed a complex network of interactions playing a role in the population dynamics of the cladoceran *Bosmina longispina maritima* in Baltic Sea plankton. The polychaete *Marenzelleria* sp. is believed to be responsible for the upward transfer of deeply buried ephippia in the sediments. In contrast, the amphipod *Monoporeia affinis* and the bivalve *Macoma balthica* may be responsible for their transfer downward in sediment. Upward transfer (carried out by the polychaete) exposes the ephippia to predation by the Mysida *Mysis mixta* which cannot prey on ephippia buried under one cm of sediment. The bivalve probably ingests part of the ephippia that it moves. Hence, the bivalve and the amphipod subtract recruits destined to the plankton and/or prey destined to the mysidan. This framework (even if based on only four actors) clearly demonstrates the complexity of interactions in the sediment cyst assemblage, which represents a real challenge for marine plankton ecologists.

5.7 Biogeographic Implications

A large percentage of microorganisms in natural systems are inactive (from 20 to 80% according Lennon and Jones (2011)), and dispersal of dormant stages may contribute to the apparent cosmopolitan distributions of taxa. Among freshwater Protista and small Metazoa at least, the capability to produce resting stages facilitates dispersal by air, dust, and moving animals (but see the critical review of Foissner (2006)).

The ability to avoid being affected by adverse conditions automatically allows cysts to pass barriers of unsuitable conditions that are also geographical. The possibility of seed transfers between separate water bodies by passive transport in the gut of fishes swallowed by migrating water birds was proposed by Darwin (in Reznick 2011, p. 334) to explain the wide distribution of freshwater organisms. More recently Figuerola and Green (2002) reported the possibility that cysts could also be exchanged between different environments in birds' gut. Indeed, the digestive enzymes of birds were found to be inefficacious either on the viability or on the hatching of cysts.

This framework has yet to be verified for marine environments, where birds cannot play the same role as in brackish waters or flooded plains. However, as already mentioned, benthic detritivores as well as fishes can be vectors of plankton species whose cysts are resistant to digestion and can survive in the gut of a fish for days, before being expelled elsewhere with feces.

Many studies demonstrate that cysts can be transported long distances far from their deposition site in the gut of vagile or swimming organisms. The amplitude of the resulting dispersal depends on the mobility of the swallowing animal, its behavior (migratory or sedentary), the time spent by cysts in its gut, and its own trophic destiny, i.e., whether it becomes the prey of other species. Consequently, cyst distribution in geographic space is the result of a complex framework.

The sculptures on the surface of the majority of marine cysts have been interpreted as being useful for many purposes (Belmonte et al. 1997). One of these is possibly to favor their clinging to floating or moving objects. This form of dispersal may account for the geographic distribution of species, including their appearance in previously unreported sites.

The majority of nonindigenous species (NIS) in the world today are attributed to dispersion via human vectors. In marine environments, such vectors include the ballast waters of ships navigating between different continents. This mechanism is believed to have been responsible for about 400 NISs present today along the marine coasts of the USA (Ruiz et al. 1997) and about 4,000 NISs (most of them planktonic) dispersed throughout the world (Carlton and Geller 1993; Ruiz et al. 2000).

As cysts, the majority of these traveling species survive the adversities in the ballast tanks, where they accumulate in the bottom sediments and rest also for long periods (Bailey et al. 2003; Wonham et al. 2005). Although a true ecological community of active stages has been described as functioning in such tanks during the period of navigation (Galil and Hulsemann 1997), cysts represent the best way to travel.

Considering that ballast waters are responsible even for transcontinental transfer, they represent for marine biogeography what birds represent for wetland biogeography.

When new species are recorded in ports, it is highly likely that they arrived with ships. The calanoid *Paracartia grani*, well known from the North Atlantic, was first reported in the Mediterranean in the port of Malaga in the late twentieth century (Rodrigues and Vives 1984). Subsequent report of the species (Lakkis and Zeidane 1990), in Lebanon ports, was associated with its capacity to produce resting eggs (Guerrero and Rodrigues 1998), supporting the hypothesis of dispersal via ships (Belmonte and Potenza 2001).

Navigation also seems to be the cause of dispersal for species living in apparently isolated habitats. The cladoceran *Podon polyphemoides* and the calanoid *Acartia clausi* have been reported in the Caspian Sea (an enclosed sea) since 1957 and 1984, respectively. Their arrival in that sea has been attributed to its navigable connection, via the Volga-Don canal, with the Azov-Black Sea system (Zaitsev and Ozturk 2001).

Navigation via the Volga-Don-Neva has also been definitively accepted as the most probable responsible for the transfer of fauna from the Ponto-Caspian system (Black, Azov, and Caspian Seas) to the Baltic Sea along the "trans-Sarmatian route." Although some species that have recently invaded the Baltic Sea from the Caspian (such as the cladoceran *Cercopagis pengoi*, first reported in the Baltic by Leppäkoski and Olenin (2000)) are known to be highly tolerant of the freshwaters of rivers, they appear to have crossed the Sarmatian plain and invaded the Baltic Sea only recently. The construction of canals and the beginning of their navigation provides the best explanation for the recent northward dispersal of many species, as passengers.

The Sarmatian route is not only a conduit for northward dispersal. In the case of the calanoid *Acartia tonsa*, southward dispersal has been proposed, apparently bucking the present climate change trend, which is expected to drive southern species northward. *A. tonsa* was first reported in the Baltic Sea by Smirnov (1935), just a few years after its first record in European Atlantic waters (Remy 1927 in Brylinsky Brylinski 1981). The species was reported in the Mediterranean basin more than 50 years later (Gaudy and Viñas 1985; Farabegoli et al. 1989), and in the Black Sea later still (Belmonte et al. 1994), allowing Belmonte and Potenza (2001) to hypothesize penetration through the Gibraltar Strait. Recently, however, it was ascertained that samples from the Black Sea studied by Belmonte et al. (1994) were actually collected in 1976 (Alexandrov, in Zaitsev and Ozturk 2001, p. 89–94); hence, the species was already in the Black Sea before its appearance in the Mediterranean Sea. Although its arrival in the Bay of Sevastopol (the Crimea, Black Sea) could be attributed to any incoming ship, the Sarmatian route, direct from the Baltic, becomes a possibility.

Hence, the river connection over the Sarmatian plain carries species in both directions, toward either the Baltic or the Black-Caspian basins. Although non-cyst-producing organisms (such as the ctenophoran *Mnemiopsis leydi*) also use this route, the resting eggs of *C. pengoi* and *A. tonsa* better enable them to survive the adverse conditions they face along this thousand-kilometer river journey.

References

Albertsson J, Leonardsson K (2000) Impact of a burrowing-feeder, *Monoporeia affinis*, on viable zooplankton resting eggs in the northern Baltic Sea. Mar Biol 136:611–619

Alekseev VR (1990) Diapausa rakoobrazhik. Ekologo physiologicheskye aspekty. Nauka, Moscow. 143 pp. (in Russian)

Alekseev VR (2007) Diapause in crustaceans: peculiarities of induction. In: Alekseev VR et al (eds) Diapause in aquatic invertebrates. Springer, Berlin, pp 29–63

Anderson DM (1998) Physiology & bloom dynamics of toxic Alexandrium species, with emphasis on life cycle transitions. In: Anderson DM et al (eds) Physiological ecology of harmful algal blooms, NATO ASI Series G41, pp 29–48

Bailey SA, Duggan IC, van Overdijk CDA, Jenkins PT, MacIsaac HJ (2003) Viability of invertebrate diapausing eggs collected from residual ballast sediment. Limnol Oceanogr 48:1701–1710

Ban S (1992) Effects of temperature, photoperiod, and population density on induction of diapause egg production in *Eurytemora affinis* (Copepoda: Calanoida) in Lake Ohnuma, Okkaido, Japan. J Crustac Biol 12:361–367

Ban S (1994) Induction of diapause egg production in *Eurytemora affinis* by their own metabolites. Hydrobiologia 292(293):185–189

Barnes RSK, Hughes RN (1999) An introduction to marine ecology, 3rd edn. Blackwell Publ, Oxford

Baumgartner MF, Tarrant AM (2017) The physiology and ecology of diapause in marine copepods. Annu Rev Mar Sci 9:387–411

Belmonte G (1992) Diapause egg production in Acartia (Paracartia) latisetosa (Crustacea, Copepoda, Calanoida). Bollettino di Zoologia 59:363–366

Belmonte G (1997) Resting eggs in the life cycle of Acartia italica and A. adriatica (Copepoda, Calanoida, Acartiidae). Crustaceana 70:114–117

Belmonte G (1998a) The egg morphology of 7 Acartiidae species. A preliminary survey for the ootaxonomy of calanoids. J Mar Syst 15:35–39

Belmonte G (1998b) Pteracartia a new genus of Acartiidae (Calanoida, Diaptomoidea) for Acartia josephinae Crisafi, 1974. J Mar Syst 15:359–368

Belmonte G (2018) Calanoida (Crustacea: Copepoda) of the Italian fauna; a review. European Zoological Journal 85:274–290

Belmonte G, Pati AC (2007) Hatching rate and diapause duration in eggs of Paracartia latisetosa (Copepoda: Calanoida). J Plankton Res 29:139–147

Belmonte G, Potenza D (2001) Biogeography of the family Acartiidae (Calanoida) in the Ponto-Mediterranean province. Hydrobiologia 453(454):171–176

Belmonte G, Puce M (1994) Morphological aspects of subitaneous and resting eggs from Acartia josephinae (Calanoida). Hydrobiologia 292(293):131–135

Belmonte G, Rossi V (1998) Resurrection and time travelling: diapause in crustaceans (and others). Trends in Ecology and Evolution 13:4–5

Belmonte G, Rubino F (2019) Resting cysts from coastal marine plankton. Oceanogr Mar Biol Annu Rev 57. (in press)

Belmonte G, Mazzocchi MG, Prusova IY, Shadrin NV (1994) *Acartia tonsa* a species new for the Black Sea fauna. Hydrobiologia 292(293):9–15

Belmonte G, Castello P, Piccinni MR, Quarta S, Rubino F, Geraci S, Boero F (1995) Resting stages in marine sediments off the Italian coast. In: Elefteriou A et al (eds) Biology and ecology of shallow coastal waters. Olsen & Olsen Publ, Fredensborg, pp 53–58

Belmonte G, Miglietta A, Rubino F, Boero F (1997) Morphological convergence of resting stages produced by planktonic organisms: a review. Hydrobiologia 335:159–165

Boero F (1994) Fluctuations and variations in coastal marine environments. Mar Biol 15:3–25

Boero F, Belmonte G, Fanelli G, Piraino S, Rubino F (1996) The continuity of living matter and the discontinuities of its constituents: do plankton and benthos really exist? Trends in Ecology and Evolution 11:177–180

Boero F, Belmonte G, Bussotti S, Fanelli G, Fraschetti S, Giangrande A, Gravili C, Guidetti P, Pati AC, Piraino S, Rubino F, Saracino OD, Schmich J, Terlizzi A, Geraci S (2004) From biodiversity and ecosystem functioning to the roots of ecological complexity. Ecol Complex 1:101–109

Bohonack AJ, Holland MD, Santer B, Zeller M, Kearns CM, Hairston NG Jr (2006) The population genetic consequences of diapause in Eudiaptomus copepods. Arch Hydrobiol 167:183–202

Braiko VD (1966) Biology of the winter eggs of *Penilia avirostris*. Doklady Biological Sciences (English traduction of Doklady Academic Nauk SSSR) 170:681–683

Brierley AS, Thomas DN (2002) Ecology of southern ocean pack ice. Adv Mar Biol 43:173–276

Brylinski JM (1981) Reports on the presence of *Acartia tonsa* Dana (Copepoda) in the area of Dunkirk and its geographical distribution in Europe. J Plankton Res 3:255–261

Buck KR, Bolt PA, Bentham W, Garrison DL (1992) A dinoflagellate cyst from the Antarctic Sea ice. J Phycol 28:15–18

Càceres CE, Tessier AJ (2003) How long to rest: the ecology of optimal dormancy and environmental constraint. Ecology 84:1189–1198

Carlton JT, Geller JB (1993) Ecological roulette: the global transport of non indigenous marine organisms. Science 261:78–82

Chesson PL, Warner RR (1981) Environmental variabilità promotes coexistence in lottery competitive systems. Am Nat 117:923–943

Clegg JS (1997) Embryos of *Artemia franciscana* survive for years of continuous anoxia: the case for complete metabolic rate depression. J Exp Biol 200:467–475

Cohen D (1966) Optimizing reproduction in a randomly varying environment. J Theor Biol 12:119–129

Cohen PA, Knoll AH, Kodner RB (2009) Large spinose microfossils in Ediacaran rocks as resting stages of early animals. Proc Natl Acad Sci 106:6519–6524

Dahms H-U (1995) Dormancy in the Copepoda ? An overview. Hydrobiologia 306(3):199–211

Dahms H-U, Li X, Zhang G, Quian P-Y (2006) Resting stages of *Tortanus forcipatus* (Crustacea, Calanoida) in sediments of Victoria harbour, Hong Kong. Estuarine Coastal and Shelf Sciences 67:562–568

Drillet G, Jeppesen PM, Højgaard JK, Jørgensen NOG, Hansen BW (2008) Strain specific vital rates in four *Acartia tonsa* cultures II: life history traits and biochemical contents of eggs and adults. Aquaculture 279:47–54

Drillet G, Hay S, Hansen BW, O'Neill FG (2014) Effects of demersal otter trawls on the re-suspension of copepod resting eggs and its potential effects on recruitment. J Fisheries and Livestock Production 2. https://doi.org/10.4172/2332-2608.1000114

Dumont HJ, Nandini S, Sarma SSS (2002) Cyst ornamentation in aquatic invertebrates. A defense against predation Hydrobiologia 486:161–167

Ellner SP, Hairston NG Jr, Kearns CM, Babai D (1999) The roles of fluctuating selection and long term diapause in microevolution of diapause timing in a fresh water copepod. Evolution 53:111–122

Engel M (2005) Calanoid copepod resting eggs – a safeguard against adverse environmental conditions in the German bight and the Kara Sea? Berichte zur Polarforshung und Meeresforshung 508:1–108

Farabegoli A, Ferrari I, Manzoni C, Pugnetti A (1989) Prima segnalazione nel Mare Adriatico del copepode calanoide *Acartia tonsa*. Nova Thalassia 10(suppl.1):207–208

Figuerola J, Green AJ (2002) Dispersal of aquatic organisms by waterbirds; a review of past research and priorities for future studies. Freshw Biol 47:483–494

Foissner W (2006) Biogeography and dispersal of micro-organisms: a review emphasizing protists. Acta Protozool 45:111–136

Foissner W, Müller H, Agatha S (2007) A comparative fine structural and phylogenetic analysis of resting cysts in oligotrich and hypotrich Spirotrichea (Ciliophora). Eur J Protistol 43:295–314

Friedrich C, De Smet WH (2000) The rotifer fauna of Arctic Sea ice from the Barents Sea, Laptev Sea, and Greenland Sea. Hydrobiologia 432:73–89

Gaines S, Roughgarden J (1985) Larval settlment rate: a leading determinant of structure in an ecological community of the marine intertidal zone. Proc Natl Acad Sci 83:3707–3711

Galil BS, Hulsemann N (1997) Protist transport via ballast water – biological classification of ballast tanks by food web interactions. Eur J Protistol 33:244–253

Garrison DL, Buck KR (1989) The biota of Antarctic pack ice in the Weddell Sea and Antarctic peninsula regions. Polar Biol 10:211–219

Gaudy R, Viñas MD (1985) Premiere signalization en Méditerranée du copépode pélagique *Acartia tonsa*. Rapports de la Commission Internationale pour l'Exploration Scientifique de la Mer Méditerranée 29:227–229

Giangrande A, Geraci S, Belmonte G (1994) Life-cycle and life-history diversity in marine invertebrates and the implications in community dynamics. Oceanogr Mar Biol Annu Rev 32:305–333

Giangrande A, Montresor M, Cavallo A, Licciano M (2002) Influence of *Naineris laevigata* (Polychaeta: Orbiniidae) on vertical grain size distribution, and dinoflagellate resting stages in the sediment. J Sea Res 47:97–108

Giannakourou A, Orlova TY, Assimakopoulou G, Pagou K (2005) Dinoflagellate cysts in recent marine sediments from Thermaikos gulf, Greece: effects of resuspending events on vertical cyst distribution. Cont Shelf Res 25:2585–2596

Gilbert JJ (1974) Dormancy in rotifers. Trans Am Microsc Soc 93:490–513

Gilbert JJ (1992) Rotifera. In: Adiyodi KG, Adiyodi RG (eds) Reproductive biology of invertebrates, Sexual differentiation and behaviour, vol V. Oxford/IBH Publishing Co, New Delhi, pp 115–136

Gilbert JJ (2004) Population density, sexual reproduction, and diapause in monogonont rotifers: new data for Brachionus and a review. J Limnol 63:32–36

Gilbert JJ (2007) Timing of diapause in monogonont rotifers. In: Alekseev VR et al (eds) Diapause in aquatic invertebrates. Theory and human use. Springer, Berlin, pp 11–27

Glippa O, Denis L, Lesourd S, Souissi S (2014) Seasonal fluctuations of the copepod resting egg bank in the middle seine estuary, France: impact on the nauplii recruitment. Estuar Coast Shelf Sci 142:60–67

Gradinger R (1999) Vertical fine structure of the biomass and composition of algal communities in Arctic pack ice. Mar Biol 133:745–754

Grice GD, Marcus NH (1981) Dormant eggs of marine copepods. Occeanogr Mar Biol Annu Rev 19:125–140

Guelorget O, Perthuisot JP (1992) Paralic ecosystems - biological organization and functionning. Vie Milieu 42:215–251

Guerrero F, Rodrigues V (1998) Existence and significance of *Acartia grani* resting eggs (Copepoda: Calanoida) in sediments of a coastal station in the Alboran Sea (SE Spain). J Plankton Res 20:305–314

Gyllström M, Hansson L-A (2004) Dormancy in freshwater zooplankton: induction, termination and the importance of benthicpelagic coupling. Aquat Sci 66:274–295

Hairston NG Jr (1996) Zooplankton egg banks as biotic reservoirs in changing environments. Limnol Oceanogr 41:1087–1092

Hairston NG Jr, De Stasio BT Jr (1988) Rate of evolution slowed by a dormant propagule pool. Nature 336:239–242

Hairston NG Jr, Fox JA (2010) Egg banks. In: Likens GE (ed) Plankton of inland waters. Elsevier, San Diego, pp 247–254

Hairston NG Jr, van Brunt RA, Kearns CN, Engstrom DR (1995) Age and survivorship of diapausing eggs in a sediment egg bank. Ecology 76:1706–1711

Hansen BW, Drillet G, Kristensen RM, Sørensen TF, Tøttrup MT (2010) Production, hatching success and surface ornamentation of eggs of calanoid copepods during a winter at 57°N. Mar Biol 157:59–68

Heikkila M, Pospelova V, Forest A, Stern GA, Fortier L, Macdonald RW (2016) Dinoflagellate cyst production over an annual cycle in seasonally ice-covered Hudson Bay. Mar Micropaleontol 125:1–24

Horner R (1985) Sea ice biota. CRC Press, Boca Raton
Horner R, Ackley SF, Dieckmann GS, Guiliksen B, Hoshia T, Legendre L, Melnikov IA, Reeburgh WS, Spindler M, Sullivan CW (1992) Ecology of sea ice biota. 1. Habitat, terminology, and methodology. Polar Biol 12:417–427
Hutchinson GE (1961) The paradox of the plankton. Am Nat 95:137–145
Ianora A (1998) Copepod life history traits in subtemperate regions. J Mar Syst 15:337–349
Ianora A, Santella L (1991) Diapause embryos in the neustonic copepod *Anomalocera patersoni*. Mar Biol 108:387–394
Ikavalko J (2001) On the presence of some selected Heterokontophyta (Chrisophyceae, Dictyochophyceae, Bicocoecida) and cysts ("archaeomonads") from sea ice – a synopsis. Nova Hedwigia 122:41–54
Ikavalko J, Thomsen HA (1997) The Baltic Sea ice biota (march 1994): a study of the protistan community. Eur J Protistol 33:229–243
Ikavalko, J., Werner, I., Roine, T., Karell, K., Granskog, M. & Ehn, J. 2004. Sea ice biota in the northern Baltic Sea in February and April 2002. Proceedings of the 17th International IAHR Symposium on Ice. Saint Petersburg, Russia, 21–25 June 2004, pp 18–23
Jiang X, Wang G, Li S (2004) Age, distribution and abundance of viable resting eggs of *Acartia pacifica* (Copepoda: Calanoida) in Xiamen Bay, China. J Exp Biol Ecol 312:89–100
Kaiser MJ, Attrill MJ, Jennings S, Thomas DN, Barnes DKA, Brierley AS, Hidding JG, Kaartokallio H, Polunin NVC, Raffaelli DG (2005) Marine ecology. In: Processes, systems, and impacts. Press, Oxford University
Kamiyama T, Itakura S, Nagasaki K (1995) Effects of irradiance on excystment of tintinnids from marine sediments. J Oceanogr 51:615–618
Kasahara S, Uye S, Onbè T (1974) Calanoid copepods eggs in sea bottom-muds. Mar Biol 26:167–171
Kerfoot WC, Weider LJ (2004) Experimental paleoecology (resurrection ecology): chasig Van Valen's red queen hypothesis. Limnol Oceanogr 49:1300–1316
Kim Y-O, Suzuki T, Taniguchi A (2002) A new species in the genus Cyrtostrombidium (Ciliophora, Oligotrichia, Oligotrichida): its morphology, seasonal cycle and resting stage. J Eukaryot Microbiol 49:338–343
Kokinos JP, Anderson DM (1995) Morphological development of resting cysts in cultures of the marine dinoflagellate *Lingulodinium polyedrum* (= L. machaerophorum). Palynology 19:143–166
Komazawa H, Endo Y (2002) Experimental studies on hatching conditions of the resting eggs of marine cladocerans and their seasonal variation in Onagawa Bay. Tohoku J Agri Res 52:57–85
Kremp A, Shull DH, Anderson DM (2003) Effects of deposit feeder gut passage and fecal pellet encapsulation on germination of dinoflagellate resting cysts. Mar Ecol Prog Ser 263:65–73
Kremp A, Rengefors K, Montresor M (2009) Species specific encystment patterns in three Baltic cold-water dinoflagellates: the role of multiple cues in resting cyst formation. Limnol Oceanogr 54:1125–1138
Kuwata A, Tsuda A (2005) Selection and viability after ingestion of vegetative cells, resting spores and resting cells of the marine diatom, *Chaetoceros pseudocurvisetus*, by two copepods. J Exp Mar Biol Ecol 322:143–151
Lakkis S, Zeidane R (1990) Associations congeneriques d'Acartia (Copepoda, Calanoida) dans les eaux cotieres Libanaises : Calcul des indices d' « Overlap » et de « Niche hypervolume ». Rapport de la Commission Internationale pour l'Exploration Scientifique de la Mer Méditerranée 32:222–223
Lennon JT, Jones SE (2011) Microbial seed banks: the ecological and evolutionary implications of dormancy. Nat Rev Microbiol 9:119–130
Leppäkoski E, Olenin S (2000) Non-native species and rates of spread: lessons from the brackish Baltic Sea. Biol Invasions 2:151–163
Levington JS (2001) Marine biology: function, biodiversity, ecology, 2nd edn. Oxford University Press, New York

Lewin R (1986) Supply-side ecology. Science 234:25–27
Lindley IA (1992) Resistant eggs of Centropagoidea (Copepoda: Calanoida): a possible preadaptation to colonization of inland waters. J Crustac Biol 12:368–371
Lutz RV, Marcus NH, Chanton JP (1992) Effects of low oxygen concentrations on the hatching and viability of eggs of marine calanoid copepods. Mar Biol 114:241–247
Madhupratap M, Nehring S, Lenz J (1996) Resting eggs of zooplankton (Copepoda and Cladocera) from the Kiel Bay and adjacent waters (southwestern Baltic). Mar Biol 125:77–87
Mansingh A (1971) Physiological classification of dormancies in insects. Can Entomol 103:983–1009
Marcus NH (1979) On the population biology and nature of diapause of *Labidocera aestiva* (Copepoda, Calanoida). Biol Bull 157:297–305
Marcus NH (1984) Recruitment of copepod nauplii into the plankton: importance of diapause eggs and benthic processes. Mar Ecol Prog Ser 15:47–54
Marcus NH (1987) Differences in the duration of egg diapause in *Labidocera aestiva* (Copepoda, Calanoida) from the woods hole, Massachussetts, region. Biol Bull 163:169–177
Marcus NH (1990) Calanoid copepod, cladoceran, and rotifer eggs in sea-bottom sediments of northern Californian coastal waters: identification, occurrence and hatching. Mar Biol 105:413–418
Marcus NH, Boero F (1998) Production and plankton community dynamics in coastal aquatic systems: the importance of benthic pelagic coupling in the forgotten role of life cycles. Limnol Oceanogr 43:763–768
Marcus NH, Lutz R, Burnett W, Cable P (1994) Age, viability and vertical distribution of zooplankton resting eggs from an anoxic basin: evidence of an egg bank. Limnol Oceanogr 39:154–158
Mauchline J (1998) The biology of calanoid copepods. Adv Mar Biol 33:1–710
McMinn A, Bolch CJS, Hallegraeff G (1992) Cobricopshaeridium Harland & Sarjeant : dinoflagellate cyst or copepod egg ? Mircopaleontology 38:315–316
Meiers K, Feeling J, Graskog M, Spindler M (2002) Abundance, biomass and composition of biota in Baltic Sea ice and underlying water (march 2000). Polar Biol 25:761–770
Mertens KN, Dale B, Ellegaard M, Jansson I-M, Godhe A, Kremp A, Louwye S (2011) Process length variation in cysts of the dinoflagellate *Protoceratium reticulatum*, from surface sediments of the Baltic–Kattegat–Skagerrak estuarine system: a regional salinity proxy. Boreas 40:242–255
Meunier A (1910) Microplankton des Mers de Barents et de Kara. In: Duc D'Orleans: Campagne Arctique de 1907. C. Bulens, Bruxelles
Meysman FJR, Middelburg JJ, Heip CHR (2006) Bioturbation: a fresh look at Darwin's last idea. Trends Ecol Evol 21:688–695
Miyashita LK, Gaeta SA, Lopez RM (2011) Life cycle and reproductive traits of marine podonids (Cladocera, Onychopoda) in a coastal subtropical area. J Plankton Res 33:779–792
Modig H, Òlafsson E (2001) Survival and bioturbation of the amphipod *Monoporeia affinis* in sulphide rich sediments. Mar Biol 138:87–92
Montresor M, Nuzzo L, Mazzocchi M (2003) Viability of dinoflagellate cysts after the passage through the copepod gut. J Exp Mar Biol Ecol 287:209–221
Mugrabe G, Barros S, Marazzo A, Valentin JL (2007) Hatching rates of resting eggs of Cladocera (Crustacea; Branchiopoda) at a tropical bay, Brazil. Braz J Biol 67:527–530
Mura G (1986) SEM morphological survey on the egg shell in the Italian anostracans (Crustacea, Branchiopoda). Hydrobiologia 134:273–286
Odum EP (1971) Fundamentals of ecology. Saunders Co., Philadelphia
Onbé T (1974) Studies on the ecology of marine cladocerans. J Faculty of Fishery and Animal Husbandry, Hiroshima University 13:83–179
Onbé T (1985) Seasonal fluctuations in the abundance of populations of marine cladocerans and their resting eggs in the Inland Sea of Japan. Mar Biol 87:83–88
Onbé T (1991) Some aspects of the biology of resting eggs of marine cladocerans. In: Wenner A, Kuris A (eds) Crustacean egg production, Crustacean issues 7. A.A. Balkema, Rotterdam, pp 41–45

Onoué Y, Toda T, Ban S (2004) Morphological features and hatching patterns of eggs in Acartia steueri (Crustacea, Copepoda) from Sagami Bay, Japan. Hydrobiologia 511:17–24

Pati AC, Belmonte G (2003) Disinfection efficacy on cysts viability of *Artemia franciscana* (Crustacea), *Hexarthra fennica* (Rotifera), and Fabrea Salina (Ciliophora). Mar Biol 142:895–904

Persson A (2000) Possible predation of cyst - a gap in the knowledge of dinoflagellate ecology? J Plankton Res 22:803–809

Persson A, Rosenberg R (2003) Grazing on marine dinoflagellate cysts by benthic deposit feeding animals. Harmful Algae 2:43–50

Philippi T, Sen J (1989) Hedging one's evolutionary bets, revisited. Trends Ecol Evol 4:41–44

Piscia R, Tabozzi S, Bettinetti R, Nevalainen L, Manca MM (2016) Unexpected increases in rotifer resting egg abundances during the period of contamination of Lake Orta. J Limnol 75. https://doi.org/10.4081/jlimnol.2016.1300

Posi ME, BelmoParnte G (2011) Ritmi di produzione di uova di diapausa in Paracartia latisetosa (Copepoda, Calanoida). Thalassia Salentina 33:83–94

Rathaille AN, Raine R (2011) Seasonality in the excystment of Alexandrium minutum and *Alexandrium tamarense* in Irish coastal waters. Harmful Algae 10:629–635

Ratkova TN, Wassmann P (2005) Sea ice algae in the white and Barents seas: composition and origin. Polar Res 24:1–16

Redden AM, Daborn GR (1991) Viability of subitaneous copepod eggs following fish predation on egg-carrying calanoids. Mar Ecol Prog Ser 77:307–310

Reid PC, John AWG (1978) Tintinnid cysts. J Mar Biol Assoc U K 58:551–557

Reid PC, John AWG (1983) Resting cysts in the ciliate class Polyhymenophorea: phylogenetic implications. J Protozool 30:710–712

Remy P (1927) Note sur un copepode de l'eau saumatre du canal de Caen a la mer, *Acartia tonsa* Dana. Ann Biol Lacustre 15:169–186

Reznick DN (2011) The "origin" then and now: an interpretive guide to the "origin of species". University Press, Princeton

Ribeiro S, Berge T, Lundholm N, Andersen TJ, Abrantes F, Ellegaard M (2011) Phytoplankton growth after a century of dormancy illuminates past resilience to catastrophic darkness. Nat Commun 2:311–317

Ricci C (2001) Dormancy patterns in rotifers. Hydrobiologia 446(447):1–11

Ricci C, Fontaneto D (2009) The importance of being a bdelloid. Ecological and evolutionary consequences of dormancy. Ital J Zoo 76:240–249

Rodrigues V, Vives F (1984) Copepodes de las aguas portuarias de Malaga. Investig Pesq 48:235–254

Rozanska M, Poulin M, Gosselin M (2008) Protist entrapment in newly formed sea ice in the coastal Arctic Ocean. J Mar Syst 74:887–901

Rubino, F. & Belmonte, G. 2019. A new cyst morphotype from recent sediments of the mar piccolo of Taranto (southern Italy, Ionian Sea).. Progress in Aqua Farming and Marine Biology. (in press)

Rubino F, Moscatello S, Belmonte M, Ingrosso G, Belmonte G (2013) Plankton resting stages in the marine sediments of the bay of Vlore (Albania). Intern J Ecol, 2013, n.ID 101682

Rubino F, Cibic T, Belmonte M, Rogelja M (2016) Microbenthic community structure and trophic status of sediments in the mar piccolo of Taranto (Mediterranean, Ionian Sea). Environ Sci Pollut Res 23:12624–12644

Ruiz GM, Carlton JT, Grosholtz ED, Hines AH (1997) Global invasions of marine and estuarine habitats by non-indigenous species: mechanisms, extent, and consequences. Am Zool 37:621–632

Ruiz GM, Rawlings TK, Dobbs FC, Drake LA, Mullady T, Huq A, Colwell RR (2000) Global spread of microorganisms by ships Ballast water discharged from vessels harbours. A cocktail of potential pathogens. Nature 408:49–50

Sacchi CF (1985) Le sel de La Palice: réflexions sur le paralin méditerranéen. Memorie di Biologia Marina ed Oceanografia 15:71–89

Santella L, Ianora A (1990) Subitaneous and diapause eggs in Mediterranean populations of *Pontella mediterranea* (Copepoda: Calanoida): a morphological study. Mar Biol 104:83–90

Santella L, Ianora A (1992) Fertilization envelope in diapause eggs of *Pontella mediterranea* (Crustacea, Copepoda). Mol Reprod Dev 33:463–469

Sazhina LI (1968) O zhimuyushchik yaitzak morskik Calanoida. Zool Zhurnal 47:1554–1556. (in Russian)

Sazhina LI (1987) Rashmioshenie, rost, produkziya morskik veslonogik rakoobrashiyk. Naukova Dumka, Kyev. (in Russian)

Schnack-Schiel SB, Dieckmann GS, Gradinger R, Melnikov IA, Spindler M, Thomas DN (2001) Meiofauna in sea ice of the Weddel Sea (Antarctica). Polar Biol 24:724–728

Seger J, Brockmann HJ (1987) What is bet-hedging? Oxford Survey of Evolutionary Biology 14:357–367

Sichlau MH, Hansen JLS, Andersen TJ, Hansen BW (2011) Distribution and mortality of diapause eggs from calanoid copepods in relation to sedimentation regimes. Mar Biol 158:665–676

Siokou-Frangou I, Zervoudaki S, Kambouroglou V, Belmonte G (2005) Distribution of mesozooplankton resting eggs in seabottom sediments of Thermaikos gulf (NW Aegean Sea, Greece) and possible effects of sediment resuspension. Cont Shelf Res 25:2597–2608

Smirnov SS (1935) Über das Auftreten von Acartia tonsa Dana (Copepoda) in Finnischen Meerbusen. Proc Natl Acad Sci URSS 3(5):237–240

Stabili L, Miglietta AM, Belmonte G (1999) Lysozyme-like and trypsin like activities in the cyst of *Artemia franciscana* Kellog 1906. Is there a passive immunity in a resting stage? J Exp Mar Biol Ecol 237:291–303

Tarazona E, García-Roger EM, Carmona MJ (2017) Experimental evolution of bet hedging in rotifer diapause traits as a response to environmental unpredictability. Oikos 126:1162–1172

Templeton AR, Levin DA (1979) Evolutionary consequences of seed pools. Am Nat 114:232–249

Thomas DN, Dieckmann GS (2002) Antarctic Sea ice – a habitat for extremophiles. Science 295:641–644

Tunnacliffe A, Lapinski J, McGee B (2005) A putative LEA protein, but no trehalose, is present in anhydrobiotic bdelloid rotifers. Hydrobiologia 546:315–321

Underwood AJ, Fairweather PG (1989) Supply side ecology and benthic marine assemblages. Trends Ecol Evol 4(1):16–20

Uye S-I (1985) Resting egg production as a life-cycle strategy of marine planktonic copepods. Bull Mar Sci 37:440–449

Valiela I (1995) Marine Ecological Processes, 2nd edn. Springer Publishing + Business Media, New York

Viitasalo S (2007) Effects of bioturbation by three macrozoobenthic species and predation by necto-benthic mysids on cladoceran benthic eggs. Mar Ecol Prog Ser 336:131–140

Viitasalo M, Katajisto T (1994) Mesozooplankton resting eggs in the Baltic Sea: identification and vertical distribution in laminated and mixed sediments. Mar Biol 120:455–465

Viitasalo S, Viitasalo M (2004) Predation by the mysid shrimps Mysis mixta and *M. relicta* on benthic eggs of *Bosmina longispina* maritima (Cladocera) in the northern Baltic Sea. Mar Ecol Prog Ser 281:155–163

Wang G, Jiang X, Wu L, Li S (2005) Differences in the density, sinking rate and biochemical composition of Centropages tenuiremis (Copepoda: Calanoida) subitaneous and diapause eggs. Mar Ecol Prog Ser 288:165–171

Werner I, Ikavalko J, Schünemann H (2007) Sea-ice algae in Arctic pack ice during late winter. Polar Biol 30:1493–1504

Williams-Howze J (1997) Dormancy in the free-living copepod orders Cyclopoida, Calanoida, and Harpacticoida. Oceanogr Mar Biol Annu Rev 35:257–321

Wonham MJ, Bailey SA, MacIsaac HJ, Lewis MA (2005) Modelling the invasion risk of diapausing organisms transported in ballast sediments. Can J Fish Aqua Sci 62:2386–2398

Zaitsev Y, Ozturk B (eds) (2001) Exotic species in the Aegean, Marmara, black, Azov and Caspian seas. Istanbul, Turkish Marine Research Foundation

Part II
The Role of Diapause in Science and Human Uses

Chapter 6
Studying Dormancy in Space Conditions

Victor R. Alekseev, Margarita A. Levinskikh, Natalia D. Novikova, and Vladimir N. Sychev

Abstract Complicated factors of space flight, combined with environmental and technical peculiarities, can affect living organisms inside spacecraft or orbital stations: microgravity, space corpuscular radiation, magnetic and electric fields, the atmosphere (gas composition, ratio, and pressure), technical noise, vibration, and various other factors. In outer space, ultraviolet radiation, a large range of very high and very low temperatures, the vacuum, and micrometeorites increase the possible harmful effects on living organisms. Dormant stages of fish, insects, crustaceans, plant seeds, bacterial spores, and fungi were tested in three types of experiments at the International Space Station (ISS): (1) inside the station (the Aquarium program), with a limited set of aquatic species; (2) outside the station in outer space but without exposure to ultraviolet radiation (the Biorisk program); and (3) in modified outer space conditions imitating the Mars surface environment (the Expose program). This chapter discusses the most important results obtained in these ISS experiments, which have been performed since 2005.

Keywords Dormant stages · Aquatic organisms · ISS · Space experiments · Astrobiology · Biorisk · Expose · UV radiation · Vacuum · Survivorship · Mars surface conditions

V. R. Alekseev (✉)
Zoological Institute of Russian Academy of Sciences, St. Petersburg, Russia
e-mail: alekseev@zin.ru

M. A. Levinskikh · N. D. Novikova · V. N. Sychev
Institute of Biomedical Problems of RAS, Moscow, Russia
e-mail: ritalev@imbp.ru; novikova@imbp.ru; vnsychev@imbp.ru

V. R. Alekseev, B. Pinel-Alloul (eds.), *Dormancy in Aquatic Organisms. Theory, Human Use and Modeling*, Monographiae Biologicae 92,
https://doi.org/10.1007/978-3-030-21213-1_6

6.1 Introduction

It has been shown recently that one or another form of dormancy—and, consequently, the peculiar stages that secure survival under conditions incompatible with life—are inherent to many organisms, from bacteria to vertebrates (Alekseev and Fryer 1998; Alekseev et al. 2007a, b). Having a similar pattern in evolutionarily distant organisms, resting stages seem to have a common molecular genetic basis that is probably due to standing adaptation. It is no great exaggeration to say that all living things have the ability to arrest vital activities, a trait that is either revealed or not, depending on the quality of the environment. It is not unusual for this stage, on the borderline between life and death, to be controlled by specific signals and/or internal factors (e.g., the biological clock). With this knowledge we may set ourselves the challenge of learning how to induce and cancel dormancy in individual species and artificial ecosystems.

This adaptation, especially if it is controlled by an external signal, bears much promise for space biology, and biological life support systems (BLSSs) in particular, and can be the focal point of research programs, including experiments performed inside and outside space vehicles.

Investigations to advance the use of animal and plant dormancy (in the forms of cryptobiosis and some other forms of biological rest) in space exploration have been conducted in several programs in astrobiology, outlined in previous publications (Alekseev et al. 2007b). These include:

- The study of dormancy from the perspective of its integration into ecological life support systems
- Dormancy and adaptation of higher vertebrates to extended body metabolism
- The microbiological safety of space flight
- Planetary and interplanetary quarantine
- The search for extraterrestrial life

This chapter gives a brief description of the program and the objects used in space experiments at the International Space Station (ISS) within the last decade.

6.2 Experiments on Crustacean Dormancy Stages Inside the Space Station

This study was conducted to elaborate a new technology for creation of an artificial ecosystem outside the Earth's biosphere (Alekseev and Sychev 2006; Alekseev et al. 2007a). Maintenance of aquatic animal and plant resting stages in space conditions will be an essential part of this biotechnology. We have examined reactivation of resting stage and life cycle parameters in two crustacean species (*Daphnia magna* and *Streptocephalus torvicornis*) during a space mission after a month of exposure of their dried resting eggs in the Russian segment (RS) of the ISS. Special attention was paid to possible negative changes caused by a set of factors at the ISS, including low gravity, radiation, magnetic/electric fields, and the biological impact of bacterial–fungal flora.

6.2.1 Aquatic Invertebrate Species Used in ISS Exposure

Resting eggs of *D. magna* and *S. torvicornis* were collected in artificial ponds for *Daphnia* cultivation used in sturgeon aquaculture at the Trudfront State Sturgeon Farm (in the Astrakhan District in the South of European Russia at 46°00″N; 47°30″E). The betony ponds have existed for more than 30 years and regularly operate in May–June during the short period of plankton feeding in sturgeon larvae. At the end of June, *Daphnia* ephippia collected from the ponds are placed in fiber bags and kept dry at outside temperatures until next spring. During winter they pass through natural cold termination and, before use in our experiments, these eggs were almost activated and ready for development. Preliminary testing in May 2005 indicated a high level of reactivation (up to 80% in *D. magna*) in the resting eggs of both species.

6.2.2 Preparation of Dormant Stages for the Space Mission

In a laboratory, resting eggs were separately fixed onto sticky plastic film mounted as special windows in carton blocks, covered with a protective paper sheet, and placed in zipped plastic envelopes to avoid crushing and unsanctioned movement during transportation to the ISS and storage in microgravity.

Each carton block contained three windows with numerous resting eggs of *D. magna* and of *S. torvicornis.* The carton blocks were then randomly split into four groups of nine replicates each: (1) for exposure at the ISS for 1 month, (2) for exposure at the ISS for 8 months, (3) for laboratory controls (kept at 20 °C in natural light), and (4) for cold controls (kept at 4 °C in darkness—their optimal conditions).

In the service module of the ISS RS, envelopes containing groups 1 and 2 were placed in a Plant-2 sample-holder in controlled light, radiation, and temperature conditions near the Lada microgreenhouse (Fig. 6.1).

After 10 days in orbit the samples were delivered back to the Earth's surface and, in parallel with the control samples, were transported by VA to the ecological physiology laboratory at the Max Planck Institute of Limnology (Ploen, Germany) for detailed analyses of the reactivation and productive potential in these two species.

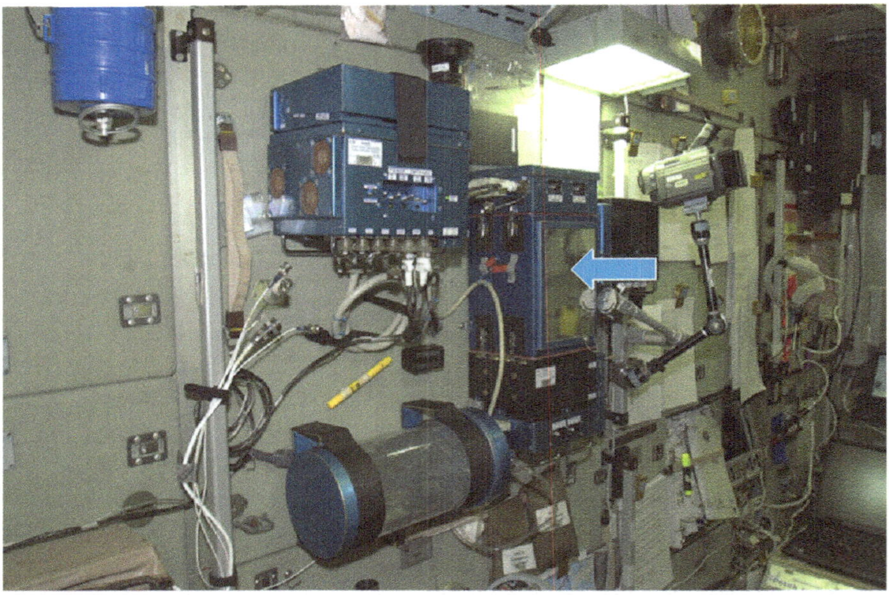

Fig. 6.1 Placement (blue arrow) of dormant eggs in the Russian segment of the International Space Station

6.2.3 Reactivation Protocol

The number of resting eggs were calculated on each plastic film and then placed into petri dishes, rinsed with cold (10° C) water, and preliminarily aerated within 1 h. The resting eggs then spent 2 days in a dark room at 10 °C for gradual swelling. After this preliminary preparation the resting eggs on plastic film were transferred to a thermostat with a fluctuating temperature and a long-day photoperiod. During the day phase (lasting 14 h) they stayed at 25 °C and at nighttime (lasting 10 h) they stayed at 15 °C. This imitated the natural light and temperature conditions in spring. The petri dishes were checked twice per day, and all reactivated animals were picked up, imaged under a microscope with a digital camera, and then used for growing experiments or dried (at 60 °C for 24 h) and weighed on a Sartorius balance with a resolution of 0.0001 mg. Usually, not just a single neonate but small groups of 5–10 individuals were used for weighing. The pictures were then used to measure the neonates' length on a computer workstation, using the Adobe measurement tool with a resolution of 0.001 mm.

The testing of reactivation lasted at least 3 days after the last neonate appeared (normally at 2–4 weeks). Resting eggs in the same dishes were then returned to a cold (10 °C) and dark room for 2 weeks. After the cold treatment, the same cycle of reactivation was repeated, and the total reactivation time was no less than 1 month. We calculated the rate of reactivation for every day and the total numbers of hatched neonates from the actual number of resting eggs. We tested *D. magna* reactivation in 12 replicates, with 16–31 (average 25) ephippia in each, for ISS-exposed embryos; 16 replicates, with 4–68 (average 29) ephippia in each, for control embryos; and 8 replicates, with 16–31 (average 27) ephippia in each, for cold-control embryos. In cysts of *S. torvicornis*, reactivation was tested in the same groups in 8 replicates (42–97, average 76), in 16 replicates (41–98, average 70), and in 8 replicates (56–87, average 74), respectively.

This protocol of reactivation was applied in the Biorisk and Expose experiments in outer space.

6.2.4 *Testing of* D. magna *Embryo Resistance to Parasite Invasion*

Spores of a *Daphnia* parasitic fungus, *Pitium daphniarum*, were applied to test the level of resistance in *D. magna* embryos. The spores were obtained from a wild culture of *P. daphniarum* via cultivation of its mycelium in a drying drop of water at a cold temperature for 3 days. The presence of the spores in the culture was checked under a microscope. Then the mycelium was cut into equal portions, which were added to petri dishes with ten *D. magna* embryos at the first stage of cleavage. We tested the resistance of the embryos to the parasite invasion in both the exposed and control groups of *D. magna* in 4–6 replicates each. For a reference outside group we used *D. magna* embryos with a suppressed reactivation rate. This reference group was kept in unfavorable conditions (at a high temperature in constant light and dry air) for 6 months.

6.2.5 *Cultivation and Testing of Life Cycle Parameters*

Daphnids and phyllopods were cultivated in groups of 4–6 individuals in 250-mL flow-through vessels (Lampert et al. 1988) at a flow rate of 1 L day^{-1} under constant food and photoperiod conditions at 25° C (Fig. 6.2). The crustaceans were fed with the green alga *Scenedesmus obliquus* (strain SAG 276-3a from the Gottigen algal collection) grown in a thermostat with modified CHU-12 medium (Muller 1972) at a constant rate of 0.7 L day^{-1} to maintain equal quality of food for all treatments.

An experiment started with placement of the neonates in conditions of a high food concentration (1 mg carbon L^{-1}) with long photoperiods (14 h of light) and short photoperiods (10 h of light). The food medium was replaced every morning, and the animals were checked twice per day. After the females had released their first clutch, we recorded the time, counted the neonates (the size of the first clutch), and determined the neonate dry mass. Females were cultivated under these conditions until they deposited their second clutch of eggs in the brood pouch, then they were dried and weighed. The neonate mass and final female mass were used to

Fig. 6.2 Cultivation equipment at the Max Planck Institute of Limnology (Ploen, Germany), used in Earth laboratory testing of *Daphnia magna* and *Streptocephalus torvicornis*

calculate the somatic growth. Several females from the control and ISS-exposed groups were used to obtain a third clutch, from which the neonates were then cultivated in the same conditions in which their mothers laid eggs. After maturation they were checked for sex, size, and dry mass.

The data obtained were not always normally distributed; hence, after a test of assumptions, we used either a *t* test or a nonparametric test (Mann–Whitney test) for assessment of significant differences. The statistical analyses were run using the Statistica-6 program.

6.2.6 Hatching of D. magna *after ISS Exposure*

Hatching of juveniles from the ISS-treated ephippia and the control group started on the third day after they were moved from 10 °C to 25 °C (the fifth day after the start of activation) and on the fifth day (the seventh day after the start of activation) for embryos kept in unfavorable conditions (Figs. 6.3 and 6.4).

On the next day, in these two variants, the maximum hatching rates were observed in the ISS-treated group and the control group, and within 7–11 days (longer in the control group) the hatching rate gradually declined until it reached zero. In the former variant the maximal hatching rate happened close to the expiry of the process, which stopped abruptly on the sixth day after the start. The average reactivation rates were 39.6% in the ISS-treated embryos and significantly higher (51.8%) in the

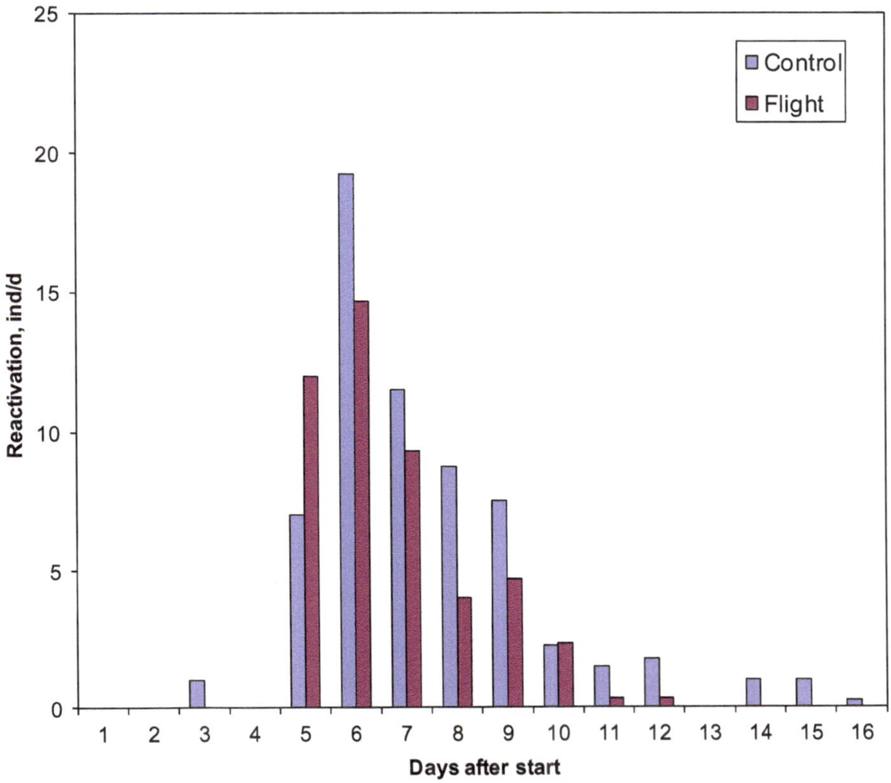

Fig. 6.3 Neonate hatching from *Daphnia magna* ephippia exposed to space flight and control conditions

control group (t test, $p = 0.035$) when calculated on the basis of the number of ephippia. As we found a variety of embryo numbers among the ephippia (about 13 eggs per 10 ephippia), the real reactivation rates were 30.6% and 39.9%, respectively. In embryos kept in unfavorable conditions, the real reactivation rate was significantly lower than those in the other variants (19.0%).

Besides the differences in the reactivation rate and in the duration of this process, the tested groups were also divided in terms of the dynamics of the neonate dry weights (Fig. 6.4). In the control group the neonate dry weights at the beginning of the hatching period (1–3 days) were significantly lower than those in the middle (4–5 days) of the process (t test, $p = 0.02$). In the ISS-treated embryos we found the opposite neonate dry weight dynamics, and the difference was also confirmed statistically (t test, $p = 0.0174$).

Actually, in the control group the neonates with a low dry weight (0.0049 ± 0.00047 mg) hatched first, whereas in the ISS-treated group the biggest embryos (weighing 0.0056 ± 0.00034 mg) appeared first. This difference was confirmed statistically (t test, $p = 0.0201$).

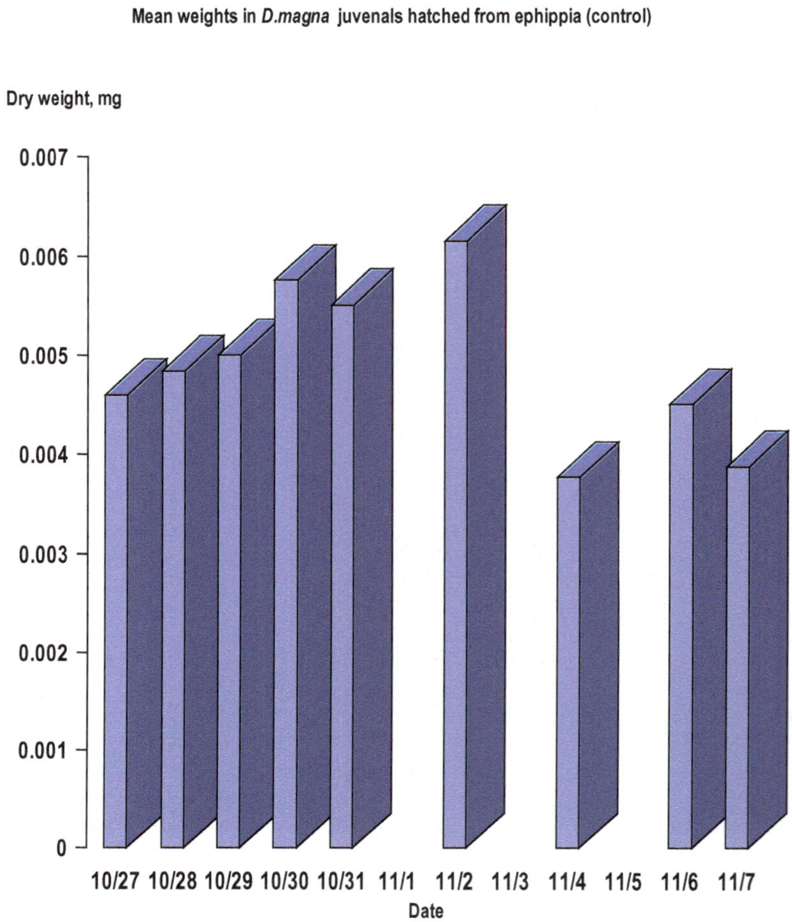

Fig. 6.4 Dry weight of *Daphnia magna* neonates hatched from ephippia in control conditions

In the middle of the hatching period we found bigger neonates (weighing 0.0056 ± 0.000418 mg) in the control group than in the ISS-treated group (weighing 0.0046 ± 0.000579 mg); this difference was also confirmed by a *t* test ($p = 0.002$).

At the same time, the difference in the average dry weights in these two groups, calculated for all periods of hatching, was not significant ($p = 0.504$), despite the relatively big sample size (34 weights in the two groups). This was easily explained by the fluctuation in the neonate dry weight in these groups (Figs. 6.4 and 6.5).

We proposed that at the ISS, the *D. magna* embryos were affected by a negative factor or by a set of factors that suppressed their vital ability (a kind of space flight–induced stress), so the weakest embryos of lower weight could not hatch and their place in the ISS group was substituted by large-sized neonates that possibly also had

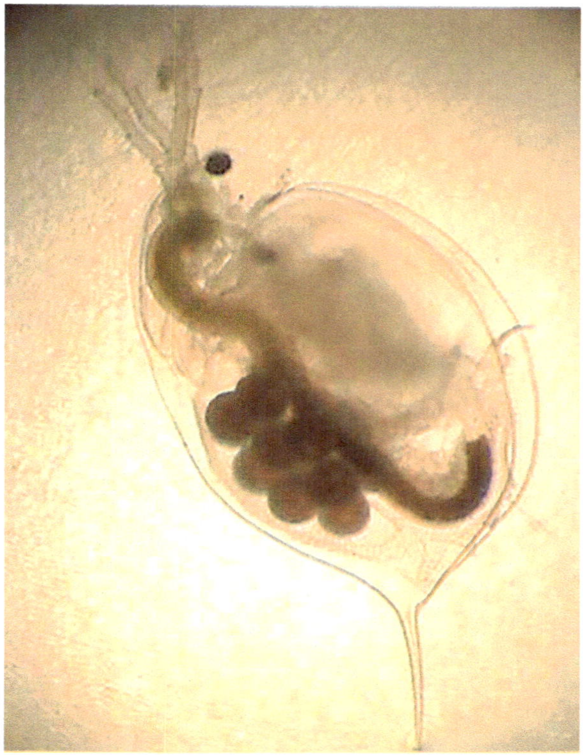

Fig. 6.5 The first *Daphnia* female obtained from the ISS-exposed resting eggs

a suppressed but still sufficient vital ability to hatch from their shells. This theory of stress signals in space flight affecting the *D. magna* dried embryos was confirmed in the following cultivation tests described in Sects. 6.2.7–6.2.11.

6.2.7 Hatching of S. torvicornis *After ISS Exposure*

Hatching of phyllopods from ISS-treated cysts and control cysts started simultaneously on the day after they were moved to conditions of 25° C (Fig. 6.6). A clear difference seen in the total duration of hatching between these variants (3 days in ISS-treated cysts versus 9 days in control cysts) was also accompanied by a significant difference in their total reactivation rates. As in the case of *D. magna*, there was significantly less reactivation (only about half) in the ISS-treated cysts of *S. torvicornis* than in the control cysts (*t* test, $p = 0.034$). Cysts of this species kept in unfavorable conditions had a reactivation efficiency close to zero. Neonate dry weights were documented only in the control group (Fig. 6.6). The biggest neonates hatched on the first day, then the average weight of the neonates declined gradually.

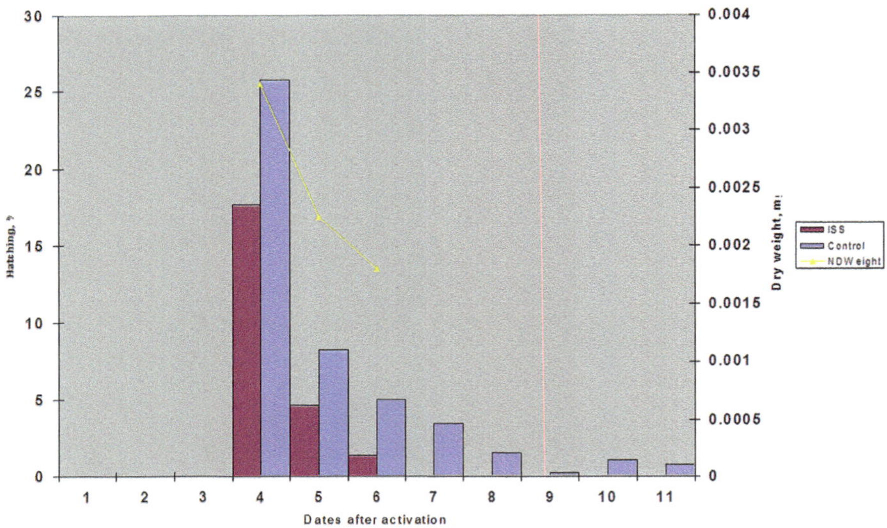

Fig. 6.6 Neonate hatching from *Streptocephalus torvicornis* resting eggs exposed to space flight (on the International Space Station (*ISS*; *red bars*)) and control conditions (*blue bars*), and neonate dry weight in ISS-exposed eggs (*yellow line*)

6.2.8 Testing of D. magna *Embryo Resistance to Parasite Invasion*

Some aquatic fungi are better known as saprophytes than as real parasites, but many of them are harmful to their hosts, drastically reducing the hosts' reproductive success and survival (Ebert 2005).

To test our hypothesis on lower vital ability in ISS-exposed *D. magna* embryos versus control embryos, we tested resistance to fungal parasite invasion with *P. daphniarum* in several groups of eggs (Fig. 6.7). Two parameters were used to characterize resistance to the parasite: (1) the intensity of infection, calculated as the number of eggs infected by the parasite, expressed as a percentage; and (2) the spread of infection, calculated as the number of variants successively infected by the parasite, expressed as a percentage.

Two groups of embryos exposed to orbit and tested in this experiment showed higher sensitivity to parasite invasion than embryos from the control group (Fig. 6.8). The lowest resistance was found in embryos with a suppressed reactivation rate caused by unfavorable storage conditions. In terms of the intensity of infection, the ISS-exposed samples were closer than the control samples to the embryos with a suppressed reactivation rate. The difference in the intensity of infection between ISS-exposed embryos and control embryos was statistically significant (*t* test, $p < 0.04$).

In this study we used the intensity of infection in each group of embryos as a reference for egg capacity. In fact, this also could be explained by different rates of

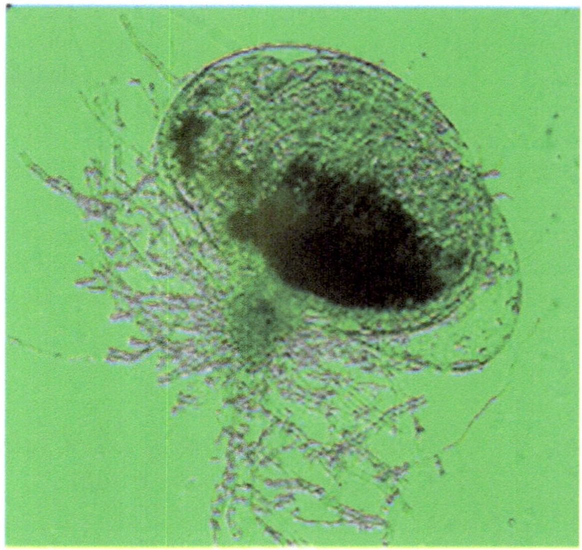

Fig. 6.7 Mycelium of the parasitic fungus *Pitium daphniarum* infecting an embryo of *Daphnia magna* exposed to space flight

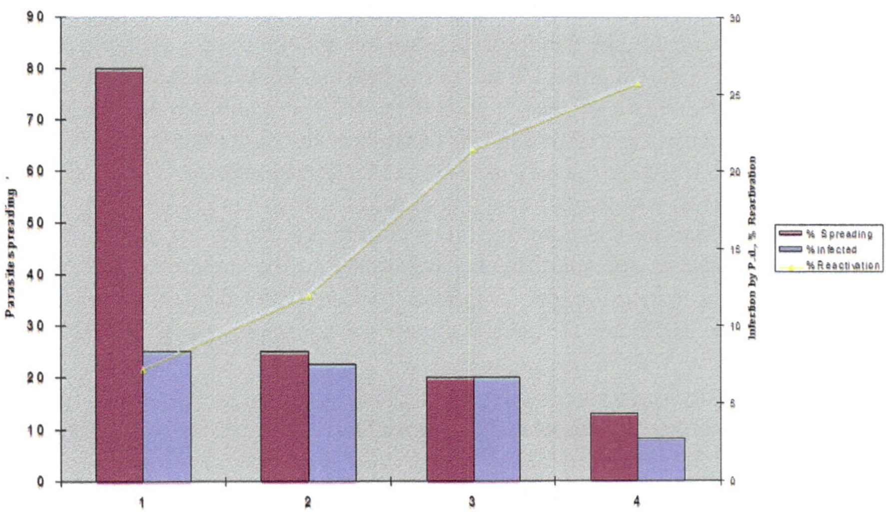

Fig. 6.8 Infection of *Daphnia magna* resting eggs by the parasitic fungus *Pitium daphniarum*: **1** eggs with a suppressed reactivation rate; **2** first group from space; **3** second group from space; **4** control group. *Red bars* denote the percentage of parasite spread (number of experiments with *P. daphniarum* found), *blue bars* denote the percentage of parasite infection, and the *yellow solid line* denotes the percentage of reactivation

egg mortality in these groups, so the fungi could have decomposed embryos that had died before the test. We hoped that in our experiments, we observed that the fungi had infected live embryos that then hatched, lived, and even moved within several hours (Fig. 6.8). Also, periodically we found recently born juveniles surrounded with fungal mycelium. Such mycelium, in accordance with our observations, should grow 2–3 days after infection, meaning that these juveniles had been infected as embryos. This allowed us to conclude that *P. daphniarum*, being a real parasite, more intensively infected weaker but live ISS-exposed eggs than those in the control group.

It is interesting to note that the resistance of *D. magna* embryos to parasite invasion was 1.43 times lower in the ISS-treated group than in the control group. A similar difference (1.5 times lower in the ISS-treated group than in the control group) could be calculated for the spread of infection in these embryos (Fig. 6.8). This possibly corresponds to the rate of the decline in the vital ability of the ISS-treated embryos, caused by space-related factors.

6.2.9 *Cultivation and Testing of Life Cycle Parameters in* D. magna *After Space Flight*

These experiments were designed to determine the effects of resting egg treatment with space factors on life history parameters in *D. magna*, such as the time to release of the first clutch, the survival to maturation, the size of the first clutch, the mass of the first-clutch neonates, the female dry mass at the age of the second-clutch release, and the possible maternal influence on sex in offspring. We found significant effects of ISS exposure for two of these parameters. Table 6.1 summarizes the results obtained for the embryos exposed to space and laboratory environments.

The first-clutch size in females born from ISS-treated embryos (11.14 ± 3.592 eggs per individual) was smaller than that in the control group (14.39 ± 2.847 eggs per individual), and this was confirmed by a *t* test ($p = 0.0477$).

A small but also significant difference was found for the maturation time, which was slightly longer in ISS-exposed animals (9.95 ± 0.284 days) than in control animals (10.31 ± 0.372 days) (Mann–Whitney test, adjusted $z = 2.09$, $p = 0.033$).

Table 6.1 Means and standard deviations (*SDs*) of life history parameters of diapausing embryos of *Daphnia magna* exposed to space on the International Space Station (*ISS*) and kept in laboratory conditions in experiments 1 and 2

	Control (laboratory environment)				ISS (space station environment)			
	First experiment		Second experiment		First experiment		Second experiment	
Parameter	Mean	SD	Mean	SD	Mean	SD	Mean	SD
Female mass (mg)	0.3837	0.0254	0.3366	0.0280	0.3918	0.0635	0.2549	0.0512
Female length (mm)	3.0937	0.1364	3.1307	0.1605	3.4970	0.1593	3.0560	0.0919
Clutch 1 (neonates per individual)	67.20	16.08	70.00	10.22	49.00	17.47	56.00	16.57
Neonate mass (mg)	0.0073	0.0005	0.0073	0.0006	0.0084	0.0016	0.0063	0.0006
Survival (%)	96.00	8.94	96.00	8.94	90.00	11.55	100.00	0.00
Maturation time (days)	9.80	0.27	10.10	0.22	10.63	0.25	10.00	0.00

Both of these parameters are very important for *Daphnia* population dynamics (Lynch 1980).They are mainly responsible for the fitness of the population or clones for environmental conditions, and they also play a key role in their productivity, both in culture conditions and in nature (Ebert 1993).

Despite our expectations, there were no confirmed differences between the ISS-treated and control groups in terms of survival to maturation, neonate mass, and female mass.

6.2.10 *Male Production in* D. magna *as a Stress-Modulated Response After Space Flight*

We traced the growth of offspring (the third clutch) of females exposed to space and kept in laboratory conditions as diapausing embryos (Fig. 6.9). A remarkable difference in the sexual structure of these offspring between the ISS and laboratory groups was found. All five replicates of offspring obtained from females exposed to space contained high proportions of males, ranging from 30% to 78% (54.9 ± 25.94%). Not a single male appeared among the offspring from the control group cultivated in the same conditions.

It is known that a shift from parthenogenesis to bisexual reproduction, especially in *Daphnia*, is a response to environmental stresses of different origins—such as a decline in the availability of food, photoperiod changes, and signal metabolites—mainly in the maternal environment (Glazier 1992). In our experiments the culture

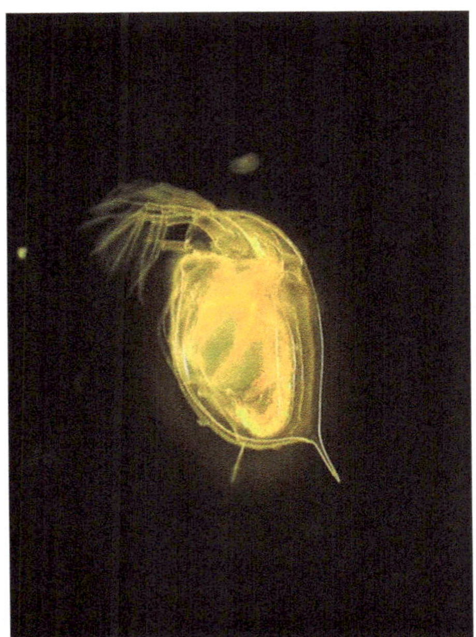

Fig. 6.9 Male of *Daphnia magna*—the second generation grown from ephippia exposed to space flight

conditions used for the ISS and control groups were equivalent, stable, and favorable for parthenogenesis within the two following generations. The only clearly seen difference between them was in the environmental history of the diapausing embryos exposed at the ISS for 1 month. It seems that the month of exposure to space conditions was accepted by the diapausing embryos as a stress signal, and this information was then transmitted via a maternal effect to their offspring, as has been shown with regard to the photoperiod and other biological information in many invertebrates, including *Daphnia* (Mousseau and Fox 1998; Alekseev and Lampert 2001).

This gives us more proof to claim that environmental stress experienced at the space station is also a reason for the other differences between the ISS-treated and control groups that were found in our previous experiments on reactivation, parasite resistance, and life cycle analyses in *D. magna*.

The sensitivity of resting eggs to space flight factors has not been quantified yet. This may be a result of space radiation, magnetic fields, or even microgravity. Also, these factors may work together and produce a synergetic effect. In such cases, these negative effects of space exposure on biological objects, including humans, may be difficult to recognize. This creates additional problems for creation, exploitation, and even transportation of artificial ecosystems outside the Earth's biosphere.

6.2.11 Cultivation and Testing of Life Cycle Parameters in S. torvicornis *After Space Flight*

S. torvicornis, grown from eggs collected in a pond, is characterized by a very high level of variability in its life cycle. That is why only a few significant differences between experiments done on samples of this species exposed to space and kept in a laboratory as diapausing cysts were found. Table 6.2 summarizes the results obtained for embryos exposed to space and laboratory environments.

The only significant effect found in the analyses of the *S. torvicornis* life cycle was an influence of the (photoperiod) day length on the length of the animals at the age of 12 days (the time of the appearance of eggs in the female brood sac). The length of the animals differed between a long day (9.838 ± 1.1546 mm) and a short day (8.676 ± 1.9123 mm) (Mann–Whitney test, $z = 2.021$, $p = 0.043$). Other parameters such as the female/male length or clutch size also showed differences between experiments in long-/short-day conditions and ISS/control environments, but huge population polymorphisms resulted in large standard deviations and masked these effects, in our judgment.

S. torvicornis is an obligatory bisexual phyllopod, so it is impossible to obtain a genetically homogenous strain, as is found in some *Daphnia*, in which obligatory parthenogenesis occurs, decreasing the genetic variation. To overcome these obstacles of population polymorphism in *S. torvicornis* resting eggs, it is necessary to keep the eggs of the species in controlled environmental conditions, including food, temperature, and day length.

Table 6.2 Means and standard deviations (*SDs*) of life history parameters of diapausing cysts of *Streptocephalus torvicornis* exposed to space on the International Space Station (*ISS*) and kept in laboratory conditions

	Control (laboratory environment)				ISS (space station environment)			
	Long day		Short day		Long day		Short day	
Parameter	Mean	SD	Mean	SD	Mean	SD	Mean	SD
Average length at age of 12 days (mm)	9.93	1.52	8.71	2.90	9.77	0.96	8.64	0.34
Female length at age of 12 days (mm)	10.41	1.20	9.48	3.76	10.32	1.40	9.89	0.67
Male length at age of 12 days (mm)	8.73	0.10	8.23	2.88	9.56	1.17	8.67	0.41
Clutch 1 (cysts per individual)	3.25	3.32	8.88	11.68	2.50	1.87	11.04	12.04
Egg production in 3 weeks (cysts per individual)	9.12	5.38	8.88	11.36	11.80	6.29	14.04	10.43
Survival (%)	85.00	19.15	100.00	0.00	84.00	16.73	85.00	19.15
Maturation time (days)	16.50	1.00	17.25	1.50	15.60	1.67	16.00	0.00

6.2.12 Conclusion of the Experiments on Dormancy at the ISS

In *D. magna* and *S. torvicornis*, we found statistically confirmed differences in reactivation efficiency between resting eggs exposed at the ISS and the control group kept in the laboratory. Embryos of *D. magna* exposed to orbit were more sensitive to fungal infection (with *P. daphniarum*) than the reference group. In culture experiments, *D. magna* juveniles obtained from ISS-exposed eggs had a longer maturation time and lower offspring productivity than those in the control group. Exposure of resting eggs to space conditions also induced male production in *D. magna* offspring.

S. torvicornis, in fact, also demonstrated some differences in life cycle parameters between the ISS-exposed and control groups of resting eggs, but most of the differences were not statistically significant, because there were very high levels of population variability in the quality of the juveniles. The only significant effect found in analyses of the *S. torvicornis* life cycle was the influence of the day length on the animal size at the age of egg appearance in the female brood sac.

The following brief generalizations should be mentioned to finalize this part of the paper:

- Even dried diapausing *D. magna* embryos seem to be sensitive to some extreme environmental factors, which may be interesting for crustacean ecology not only in space but also in Earth environments.
- Discovery of similar responses of resting eggs in other species or plant seeds could be important for transportation of ecosystems on long-term space missions and subsequent cultivation at planetary stations.
- To understand the complicated and profound effects of space flight factors on live systems, including humans, wide interdisciplinary collaboration is required.

6.3 Biorisk Experiment on Dormancy in Outer Space

Planetary protection is one of the key problems of deep space exploration. It is essential to assess the risks associated with possible transfer of biological material across interplanetary space during space exploration.

The Biorisk equipment has been developed at the Institute of Biomedical Problems of the Russian Academy of Sciences (Moscow, Russia) to perform space experiments on long-term exposure of resting stages of organisms on the outside of the ISS RS. The Biorisk device consists of three containers in which 24 plastic petri dishes, measuring 65 mm in diameter, are placed with the studied organisms inside them (Fig. 6.10). Each dish contains an air-permeable bacterial filter that is impenetrable by the microorganisms. The containers are fastened onto a special platform on the outside of the station (Fig. 6.10). Before each container is fastened, its covering is opened. The experiment cyclogram allows for container exposure on the outside of the ISS RS for 6, 12, or 23 months.

In the first experiment with Biorisk, which was started in January 2005, the test organisms were prokaryotes (*Bacillus*) and eukaryotes (fungi of the genera *Penicillium*, *Aspergillus*, and *Cladosporium*). The durations of exposure to open space were 7, 12, and 18 months. In this experiment, microorganisms kept on the outer surface of the ISS for a long time (1.5 years) retained lower viability under the conditions of the space environment than control microorganisms.

In the second experiment in the framework of the Biorisk program, the resting stages of various taxonomic groups (from bacteria to vertebrate animals and higher plants) served as biological test objects (Table 6.3).

On February 16, 2007, test objects were put into three containers of the Biorisk equipment, the number and composition of the objects being the same in all containers. Spores of bacteria and mold fungi were applied to metal plates prepared from materials similar to those used for constructing the envelope of the ISS RS.

a **b**

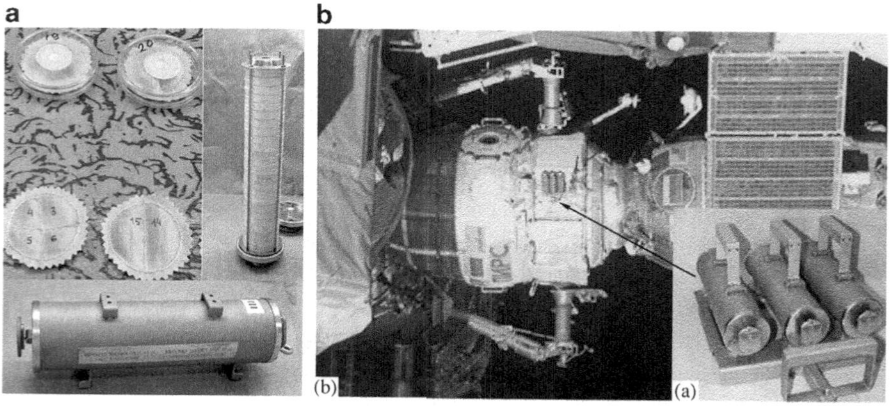

Fig. 6.10 Biorisk experimental device and samples: **a** placement of Biorisk device on the International Space Station; **b** the Russian Zvezda module (black arrow). (After Baranov et al. 2009)

Table 6.3 Test objects used in the Biorisk-MSN-2 experiment

Taxon	State	Common/genus/ family name	Species	Strain/line/variety
Higher plants	Air-dried seeds	Mustard	*Brassica rapa*	Improved basic
		Mustard	*Brassica rapa*	High-anthocyanin
		Mustard	*Brassica rapa*	Anthocyaninless
		Mustard	*Brassica rapa*	Rosette
		Mustard	*Brassica rapa*	Somatic variegation
		Mustard	*Brassica rapa*	AstroPlants
		Mustard	*Brassica juncea*	Red Giant
		Thale cress	*Arabidopsis thaliana*	K-310
		Thale cress	*Arabidopsis thaliana*	Columbia
		Apple of Peru	*Nicandra physaloides*	Black Pod
		Tomato	*Lycopersicon esculentum*	Micro-Tom
		Garden radish	*Raphanus sativus*	Cherry Bomb
		Barley	*Hordeum vulgare*	K-305
		Barley	*Hordeum vulgare*	Haruna Nijo
		Rice	*Oryza sativa*	Kasalath
		Rice	*Oryza sativa*	Murasaki Mai
Vertebrates	Dried fish eggs	Killifish	*Nothobranchius guentheri*	–
Insects	Cryptobiotic larvae	Chironomidae	*Polypedilum vanderplanki*	–
Lower crustaceans	Resting eggs	*Daphnia*	*Daphnia magna*	–
		Daphnia	*Daphnia pulicaria*	–
		Brine shrimp	*Artemia salina*	–
		Triops	*Triops cancriformis*	–
		Fairy shrimp	*Streptocephalus torvicornis*	–
		Ostracoda	*Eucypris ornata*	–
Mold fungi	Spores	*Aspergillus*	*Aspergillus sydowii*	#9-6
		Aspergillus	*Aspergillus versicolor*	#4-3-4
		Penicillium	*Penicillium aurantiogriseum*	#9-9
		Penicillium	*Penicillium expansum*	#4-3-3
Bacteria	Spores	*Bacillus*	*Bacillus subtilis*	#20
			Bacillus licheniformis	#24
			Bacillus subtilis	#25
			Bacillus subtilis	#2335/105

The first container was exposed on the outside of the Russian segment of the International Space Station from July 6, 2007, to August 15, 2008

Plant seeds, resting eggs of lower Crustacea, bloodworms, and dried eggs of killifish were placed into cotton cloth bags and then into plastic petri dishes (Fig. 6.10). Afterward, the equipped Biorisk was transferred to the Energia block to be delivered to the ISS RS on April 15, 2007, by the Soyus-TMA-10 spaceship. The Biorisk device was fastened onto the outside of the ISS by the Russian cosmonauts F.N. Yurchishin and O.V. Kotov on June 6, 2007, during extravehicular activity (EVA). Because of EVA rescheduling in the ISS RS, the first container was removed 13 months (instead of 6 months) after that (on July 15, 2008) by the astronauts S.A. Volkov and O.D. Kononenko. The first container was returned to Earth on October 24, 2008. At 5 p.m. on October 24, 2008, it was delivered to the Institute of Biomedical Problems. Examination of the metal container and withdrawal of the petri dishes containing the test objects were conducted immediately. An external visual examination showed that the plastic petri dishes were considerably deformed. When they were opened, the metal plates with the bacterial and fungal spores were found to be fused to the plastic, whereas the cotton cloth bags containing seeds and resting eggs of crustaceans and fish remained visibly unchanged.

The resting eggs of the lower Crustacea and fish were examined at the Zoological Institute of the Russian Academy of Sciences (St. Petersburg, Russia). The phyllopod resting eggs remained viable, although the rate of egg activation in the experiment (on *Artemia salina*) was significantly lower (by a factor of about five) than that in the control. The survival of resting eggs of the tadpole shrimp *Triops cancriformis* ranged from about 5% to about 8%. Although only a small proportion of the experimental eggs were capable of reactivating, the growth of the hatched tadpole shrimps did not differ from that in the control group. The ostracod *Eucypris ornata* was also shown to retain its viability and capability for successful reproduction. The reactivation of the *D. magna* eggs started after a significant lag (on November 25, 2008) as compared with the reactivation of the phyllopod and ostracod eggs. The flight sample of reactivated *D. magna* could be divided into three groups of the same size, differing in their degree of damage: (1) individuals that died immediately after birth, (2) those surviving until sexual maturity but incapable of reproduction, and (3) those capable of reproducing several viable fry generations. Six seeds of the mustard *Brassica rapa* anthocyaninless line and three seeds of the radish *Raphanus sativus* variety Cherry Bomb were soaked on October 25. On October 28, 2008, all of the seeds germinated.

Regarding the exposure characteristics on the outside of the ISS RS, it should be noted that in contrast to the first Biorisk-MSN experiment, the container heating during exposure was much more vigorous in the second experiment. Unfortunately, this equipment had no temperature sensor; however, it may be suggested that the heating of the plastic dishes reached 95 °C, because in the model experiment, the plates were deformed at this temperature. In a vacuum, heat transmission occurs when materials differing in thermal conductivity come into contact. Cotton cloth bags may be expected to be good heat insulators, because of which the heating of the test objects within these bags was not too vigorous.

Unlike the other test objects, the bacterial and fungal spores underwent heating, which was apparently comparable to that of the petri dishes, because these spores were applied to metal plates that were in direct contact with the plastic of the dishes.

Because of this, the bacterial spore survival was much lower in this experiment (that of the *Bacillus subtilis* spores ranged from 0.002% to 0.02% of the baseline level).

6.3.1 Conclusion of This Biorisk Experiment

The goal of the experiments was to accumulate data about potential manifestations or limits of phenotypic adaptation and genotypic changes in bacterial–fungal associations regularly found on spacecraft construction materials. The primary purpose of the first experiment in the series was to verify the viability of test species within microbial–construction material systems after long-term exposure (comparable to the duration of a Mars mission) in open space. The secondary objective was to evaluate space flight effects on the biospecimens under study. As is known, not only microbes but also some highly organized organisms have the ability to enter a dormant phase under adverse environmental conditions. It is no exaggeration to say that many living creatures may reach a state in which their vital functions are temporarily stopped to survive unfavorable environmental effects, even in outer space conditions. These states differ in terms of the mechanisms that allow various organisms to reach dormancy, as well as in the levels of dormancy that ensure their survival. In diapause and cryptobiosis, organisms can survive a wide range of extreme chemical and physical effects (Danilevsky 1961), thus making possible the transfer of living matter across space.

In view of this, the second Biorisk experiment became a new step in understanding the survival limits of microorganisms, as well as more highly organized specimens—including dormant stages of higher plants, insects, and lower crustaceans—after their exposure to outer space for a time similar to the duration of a flight to Mars.

6.4 Expose Program—Outer Space Biological Experiment at the ISS to Support a Mars Mission

The multipurpose Expose device (Fig. 6.11) was prepared for long-term astrobiological experiments in outer space at the ISS. It was created by the European Space Agency (ESA) to test degradation of biological materials (amino acids) and spores of primitive organisms (bacteria, fungi) in space (within meteorites) and in the conditions on the surface of Mars (de Vera et al. 2019).

In contrast to Biorisk, Expose was able to record or even regulate different types of environmental data (radiation, temperature) during exposure of biological samples to space, including ultraviolet radiation (Novikova et al. 2018; de Vera et al. 2019).

The first device in this series, Expose-E, was prepared for delivery to the ISS by the US National Aeronautics and Space Administration (NASA) (Rabbow et al. 2009). For certain reasons, NASA was not able to transport it to the orbiter, and so Expose-E became an Earth-treated pilot project.

Fig. 6.11 Expose outer space device

The next ESA project, Expose-R, was transported to the ISS in Russian cargo and placed outside the ISS on the Russian Zvezda module (Rabbow et al. 2009). Along with the aforementioned set of objects prepared by the ESA, it contained a biodiversity block with seeds of higher plants and dormant stages of several aquatic invertebrates that had survived in Biorisk outer space experiments (Baranov et al. 2009; Orlov et al. 2017). All of these objects were placed in small plastic bags (measuring 10 × 10 mm and 20 × 20 mm) in three layers.

6.4.1 Expose-R and Expose-R2 Experiments on Dormancy

After more than 1.5 years of outer space exposure, the spores of microorganisms and fungi, as well as two species of plant seeds (*Arabidopsis thaliana* and tomato) and dormant crustacean eggs (*Artemia franciscana*, *E. ornata*, *D. magna*, and *S. torvicornis*) were analyzed for viability and a set of biological properties. The experiment provided evidence that not only bacterial and fungal spores but also seeds (dormant forms of plants), as well as dormant eggs of some crustaceans, have the capability to survive long-term exposure to outer space, including ultraviolet radiation, with minimal protection in the first layer of experimental samples. Dormant eggs of two crustaceans (*A. franciscana* and *E. ornata*) showed increasing rates of reactivation in the second and third layers of samples (Fig. 6.12), which reflected the decline in the penetration of ultraviolet radiation. Two other species of aquatic crustaceans (*D. magna* and *S. torvicornis*) did not survive in any of the three layers of samples. Even bacteria and fungi in the first layer, exposed to direct ultraviolet radiation, were killed (de Vera et al. 2019). At the same time, the microbes showed very high rates of survivorship (up to 90% in comparison with control samples) in darkness, which was never was found in the Biorisk experiments.

Expose-R demonstrated that the viability of *Artemia* cysts and dormant Ostracoda eggs increased as the ultraviolet radiation intensity decreased. Dormant *Daphnia* and *Streptocephalus* eggs could not survive exposure to open space in the Expose-R

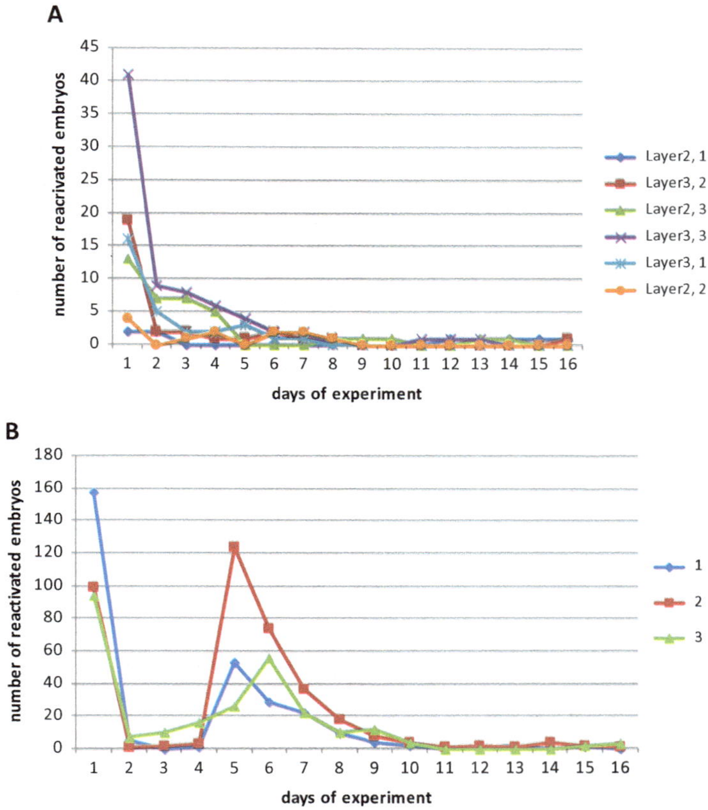

Fig. 6.12 Hatching of *Artemia franciscana* dormant cysts: **a** after 22 months of exposure to outer space within the Expose-R device (layers 2 and 3); **b** in Earth laboratory control conditions. Each variant was done in three replicates (1, 2, 3); the numbers of cysts were 600 per sample in the control group and 150 per sample in the International Space Station group

conditions. The main factor that was harmful to the last two species was ultraviolet radiation; as in our previous Biorisk long-term experiments, about 30% of the resting eggs of these species survived, successfully passed embryonal development, and established normal bisexual breeding populations, at least in two succeeding generations.

To overcome such harmful (lethal!) effects of solar light on live organisms, the construction of the Expose-R2 windows was modified. They were covered with calibrated glass filters for creation of different levels of ultraviolet radiation. In many bacteria, fungi, and plant species, it was again shown that survivorship of dormant stages in outer space increased as ultraviolet radiation levels decreased (Novikova et al. 2018). Expose-R2 was also launched with dormant stages of two invertebrates (the mosquito *Polypedilum vanderplanki* and the tadpole shrimp *T. cancriformis*), but the results of these experiments have not been published yet.

6.5 Summary of Outer Space Experiments on Dormancy

To sum up, our observations on the survivability of biological objects in the space environment have both theoretical and practical applications.

The Biorisk study was the first to demonstrate that bacterial and fungal spores, as well as higher plants and arthropods, can survive extended (up to 23-month) exposure to the harsh environment of outer space. This suggests that terrestrial and aquatic organisms could be transported to other planets by our space vehicles, particularly manned spacecraft, despite stringent waste management procedures.

Comparison of the results obtained in the Biorisk and Expose experiments, in which microbial and fungal spores (as well as dormant stages of plant and aquatic animals) were exposed to outer space factors for 23 and 22 months, respectively, allows the following conclusion to be drawn: that cosmic ultraviolet radiation is the most powerful hostile factor that affects living matter in the space environment. In the Expose-R experiment, practically all of the organisms in the top layer of the device died.

Dormancy of tropical organisms (dried eggs of cyprinodont fish) and summer-dormant eggs (ephippia of *Daphnia pulicaria*) did not protect organisms long term and cannot be used in space experiments. Dormant eggs of two crustaceans (*D. magna* and *S. torvicornis*) were sensitive to stress signals inside the ISS but did not survive the solar radiation outside the ISS. *A. franciscana* and *E. ornata* showed the highest resistance among multicellular organisms and have good potential for astrobiological experiments in outer space environments.

The dormancy concept is of great interest in regard to our search for extraterrestrial life. Our recent discoveries regarding long-term survivorship of dormant stages of algae, bacteria, fungi, and even invertebrates in outer space raise our hopes of finding life beyond the Earth. This implies the possibility of interplanetary transfer of life-forms by meteorites, comets, and human-mediated transportation.

Finally, we can again postulate that heavenly bodies, where terrestrial forms can hardly exist, may be inhabited by organisms that remain dormant for the major portion of their life cycles. The conditions on Mars, as were imitated in the Expose program, revealed the strong possibility (up to 90% for bacteria) to survive in a light-protected environment (in the ground or in permanent shadow), even for Earth organisms. On one hand, this increases the possibility of biological invasion of an intact planet by Earth life-forms, but, on the other hand, it shows that there is a chance of discovering native life on Mars, somewhere in permanent shadow (deep holes, grottos, and caves) or under the surface near the polar ice-melting area.

Acknowledgements This study was partly supported by joint Russian--Japan project N 05-04-50914a, Russian Foundation for Basic Research (RFBR) project 17-04-00027-a, and the Russian Academy of Science (topics 65.4 and 65.5). Victor R. Alekseev received a fellowship from the Max Planck Society.

The authors gratefully acknowledge the assistance of Professor Winfried Lampert (Max Planck Institute of Limnology, Ploen, Germany) for valued discussions, Vladimir Shorin (Senior Specialist in the Fishery Department, Ikrjanoye Municipal Administration, Astrakhan District, Russia) for help in obtaining the *Daphnia* resting eggs used in this study, and Maren Volquardsen (Max Planck Institute of Limnology, Ploen, Germany) for help with algal and *Daphnia* cultures.

References

Alekseev V, Fryer G (1998) Diapause in the crustaceans. Kluwer, The Netherlands

Alekseev V, Lampert W (2001) Maternal control of resting-egg production in *Daphnia*. Nature 414:899–901

Alekseev V, Sychev V (2006) Effects of space station conditions on resting egg survivorship and life cycle parameters in *D. magna*. Mat. 36 COSPAR Scientific Assembly, Beijing

Alekseev VR, de Stasio BT, Gilbert JJ (eds) (2007a) Diapause in aquatic invertebrates: theory and human use, Monographiae Biologicae 84. Springer, Dordrecht. 257 p

Alekseev V., Novikova N., Sychev V. (2007b) Studying the phenomenon dormancy: why it is important for space exploration? In: Alekseev, V., De Stasio B. Gilbert J. (eds) Diapause in aquatic invertebrates. Dordrecht Springer 207–216

Baranov V, Novikova N, Polikarpov N, Sychev V, Levinskikh M, Alekseev V, Okuda T, Sugimoto M, Gusev O, Grigor'ev A (2009) The Biorisk experiment: 13-month exposure of resting forms of organism on the outer side of the Russian segment of the International Space Station: preliminary results. Dokl Biol Sci 426:267–270. https://doi.org/10.1134/S0012496609030223

Danilevsky AS (1961) Photoperiodism and seasonal development of insects (Fotoperiodizm i sezonnoe razvitie ou nasekomukh). Leningrad University Press, Leningrad

de Vera J-P, Alawi M, Backhaus T, Baqué M, Billi D, Böttger U, Berger T, Bohmeier M, Cockell C, Demets R, Noetzel R d l T, Edwards H, Elsaesser A, Fagliarone C, Fiedler A, Foing B, Foucher F, Fritz J, Hanke F, Herzog T, Horneck G, Hübers H-W, Huwe B, Joshi J, Kozyrovska N, Kruchten M, Lasch P, Lee N, Leuko S, Leya T, Lorek A, Martínez-Frías J, Meessen J, Moritz S, Moeller R, Olsson-Francis K, Onofri S, Ott S, Pacelli C, Podolich O, Rabbow E, Reitz G, Rettberg P, Reva O, Rothschild L, Sancho LG, Schulze-Makuch D, Selbmann L, Serrano P, Szewzyk U, Verseux C, Wadsworth J, Wagner D, Westall F, Wolter D, Zucconi L (2019) Limits of life and the habitability of Mars: the ESA space experiment BIOMEX on the ISS. Astrobiology 19((2)):145–157. https://doi.org/10.1089/ast.2018.1897

Ebert D (1993) The tradeoff between offspring size and number in *Daphnia magna:* the influence of genetic, environmental and maternal effects. Arch Hydrobiol Suppl 90:453–473

Ebert D (2005) Ecology, epidemiology, and evolution of parasitism in *Daphnia*. National Center for Biotechnology Information, Bethesda

Glazier DS (1992) Effects of food, genotype, and maternal size and age on offspring investment in *Daphnia magna*. Ecology 73:910–926

Horneck G, Rettberg P, Schott J-U, Panitz C, L'Afflitto A, von Heise-Rotenburg R, Willnecker R, Baglioni P, Hatton J, Dettmann J, Demets R, Reitz G, Rabbow E (2009) EXPOSE, an astrobiological exposure facility on the International Space Station – from proposal to flight. Orig Life Evol Biosph 39(6):581–598. https://doi.org/10.1007/s11084-009-9173-6

Lampert W, Schmitt RD, Muck P (1988) Vertical migration of freshwater zooplankton: test of some hypotheses predicting a metabolic advantage. Bull Mar Sci 43:620–640

Lynch M (1980) The evolution of cladoceran life histories. Q Rev Biol 55:23–42

Mousseau TA, Fox CW (1998) Maternal effects as adaptations. Oxford University Press, New York, pp 1–375

Muller H (1972) Wachstum und Phosphatbedarf von Nitzschia actinastroides (LEMM.) v. Goor in statischer und homokontinuierlicher Kultur unter Phosphatlimitierung. Arch Hydrobiol Suppl 38:399–484

Novikova N, Poddubko S, Deshevaya E, Sychev V (2018) Survival of bacterial and fungal spores in the expose-r2 experiment. Int J Biochem 7(4):5616–5622

Orlov OI, Novikova ND, Polikarpov NA, Levinskikh MA, Deshevaya EA, Sugimoto M, Alekseev VR, Okuda T, Gusev OA, Sychev VN (2017) Planetary protection challenges in space exploration missions and ways of their resolution with account of Russian exobiology experiments. REACH 6:25–33. https://doi.org/10.1016/j.reach.2017.07.001

Chapter 7
Resistance of Dormant Eggs of Cladocera to Anthropogenic Pollutants

Egor Zadereev, Tatiana S. Lopatina, and Natalia Oskina

Abstract Many Cladocera species under unfavourable conditions produce resting eggs that can survive for years in deep diapause. Resting eggs form egg banks at the bottom sediments, which serve as a source of genetic diversity and replenish the population after periods of decline. Despite the obvious importance of resting eggs for the ecosystem functioning, studies assessing the sensitivity of resting eggs to different toxicants are scarce. We reviewed published data on the sensitivity of resting eggs to the effect of heavy metals, organic pollutants and ionizing radiation. Analysis shows that the effects of toxicants of different types on resting eggs will have different environmental consequences. Egg banks may suffer from prolonged contact of dormant eggs with heavy metals. However, the ecological relevance of these effects is low, since the effective concentrations of toxicants must be very high. In addition, the effect of heavy metals on resting eggs is not transmitted to hatchlings from exposed eggs. Taking into account high toxicity of heavy metals to active animals, we assume that the toxic effect of heavy metals is critical for active zooplankters and relatively safe for resting stages. Accumulation of artificial radionuclides in bottom sediments can have a significant impact on aquatic ecosystems through chronic effects both on survival of resting eggs and on the life cycle parameters of animals hatched from irradiated eggs. Resting eggs during reactivation are more sensitive to the effect of ionizing radiation. Pesticides and complex chemical compounds produce similar toxic effects on both resting eggs and life history parameters of hatchlings from exposed resting eggs. The bottom line is that to predict the effect of contamination of bottom sediments by different pollutants on the ecosystem structure and functioning, it is highly important to investigate the viabil-

E. Zadereev (✉) · N. Oskina
Institute of Biophysics, Krasnoyarsk Research Center SB RAS,
Akademgorodok, Krasnoyarsk, Russia

Siberian Federal University, Krasnoyarsk, Russia
e-mail: egor@ibp.ru

T. S. Lopatina (✉)
Institute of Biophysics, Krasnoyarsk Research Center SB RAS,
Akademgorodok, Krasnoyarsk, Russia

V. R. Alekseev, B. Pinel-Alloul (eds.), *Dormancy in Aquatic Organisms. Theory, Human Use and Modeling*, Monographiae Biologicae 92,
https://doi.org/10.1007/978-3-030-21213-1_7

ity of the resting eggs under a wide range of concentrations of various toxicants and different durations of direct contact of eggs with toxicants.

Keywords Resting eggs, Resistance, Heavy metals, Ionizing radiation, Toxicity

7.1 Introduction

Cladocerans play a vital role in aquatic food webs, transferring organic carbon from primary producers to fish or invertebrate predators. Under favourable conditions, they reproduce by parthenogenesis, which allows them to increase in numbers quickly and control the lower food web level (Lampert et al. 1986). Under adverse conditions, many species of Cladocera produce resting eggs (Alekseev et al. 2007). The production of resting eggs is the strategy to survive either seasonal or occasional unfavourable conditions (Radzikowski 2013). Usually in natural habitats, cladocerans produce diapausing egg at the end of the growing season (Alekseev et al. 2007). Diapausing eggs produced in different years are accumulated at the bottom sediments, where they form the egg bank. The bank of dormant eggs is important for ecology and evolution of zooplankton. It is a source of genetic diversity; natural populations can recover from the egg bank, which supports the stability of zooplankton communities (Brendonck and De Meester 2003; Hairston 1996).

Resting eggs accumulated in the sediments can be exposed to various unfavourable factors. However, dormant organisms can stay alive in the resting form for a long time and demonstrate high tolerance to different unfavourable factors. The resistance of dormant forms to abiotic factors, such as desiccation or freezing and salinity stress, has been well characterized in recent decades. Radzikowski (2013) reviewed the survival of planktonic animals under adverse abiotic factors and reported longevity of dormant forms of selected aquatic invertebrates. For example, dry resting eggs of *Daphnia magna* can tolerate extreme temperatures of -84 °C and 110 °C during several hours. Nielsen et al. (2012) demonstrated that resting eggs of freshwater zooplankton are able to survive without decline of hatching success exposure to salt concentration of 13.5 g/l of sea salt for a period of 22 months. The maximal reported age of viable dormant forms of *D. pulicaria* and *D. galeata mendotae* was 125 years (Radzikowski 2013) and of the copepod *Diaptomus sanguieneus* was 322 years (Hairston Jr. et al. 1995).

In addition to natural factors that affect resting eggs in the bottom sediments, they can be influenced by anthropogenic factors. Taking into account high resistance of resting eggs to adverse conditions, it is not surprising that they tolerate high concentrations of toxic substances (e.g. Jiang et al. 2007; Alekseev et al. 2010; Raikow et al. 2007a, b). At the same time, bottom sediments accumulate compounds that enter the water body, including toxic ones. As a result, concentrations of toxicants in the sediments may exceed those in the water column and pose a certain threat to bottom biota (e.g. Avila-Perez et al. 1999; Jiang et al. 2007).

Some studies also demonstrated that toxic compounds not only directly affected resting eggs (e.g. the efficiency of reactivation) but also influenced life cycle parameters of active animals hatched from eggs exposed to toxicants (e.g. Navis et al. 2013; Rogalski 2015). It is very important to understand whether the contact with a toxicant during the diapause will affect the parameters of growth and reproduction of animals hatched from exposed eggs.

In this chapter, we review published studies and present our results on the effects of various anthropogenic pollutants on the resting eggs of zooplankters. Pollutants that we cover in this review are heavy metals, pesticides and other complex chemical compounds, and radioactive materials. After a brief review, we discuss environmental relevance of the effects observed and suggestions for future studies.

7.2 The Effects of Heavy Metals and Organic Pollutants on Resting Eggs and Hatchlings from Exposed Eggs

Heavy metals are of high ecological importance. They demonstrate variable toxicity to active Cladocera species, and, hence, they are very often used as model toxicants in toxicological research (Wong and Wong 1990; Wong et al. 1991). Alekseev et al. (2010) showed that resting eggs of *Moina macrocopa* remained viable after 100 days of incubation in the solution with a copper concentration of 500 mg/l. Experiments with resting eggs of *Acartia pacifica* showed reduced hatching success of resting eggs after exposure to 348 mg/kg of copper and 6.8 mg/kg of cadmium (Jiang et al. 2007). Zinc concentrations of up to 5 mg/l did not affect the survival of resting eggs of *Artemia parthenogenetica* (Sarabia et al. 2008).

In experiments with resting eggs of *M. macrocopa*, we observed resistance of resting eggs to higher concentrations of heavy metals (up to 60–70 g/l) than was demonstrated previously (Lopatina et al. 2017; Oskina et al. 2018). First, we tried to determine the critical concentrations of heavy metals that are lethal for resting eggs after one-month exposure of eggs to a wide range of concentrations of heavy metals dissolved in water, but survival and hatching success of resting eggs were not affected. We used concentrations of heavy metals (from the background level to 60–70 g/l) that far exceeded toxic thresholds of these metals for active animals. When the exposure to toxicants was prolonged, we observed some adverse effects. In the experiment with contaminated artificial sediment, hatching of resting eggs was completely suppressed after 8-month exposure to high concentrations of copper (34 and 67 g/kg). None of the other treatments with sediment contaminated by different concentrations (from the background level to 60–70 g/l) of cadmium, zinc or nickel affected survival of resting eggs.

Our experiments showed that resting eggs could tolerate high concentrations of heavy metals. Recently, it was shown that resting eggs of cladoceran *Daphnia* are good biosorbent of heavy metals (Sacmaci et al. 2014). Thus, most probably protective structures of the ephippium are able to absorb high concentrations of ions of heavy metals and protect enveloped embryos from toxic exposure. However, the toxic effect of heavy metals on reactivation of eggs was manifested when we extended the duration of contact between eggs and toxicants. With the increased exposure time, the viability of resting eggs was reduced under the effect of copper, which was the most toxic heavy metal of the selected toxicants. If the viability of resting eggs is determined by the diffusion of a toxicant through the protective wall of the ephippium, the toxic effect of copper should be manifested first, as this metal is the most toxic of the selected heavy metals. We assume that longer contact with other heavy metals will also result in unfavourable effects. Indeed, Rogalski (2015) demonstrated that the concentration of heavy metals in the bottom sediments of the lake had a significant effect on the hatching success of *Daphnia* resting eggs extracted from the sediment core. However, at least for the period of up to 8 months, resting eggs protected offspring from the effect of heavy metals.

Raikow et al. (2006) examined the acute toxicity of the biocide SeaKleent (menadione), which is used for chemical treatment of residual material in ship ballast tanks, on resting eggs of several zooplankton species. SeaKleen was toxic to resting eggs of all the tested taxa. *D. mendotae* resting eggs encased in protective ephippia were the least sensitive, as indicated by 90% mortality of organisms after 24-h exposure to an 8.7 g/l concentration of the toxicant. SeaKleen induced teratogenic effects in *D. mendotae* and *Artemia* sp. In a study of the acute toxicity of glutaraldehyde and sodium hypochlorite for the resting eggs of *D. mendotae* and *Artemia* sp., Raikow et al. (2007a, b) demonstrated that glutaraldehyde was toxic to resting eggs of *Artemia* sp., while *D. mendotae* displayed erratic responses to glutaraldehyde. Sodium hypochlorite was similarly toxic to resting eggs of *Artemia* sp. and *D. mendotae*, which displayed LC_{90} of 86.5 ± 3.0 and 78.3 ± 1.6 mg/l, respectively.

Most et al. (2015) demonstrated the stimulating effect of organic pollutants on the hatchability of resting eggs of *D. longispina*. However, the offspring hatched from the eggs showed increased mortality and developmental anomalies associated with the presence of toxicants in the water. These results are difficult to interpret due to short exposure times and the presence of toxicants not only during the resting period but also at the time of reactivation of eggs. We proposed (Lopatina et al. 2017; Oskina et al. 2018) that variable resistance of resting eggs to heavy metals could be explained by methodological differences between studies. In preliminary experiments, we observed high mortality of juveniles hatched from resting eggs exposed to high concentrations of heavy metals. However, when resting eggs were thoroughly washed, no mortality of hatchlings was observed. We assumed that mortality of animals was associated with the diffusion of heavy metals bound to the surface of resting eggs to the culture medium. When we excluded the effect of water pollution by toxicants during the reactivation of eggs, we observed their high resistance to heavy metals. In most of the studies that we cited before, the method sections did not specify whether the resting eggs had been washed after exposure to toxicants. In general, standardization of techniques is considered as one of the key problems in studying the sensitivity of resting eggs of zooplankton to the effect of toxic compounds (Luc Brendonck, pers. communication).

We also found (Lopatina et al. 2017, Oskina et al. 2018) that parameters of the life cycle (life span, juvenile growth rate, number of hatched clutches, fecundity) of females hatched from resting eggs that had been exposed to heavy metals in a wide range of concentrations did not differ from those of the group of control animals. Taking into account high susceptibility of active cladocerans to the toxic effect of heavy metals, we assumed that when a toxicant penetrated through the protective ephippia, it caused mortality of encapsulated eggs. When resting eggs remained viable, hatched offspring did not differ from the control group.

The effects of heavy metals on performance of hatchlings from resting eggs exposed to toxicants should be different from the effects of toxic compounds that can modify life cycle parameters. For example, exposure to pesticides (carbaryl and fenoxycarb) affected both the reactivation of exposed resting eggs of *D. magna* and the survival and life cycle parameters of hatchlings from these eggs (Navis et al. 2013). Later, in the study of the sensitivity of resting eggs of *D. magna* at different stages of embryonic development to fenoxycarb, the authors showed that ephippium did not protect eggs against the pesticide (Navis et al. 2015). The concentration of the pesticide in the embryo tissues in the resting egg did not differ from that in an embryo devoid of the protective shell. Hence, the lethal concentration of fenoxycarb for resting eggs was only twice as high as that for living *Daphnia*. Thus, studies demonstrate that sensitivity of resting eggs to different toxicants varies considerably. Estimating the toxicity of heavy metals, organic substances and low-molecular-weight compounds, Alekseev et al. (2010) suggested that low-molecular-weight compounds were more toxic to resting eggs. Assessment of the toxicity of various compounds for resting eggs should probably require a detailed study of the structure of the ephippium and modelling of the interactions of toxicants with these protective structures.

We can summarize that resting eggs are quite resistant to the effect of heavy metals. Studies demonstrated that the survival of eggs was affected by both the duration of contact with heavy metals and their concentrations (Jiang et al. 2007; Rogalski 2015). The effect of heavy metals on resting eggs is mainly manifested as increased mortality of eggs. Performance of hatchlings from resting eggs exposed to heavy metals is usually not affected. Effects of pesticides and complex chemical compounds are observed at lower concentrations. Not only do such toxicants decrease the survival of resting eggs, but they also affect the parameters of the life cycle of hatchlings.

7.3 The Effect of Ionizing Radiation on Resting Eggs and Hatchlings from Exposed Eggs

Data on the effect of ionizing radiation on aquatic invertebrates are fragmentary and incomplete (Dallas et al. 2012). Over the past century, the level of background radioactivity has increased due to various anthropogenic activities (United Nations

Scientific Committee on the Effect of Atomic Radiation 2010). Artificial radionuclides enter aquatic ecosystems due to discharges from nuclear-power facilities, washouts from water catchment areas and nuclear fall-outs (Van der Stricht and Janssens 2010). Bottom sediments accumulate various anthropogenic pollutants including radionuclides (e.g. Kansanen et al. 1991), which expose benthic dwellers to adverse effects.

Numerous experimental data demonstrated adverse effects of ionizing radiation on active cladocerans (e.g. Alonzo et al. 2008; Massarin et al. 2010; Sarapultseva and Gorski 2013). Recent studies have also demonstrated the transgenerational effects of parental exposure to ionizing radiation on survival and reproduction of exposed females of *D. magna* and their offspring (Sarapultseva and Dubrova 2016).

Studies of the effect of ionizing radiation on resting eggs of cladocerans are scant. The ability of resting eggs of *Daphnia* sp. to endure the conditions of outer space was tested at the International Space Station (ISS) (Novikova et al. 2011). Despite the high-temperature differences and the impact of ionizing radiation (absorbed dose was equal to 2–3 Gy), part of the resting eggs retained the ability for reactivation (Novikova et al. 2011). It was noted that radiosensitivity correlated positively with the rates of metabolic processes, which resulted in the tolerance to the effect of radiation of dormant eggs of aquatic invertebrates (Eisler 1994) or animals in cryptobiosis (Watanabe 2006). Thus, we should expect relatively high tolerance of resting eggs to the effect of radiation. Some anhydrobiotic invertebrates show extremely high tolerances against radiation. A tardigrade, *Macrobiotus areolatus*, tolerates exposure to 5500 Gy of X-ray (May et al. 1964). LD_{50} for dry Artemia cysts can be as high as 5000 Gy of Co_{60} gamma radiation (Iwasaki 1964).

Active invertebrates can also tolerate high doses of radiation. Some bdelloid rotifers were able to tolerate doses of up to 600 Gy (Gladyshev and Meselson 2008). Gladyshev and Meselson (2008) suggested that the radiation resistance of bdelloid rotifers was a consequence of their ability to survive the desiccation. The damage incurred in desiccation episodes includes the DNA breakage that is repaired upon rehydration. Thus, these species have developed effective mechanisms of DNA repair that can be activated to minimize the unfavourable effects of radiation.

Recently, we determined (Zadereev et al. 2017) the dose-response relationship for the effect of gamma radiation on hatching success of resting eggs of *M. macrocopa*. The absorbed dose of 200 Gy was lethal to resting eggs. Gamma irradiation in the range of doses from the background level to 100 Gy had no effect on the survival of resting eggs. Even though survival and hatching of resting eggs was not affected by doses below 100 Gy, the effect of radiation was manifested in hatched neonates at individual and population levels. The number of clutches and net reproductive rate (R_0) of hatchlings from eggs exposed to radiation were the strongly affected parameters in experiments with individual females. The number of clutches per female was drastically reduced for females hatched from eggs exposed to 80–100 Gy. The effective dose that induced 50% R_0 reduction was 50 Gy.

Life cycle parameters of active animals had different sensitivities to the irradiation of resting eggs. Life span and growth rate of juvenile females were relatively non-sensitive parameters. The proportion of males in the progeny was too variable a parameter to be used as a sensitive endpoint. Reproduction parameters (the net reproductive rate, fecundity) of active animals were the most sensitive parameters of life cycle in response to the effect of ionizing radiation on animals during the resting stage. These findings are similar to the results of several experiments with active animals exposed to ionizing radiation. For example, the size of clutches hatched by *D. magna* decreased at radiation doses greater than 0.1 Gy. At the same time, neither survival nor somatic growth rate of females were affected by radiation doses of 1–2 Gy (Gilbin et al. 2008). Sarapultseva and Dubrova (2016) also demonstrated that the fecundity of *Daphnia* significantly decreased at a dose of 0.1 mGy or higher while the survival was significantly compromised only for *Daphnia* exposed to 1 and 10 mGy of acute γ-rays. Moreover, Alonzo et al. (2016) modelled population responses to chronic external gamma radiation in 12 laboratory species (including aquatic and soil invertebrates, fish and terrestrial mammals) and found that the net reproductive rate showed the lowest EDR_{10} (effective dose rate inducing 10% effect) in all species.

While diapausing eggs can tolerate high doses of radiation, there is a knowledge gap on the sensitivity of resting eggs to the effect of radiation during hatching. When placed in favourable conditions, the resting egg hatches in several days. During this short period, the embryo should be very sensitive to the effect of radiation, as aquatic organisms exhibit increased sensitivity to radiation during active metabolism (Donaldson and Foster 1957). Indeed, we observed higher sensitivity of resting eggs when we exposed them to gamma radiation during the reactivation. The survival rate of resting eggs was decreased at the dose of 100 Gy, the growth rate of juveniles sharply decreased for hatchlings from eggs exposed to doses 24 Gy or higher, and reproduction parameters dramatically decreased for animals hatched from eggs exposed to doses of 64 Gy or higher (Lopatina et al. 2018). We can compare sensitivity of resting eggs of *M. macrocopa* to the effect of gamma radiation during deep dormancy and during reactivation. Neither the survival of resting eggs nor life span or growth rate of hatchlings from eggs irradiated in the deep dormancy were affected by the doses of up to 100 Gy (Zadereev et al. 2016; Zadereev et al. 2017). The reproductive parameters of females that hatched from irradiated eggs dramatically decreased at the exposure doses of 80 and 100 Gy. Thus, hatchlings from resting eggs of *M. macrocopa* exposed to gamma radiation during the reactivation demonstrated higher sensitivity (Lopatina et al. 2018).

There are fragmentary data on the irradiation of resting stages of zooplankton during reactivation. For example, wet cysts of *Artemia* (initial stage of reactivation) were more than twice radiosensitive to an irradiation dose of 500 Gy compared with dry cysts (Donaldson and Foster 1957). Similar results were obtained for other aquatic animals. For example, in fish of the genus *Oryzias*, chromosomal aberrations were found when eggs at their early stages of development (immediately after fertilization) were exposed to beta- (more than 0.19 Gy) and gamma (more than 0.58 Gy) radiation (Suyama et al. 1981). The mortality of eggs of mollusc *Milax nigricans* in the stage of two blastomeres increased at a dose of gamma radiation equal to 0.2 Gy and became absolute at an exposure dose of 0.8 Gy (Zampi and Focardi 1973). Embryos of the *Milnesium cf. tardigradum* exhibited higher sensitivity to radiation exposure during their early stage of development compared to the middle and late stages of development; egg mortality increased at radiation doses above 50 Gy (Beltrán-Pardo et al. 2013).

A recent study demonstrated the transgenerational effect of parental exposure to ionizing radiation on survival and fertility of directly exposed *D. magna* females and their first-generation progeny (Sarapultseva and Dubrova 2016). At the same time, F2 progeny of irradiated F0 *Daphnia* substantially recovered (Sarapultseva and Dubrova 2016). We also studied both maternal and transgenerational effects of exposure of resting eggs to ionizing radiation in population experiments (Zadereev et al. 2017). We performed experiments with populations of *M. macrocopa* that comprised at least four generations of females that were hatched from irradiated resting eggs. We found that population performance was affected by the irradiation of the resting stage of animals that initiated the population. However, the population response was detected at a higher level than individual responses. Only populations that were initiated from hatchlings from resting eggs exposed to 100 Gy were of smaller size and with fewer juvenile and parthenogenetic females compared to control populations (Zadereev et al. 2017).

Polikarpov (1998) proposed a conceptual model of biological responses to different dose rates of ionizing radiation: the dose rates that produce damage to populations and communities should be higher than the dose rates that produce detectable effects in individual organisms. Responses at the individual and population levels to the acute exposure of the resting eggs to ionizing radiation (Zadereev et al. 2017) were consistent with Polikarpov's conceptual model (1998). The estimated ED_{50} for the R_0 at the individual level was 50 Gy. However, populations initiated from resting eggs exposed to 50 Gy did not differ in size from the control population; populations initiated from resting eggs exposed to 80 Gy demonstrated reduced numbers and altered structure with fewer juvenile and parthenogenetic females in the initial stage of development, but later the population size recovered. Population responses were most probably masked by the individual variability and recovery of the subsequent progeny of F0 females irradiated in the resting stage.

We can conclude that resting eggs of Cladocera irradiated at different stages of development are resistant to relatively high doses, which corresponds to the accepted notion that invertebrates in aquatic habitats show low vulnerability to the radiation factor (Eisler 1994). Even though the survival of resting eggs is not affected by relatively high doses of radiation, the reproductive performance of hatchlings is suppressed. This effect is most probably not transgenerational, as we observed recovery of subsequent generations of females irradiated at the resting stage. At the same time, resting eggs demonstrated higher sensitivity when they were exposed to ionizing radiation during reactivation. Thus, contamination of bottom sediments by radioactive materials could affect zooplankton communities through either adverse chronic effects on resting eggs, which will be transmitted to hatchlings, or acute irradiation effects on resting eggs during reactivation.

7.4 Environmental Context

7.4.1 *Heavy Metals and Organic Pollutants*

To discuss the environmental significance of the data on the sensitivity of resting eggs to different pollutants, we need to estimate whether critical concentrations or doses observed in laboratory experiments are observed in natural habitats, too. We observed high resistance of resting eggs to heavy metals over a relatively long time (up to 8 months). Most probably, such concentrations are unattainable in natural habitats, even in the case of severe pollution. For example, copper concentrations in bottom sediments reached only 5 g/kg in contaminated swampy areas around the mines in China (Deng et al. 2004), 0.5 g/kg in the places where the city run-off enters the water body (Vesk and Allaway 1997) and 5 g/kg in the mining region of Mexico (Razo et al. 2004). While we demonstrated (Oskina et al. 2018) that the critical copper concentration that induced mortality of resting eggs after 8-month exposure was 30 g/kg, for cadmium, zinc or nickel, even concentrations of 50 g/kg had no effect on the survival of resting eggs.

The active concentrations of toxicants will be lower when we consider the long-term storage of dormant eggs in bottom sediments (e.g. for several years). For example, Rogalski (2015) observed that hatching rate of *Daphnia* hatched from diapausing eggs isolated from sediments from four lakes that experienced varying levels of metal contamination (Cu, Cd, Zn, Hg, Pb) was negatively influenced by metal contamination and sediment age. It means that negative effects of heavy metals in bottom sediments on resting eggs will be significant only for eggs that have been kept for a long time in the egg bank. However, long-term storage reduces the viability of resting eggs. In addition, the contribution to the population dynamics of eggs, buried deep in sediments, is low. The age of resting eggs that actively contribute to the zooplankton community in lakes varies between 10 and 20 years (Hairston 1996; Hairston Jr. et al. 2000). Resting eggs of zooplankton can survive longer (e.g. up to 300 years (Hairston Jr. et al. 1995)). However, it is highly improbable that eggs buried deep in sediments will contribute to the population dynamics. In addition, Rother et al. (2010) demonstrated that the contribution of hatched animals to the spring population dynamics can be negligible in situations with an overwintering population present and will only become relevant if densities are extremely low. Thus, ecological significance of such toxic effects will be low.

In addition, in the case of high concentrations of heavy metals in bottom sediments, there is a realistic assumption that elevated concentrations of toxicants will be also observed in water. Then, toxic effects of dissolved toxicants on the newborn juveniles hatched from resting eggs will be more valid. Sarabia et al. (2008) when they studied the effect of zinc on resting eggs of Artemia made a similar conclusion. Authors suggested that in natural habitats chronic toxicity of heavy metals dissolved in water to active animals will be more important than the effect of toxic bottom sediments on resting eggs. However, we should not underestimate evolutionary implications of the effect of heavy metals and other toxicants on complex interactions between active zooplankton community and the resting egg bank (e.g. Gutierrez et al. 2017; Turko et al. 2016; Aránguiz-Acuña and Pérez-Portilla 2017; Rogalski 2017). For example, metals in freshwater environments can modulate diapause adaptive efficacy and the selective process in egg banks (e.g. Aránguiz-Acuña and Pérez-Portilla 2017).

Toxic effects of pesticides or complex chemical compounds are observed at lower concentrations of toxicants than the effects of heavy metals. Organic pollutants not only directly affect survival of resting eggs but also modify life history parameters of hatchlings from exposed resting eggs. Thus, effects of such toxicants can be manifested at lower concentrations and will have higher environmental significance.

7.4.2 Ionizing Radiation

To estimate activities of radionuclides observed in natural and manmade lakes affected by various nuclear events, we can mention lake Inba affected by the Fukushima Daiichi Nuclear Power Plant accident (Cao et al. 2017) and Lake Maggiore (Italy, Switzerland), which received some ^{137}Cs as a consequence of atmospheric nuclear fallouts (Putyrskaya et al. 2009). The maximal dose rate for the sediments of these habitats was estimated to be 6 mGy/h. Thus, it would take about 1900 years to accumulate the dose of 100 Gy, which would induce individual and population effects in hatchlings from resting eggs deposited in such sediments. For areas with a high level of radioactive contamination (Lake Karachai, lakes of the Southern Urals, the cooling pond of the Chernobyl NPP, and the marine cooling area of the Leningrad NPP), the estimated dose rates vary from 0.002 to 800 Gy per day (Kryshev and Sazykina 1998). The dose of 100 Gy in such sediments will be accumulated in 137 years (for a dose rate of 0.002 Gy per day) or in several hours (for a dose rate of 800 Gy per day).

We have already mentioned that the age of resting eggs that contribute to the zooplankton in lakes lies within the range of 10–20 years (Hairston 1996; Hairston Jr. et al. 2000). This time is long enough for resting eggs to accumulate critical doses in highly polluted lakes. We also observed that during reactivation, resting eggs were more sensitive to the effect of ionizing radiation than eggs in deep diapause. We can summarize that environmentally relevant unfavourable effects of ionizing radiation on the resting eggs in sediments can be manifested in areas with radioactive contamination that is high enough to produce either a cumulative dose or an acute dose during reactivation of resting eggs that will affect hatching of resting eggs and/or life cycle parameters of hatchlings (Fig. 7.1).

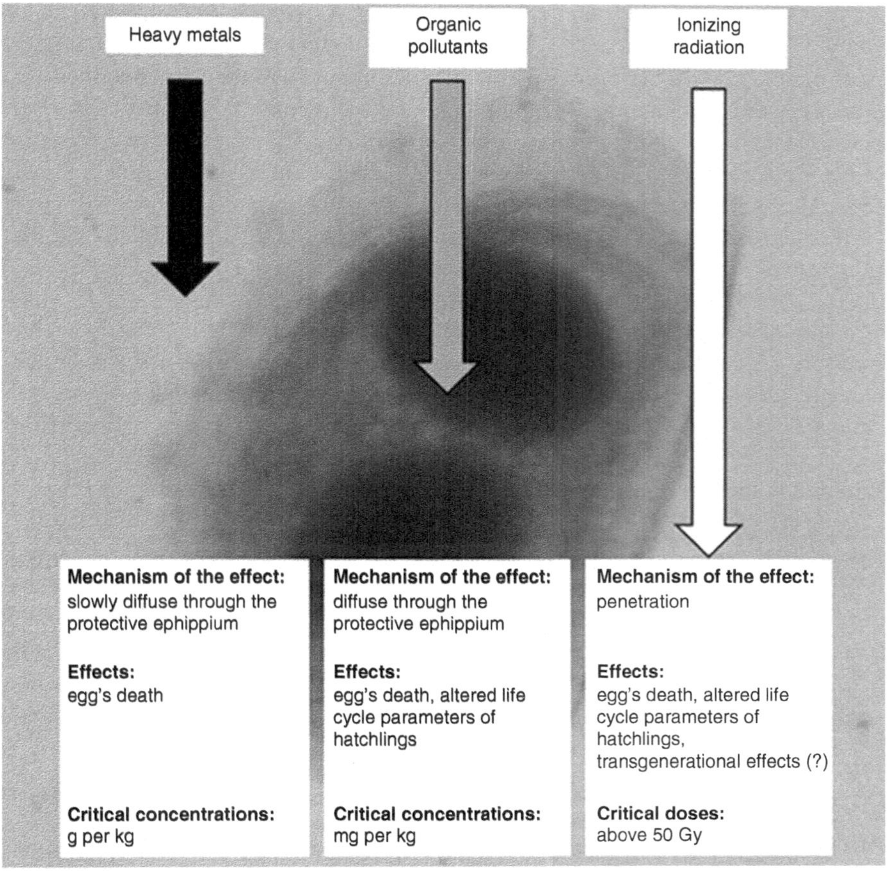

Fig. 7.1 The schematic diagram of the effects of different anthropogenic pollutants on resting eggs of Cladocera

7.5 Conclusion

Our review demonstrated that in spite of the high significance of resting eggs of zooplankton for the ecology and evolution of natural communities, studies on the negative effects of sediments contaminated with anthropogenic pollutants on resting eggs are fragmentary. We observed a recent increase in the number of papers addressing sensitivity of resting eggs to various toxic compounds. This interest was mostly stimulated by the understanding of the role that resting stages play in natural habitats and by the advances in paleolimnological research. However, many of the studies focused on the effect of a specific toxicant on resting eggs lack dose-response curves or estimates of delayed responses.

We have covered a substantial gap in this field of research in our recent detailed study of the effects of several heavy metals and gamma radiation on dormant eggs of *M. macrocopa*. We determined critical concentrations of toxicants and the lethal dose of gamma radiation for resting eggs of *M. macrocopa.* We also studied the performance of hatchlings from resting eggs exposed to either toxic or radioactive effect of pollutants. However, this is just a small contribution to this complex research area.

It is highly important to investigate mechanisms that are responsible for the viability of resting eggs under a wide range of concentrations of various toxicants and different durations of direct contact of eggs with toxicants. Detailed studies of the structure of the ephippium and interactions of toxicants with these protective structures are needed. The exposure of resting eggs to contaminated bottom sediments can affect the performance of hatchlings. Individual and population level responses should be estimated both in laboratory experiments and in natural habitats. This information is necessary to predict the effect of contamination of natural habitats by different pollutants on the ecosystem structure and functioning.

Acknowledgements The research was partially supported by the Russian Foundation for Basic Research (Project No. 15-04-05199) and the Krasnoyarsk Regional Science Foundation (Project No. 16-44-243041). We are grateful to Helen Krasova for linguistic improvements.

References

Alekseev VR, de Stasio BT, Gilbert JJ (2007) Diapause in aquatic invertebrates: theory and human use. Springer, Netherlands, p 257

Alekseev V, Makrushin A, Hwang JS (2010) Does the survivorship of activated resting stages in toxic environments provide cues for ballast water treatment? Mar Pollut Bull 61:254–258

Alonzo F, Gilbin R, Zeman FA, Garnier-Laplace J (2008) Increased effects of internal alpha irradiation in *Daphnia magna* after chronic exposure over three successive generations. Aquat Toxicol 87:146–156

Alonzo F, Hertel-Aas T, Real A, Lance E, Garcia-Sanchez L, Bradshaw C, Batlle JV, Oughton DH, Garnier-Laplace J (2016) Population modelling to compare chronic external radiotoxicity between individual and population endpoints in four taxonomic groups. J Environ Radioact 152:46–59

Aránguiz-Acuña A, Pérez-Portilla P (2017) Metal stress in zooplankton diapause production: post-hatching response. Ecotoxicology 26:329–339

Avila-Perez P, Balcazar M, Zarazua-Ortega G, Barcelo-Quintal I, Dıaz-Delgado C (1999) Heavy metal concentrations in water and bottom sediments of a Mexican reservoir. Sci Total Environ 234:185–196

Beltrán-Pardo E, Jönsson KI, Wojcik A, Haghdoost S, Harms-Ringdahl M, Bermúdez-Cruz MR, Villegas JEB (2013) Effects of ionizing radiation on embryos of the tardigrade *Milnesium cf. tardigradum* at different Stages of development. PLoS ONE. https://doi.org/10.1371/journal.pone.0072098

Brendonck L, De Meester L (2003) Egg banks in freshwater zooplankton: evolutionary and ecological archives in the sediment. Hydrobiologia 491:65–84

Cao L, Ishii N, Zheng J, Kagami M, Pan S, Tagami K, Uchida S (2017) Vertical distributions of Pu and radiocesium isotopes in sediments from lake Inba after the Fukushima Daiichi nuclear power plant accident: source identification and accumulation. Appl Geochem 78:287–294

Dallas LJ, Keith-Roach M, Lyons BP, Jhaa AN (2012) Assessing the impact of ionizing radiation on aquatic invertebrates: a critical review. Radiat Res 2012(177):693–716

Deng H, Ye ZH, Wong MH (2004) Accumulation of lead, zinc, copper and cadmium by 12 wetland plant contamination sites in China. Environ Pollut 132:29–40

Donaldson LR, Foster RF (1957) Effects of radiation on aquatic organisms, National Academy of Sciences. The effects of atomic radiation on oceanography and fisheries. Publication 551. National Academy of Sciences-National Research Council, Washington, DC, pp 96–102

Eisler R (1994) Radiation hazards to fish, wildlife, and invertebrates: a synoptic review, Contaminant hazard reviews report. US National Biological Service. U.S. Deparment of the Interior, Washington, DC. №29. 147 p

Gilbin R, Alonzo F, Garnier-Laplace J (2008) Effects of chronic external gamma irradiation on growth and reproductive success of *Daphnia magna*. J Environ Radioact 99:134–145

Gladyshev E, Meselson M (2008) Extreme resistance of bdelloid rotifers to ionizing radiation. PNAS 105:5139–5144

Gutierrez MF, Battauz Y, Caisso B (2017) Disruption of the hatching dynamics of zooplankton egg banks due to glyphosate application. Chemosphere 171:644–653

Hairston NG (1996) Zooplankton egg banks as biotic reservoirs in changing environments. Limnol Oceanogr 41:1087–1092

Hairston NG Jr, Van Brunt RA, Kearns CM, Engstrom DR (1995) Age and survivorship of diapausing eggs in a sediment egg bank. Ecology 76:1706–1711

Hairston NG Jr, Hansen AM, Schaffner WR (2000) The effect of diapause emergence on the seasonal dynamics of a zooplankton assemblage. Freshw Biol 45:133–145

Iwasaki T (1964) Sensitivity of Artemia eggs to the gamma irradiation 1. Hatchability of encysted dry eggs. J Radiat Res 5:69–75

Jiang XD, Wang GZ, Li SJ, He JF (2007) Heavy metal exposure reduces hatching success of *Acartia pacifica* resting eggs in the sediment. J Environ Sci 19:733–737

Kansanen P, Jaakkola T, Kulmala S, Suutarinen R (1991) Sedimentation and distribution of gamma-emitting radionuclides in bottom sediments in southern Lake Päijänne Finland, after the Chernobyl accident. Hydrobiology 222:121–140

Kryshev II, Sazykina TG (1998) Radioecological effects on aquatic organisms in the areas with high levels of radioactive contamination: environmental protection criteria. Radiat Prot Dosim 75:187–191

Lampert W, Fleckner W, Rai H, Taylor BE (1986) Phytoplankton control by grazing zooplankton: a study on the spring clear-water phase. Limnol Oceanogr 31:478–490

Lopatina TS, Bobrovskaya NP, Oskina NA, Zadereev ES (2017) Comparative study of the toxic effect of cadmium and nickel on the active and resting stages of Cladoceran *Moina macrocopa*. J Siberian Federal Univ Biol 10:358–372

Lopatina TS, Zadereev ES, Oskina NA, Petrichenkov MV (2018) The sensitivity of resting eggs of the cladoceran *Moina macrocopa* to the effect of ionizing radiation during the reactivation. Dokl Biochem Biophys 480:169–172

Massarin S, Alonzo F, Garcia-Sanchez L, Gilbin R, Garnier-Laplace J, Poggiale JC (2010) Effects of chronic uranium exposure on life history and physiology of *Daphnia magna* over three successive generations. Aquat Toxicol 99:309–319

May RM, Maria M, Guimard J (1964) Action differentielle des rayons x et ultraviolets sur le tardigrade *Macrobiotus areolatus*, al'etat actif et desseche. Bull Biol Fr Belg 98:349–367

Most M, Chiaia-Hernandez AC, Frey MP, Hollender J, Spaak P (2015) A mixture of environmental organic contaminants in lake sediments affects hatching from Daphnia resting eggs. Environ Toxicol Chem 34:338–345

Navis S, Waterkeyn A, Voet T, De Meester L, Brendonck L (2013) Pesticide exposure impacts not only hatching of dormant eggs, but also hatchling survival and performance in the water flea *Daphnia magna*. Ecotoxicology 22:803–814

Navis S, Waterkeyn A, Putman A, De Meester L, Vanermen G, Brendonck L (2015) Timing matters: sensitivity of *Daphnia magna* dormant eggs to fenoxycarb exposure depends on embryonic developmental stage. Aquat Toxicol 83:159–176

Nielsen DL, Smith D, Petrie R (2012) Resting egg banks can facilitate recovery of zooplankton communities after extended exposure to saline conditions. Freshw Biol 57:1306–1314

Novikova N, Gusev O, Polikarpov N, Deshevaya E, Levinskikh M, Alekseev V, Okuda T, Sugimoto M, Sychev V (2011) Survival of dormant organisms after long-term exposure to the space environment. Acta Astronaut 68:1574–1580

Oskina N, Lopatina T, Zadereev E (2018) High resistance of resting eggs of cladoceran *Moina macrocopa* to the effect of heavy metals. Bull Environ Contam Toxicol. https://doi.org/10.1007/s00128-018-2473-7

Polikarpov GG (1998) Conceptual model of responses of organisms, populations and ecosystems to all possible dose rates of ionizing radiation in the environment. Radiat Prot Dosim 75:181–185

Putyrskaya V, Klemt E, Rollin S (2009) Migration of 137 Cs in tributaries, lake water and sediment of Lago Maggiore (Italy, Switzerland) - analysis and comparison with Lago di Lugano and other lakes. J Environ Radioact 100:35–48

Radzikowski J (2013) Resistance of dormant stages of planktonic invertebrates to adverse environmental conditions. J Plankton Res 35:707–723

Raikow DF, Reid DF, Maynard EE, Landrum PF (2006) Sensitivity of aquatic invertebrate resting eggs to Seakleen (menadione): a test of potential ballast tank treatment options. Environ Toxicol Chem 25(2):552–559

Raikow DF, Landrum PF, Reid DF (2007a) Aquatic invertebrate resting egg sensitivity to glutaraldehyde and sodium hypochlorite. Environ Toxicol Chem 26:1770–1773

Raikow DF, Reid DF, Blatchley ER, Jacobs G, Landrum PF (2007b) Effects of proposed physical ballast tank treatments on aquatic invertebrate resting eggs. Environ Toxicol Chem 26:717–725

Razo I, Carrizales L, Castro J, Diaz-Barriga F, Monroy M (2004) Arsenic and heavy metal pollution of soil, water and sediments in a semi-arid climate mining area in Mexico. Water Air Soil Pollut 152:129–152

Rogalski MA (2015) Tainted resurrection: metal pollution is shown in the Daphnia egg banks. Ecology 96:1166–1173

Rogalski MA (2017) Maladaptation to acute metal exposure in resurrected clones after decades of increasing contamination. Am Nat 189(4):443–452

Rother A, Pitsch M, Huelsmann S (2010) The importance of hatching from resting eggs for population dynamics and genetic composition of Daphnia in a deep reservoir. Freshw Biol 55:2319–2331

Sacmaci S, Yilmaz Y, Kartal S, Kaya M, Duman F (2014) Resting eggs as new biosorbent for preconcentration of trace elements in various samples prior to their determination by FAAS. Biol Trace Elem Res 159:254–262

Sarabia R, Del Ramo J, Varó I, Díaz-Mayans J, Torreblanca A (2008) Sublethal zinc exposure has a detrimental effect on reproductive performance but not on the cyst hatching success of *Artemia parthenogenetica*. Sci Total Environ 398:48–52

Sarapultseva EI, Dubrova YE (2016) The long-term effects of acute exposure to ionizing radiation on survival and fertility in *Daphnia magna*. Environ Res 150:138–143

Sarapultseva EI, Gorski AI (2013) Low-dose g-irradiation affects the survival of exposed Daphnia and their offspring. Dose-Response 11:460–468

Suyama I, Etoh H, Maruyama T, Kato Y, Ichikawa R (1981) Effects of ionizing radiation on the early development of Oryzias eggs. J Radiat Res 25:125–133

Turko P, Sigg L, Hollender J, Spaak P (2016) Rapid evolutionary loss of metal resistance revealed by hatching decades-old eggs. Evolution 70:398–407

United Nations Scientific Committee on the Effect of Atomic Radiation (2010) Sources and effect of ionizing radiation. United Nations, New York. Available online at: http://www.unscear.org/docs/reports/2008/09-86753_Report_2008_Annex_B.pdf

Van der Stricht S, Janssens A (2010) Radioactive effluents from nuclear power stations and nuclear fuel reprocessing sites in the European union, 2004e08. RP 164. Publications Office of the European Union, Luxembourg. http://dx.doi.org/10.2833/27366. Available online at: https://ec.europa.eu/energy/sites/ener/files/documents/164.pdf

Vesk PA, Allaway WG (1997) Spatial variation of copper and lead concentrations of water hyacinth plants in a wetland receiving urban run-off. Aquat Bot 59:33–44

Watanabe M (2006) Anhydrobiosis in invertebrates. Appl Entomol Zool 41:15–31

Wong CK, Wong PK (1990) Life table evaluation of the effects of cadmium exposure on the freshwater cladoceran, *Moina macrocopa*. Bull Environ Contam Toxicol 44:135–141

Wong CK, Wong PK, Tao H (1991) Toxicity of nickel and nickel electroplating water to the freshwater cladoceran *Moina macrocopa*. Bull Environ Contam Toxicol 47:448–454

Zadereev ES, Lopatina TS, Zotina TA, Dementyev DV, Oskina NA, Petrichenkov MV (2016) The effect of γ-radiation on resting eggs and life cycle of cladoceran *Moina macrocopa*. Dokl Biochem Biophys 466:61–65

Zadereev E, Lopatina T, Oskina N, Zotina T, Petrichenkov M, Dementyev D (2017) Gamma irradiation of resting eggs of *Moina macrocopa* affects individual and population performance of hatchlings. J Environ Radioact 175-176:126–134

Zampi M, Focardi S (1973) Effects of ionizing radiation on the eggs of *Milax nigricans* (Schultz) Mollusca Gastropoda Pulmonata. Bolletino di zoologia 40:291–303

Chapter 8
Mechanisms Facilitating Dispersal of Dormant Eggs in a Planktonic Crustacean

Miroslaw Slusarczyk, Bernadette Pinel-Alloul, and Barbara Pietrzak

Abstract While plants are famously known to use their dormant forms for both temporal and spatial dispersal, animals are generally recognized to use them mostly for periodic survival in unfavourable conditions, despite a plethora of animals (sedentary or inhabiting isolated sites) regularly using their dormant forms for spatial dispersal, too. Freely moving planktonic crustaceans of the genus *Daphnia*, inhabiting island-like freshwater habitats, are one of the model examples. While they have low chance to cross terrestrial barriers and reach neighbourhood waterbodies in an active form, they have developed adaptations that facilitate spatial dispersal of their dormant forms. In this chapter, we broadly summarize the ultimate reasons of spatial dispersal of freshwater organisms and the modes and routes of their dislocation. We further focus on the morphological, behavioural and life history adaptations that facilitate spatial dispersal of the dormant forms of a model planktonic crustacean, *Daphnia*. Finally, we evaluate the risks and costs of passive dispersal, assess its effectiveness and describe non-adaptive consequences.

Keywords Dispersal mechanisms, *Daphnia*, Dormant eggs, Ephippial case, Oviposition behaviour

M. Slusarczyk (✉) · B. Pietrzak (✉)
Faculty of Biology, Department of Hydrobiology, University of Warsaw at Biological and Chemical Research Centre, Warszawa, Poland
e-mail: m.slusarczyk@uw.edu.pl; b.pietrzak@uw.edu.pl

B. Pinel-Alloul (✉)
GRIL, Département de sciences biologiques, Université de Montréal, Montréal, QC, Canada
e-mail: bernadette.pinel-alloul@umontreal.ca

V. R. Alekseev, B. Pinel-Alloul (eds.), *Dormancy in Aquatic Organisms. Theory, Human Use and Modeling*, Monographiae Biologicae 92,
https://doi.org/10.1007/978-3-030-21213-1_8

8.1 Spatial Dispersal of Freshwater Organisms: Causes and Routes

In geological and geographical scales, freshwaters are short living and small-sized habitats. This imposes selective forces on their inhabitants and promotes both dormancy and dispersal. While these are plants that are widely recognized to use their dormant seed stage for dispersal, animal dormant eggs are functionally analogous, and routes of their dispersal are similar.

8.1.1 Ultimate Causes Behind Spatial Dispersal of Freshwater Organisms

Living creatures are relatively fragile structures, yet they have been present on Earth for more than 3 billion years. Their long existence is owed not to their extraordinary endurance, but to their endless replication and their spatial or temporal avoidance of unfavourable conditions (Venable and Lawlor 1980; Mc Peek and Kalisz 1998; Buoro and Carlson 2014). Organisms with effective mechanisms of dispersal may survive by shifting between habitats as long as environmental variations are not correlated between neighbouring locations. All habitat are exposed to some spatial or temporal variation in abiotic conditions. Even in relatively stable abiotic setting, biotic interactions may inevitably cause spatial or temporal fluctuations in environmental conditions. Uneven distribution of resources, interspecific competition (Holt and Barfield 2002), predation (Weisser 2002) or parasitism (Boulinier et al. 2002) may affect fitness of organisms considerably and drive spatial dispersal (Gandon and Michalakis 2002). Moreover, spatial dispersal might be beneficial even in invariable locations due to kin interactions (including sib competition or parent-offspring conflict (Hamilton and May 1977)) or inbreeding depression (Bengtsson 1978; Perrin and Mazalov 1999).

Aquatic organisms inhabiting freshwater sites need to disperse more desperately than their marine counterparts do. Most freshwater lentic habitats are highly isolated, ephemeral habitats in geological and evolutionary timescales that emerge and vanish frequently. Majority of lakes formed in the northern temperate zone after the last glaciation has disappeared within the last ten thousand years (Marszelewski 2005). Shallow temporary waters persist even shorter than large deep lakes. Thus, most if not all species inhabiting freshwater bodies need effective mechanisms of overland dispersal to thrive in time.

Environmental deteriorations may have abiotic or biotic origin, operate at local or regional scale and concern all organisms – mobile and immobile ones. While mobile organisms may seek suitable locations in the active form, the sedentary species or those living in isolated habitats frequently use their dormant forms for passive dispersal (Bohonak and Jenkins 2003; Panov et al. 2004; Panov and Cáceres 2007).

8.1.2 Importance of Dormant Forms in Animal Dispersal in Freshwaters

Most freshwater animals are able to move actively within aquatic habitats. Some of them may also travel between waterbodies along permanent or temporal waterway connections (Michels et al. 2001; Shurin and Havel 2002; Leuven et al. 2009). Yet, many lakes and ponds are island-like habitats, permanently isolated from each other by impenetrable terrestrial barriers. They are not lifeless, though, but inhabited by various aquatic species that colonize them quickly (Louette and de Meester 2005; Juračka et al. 2016). Some animals may actively cross short terrestrial barriers between aquatic habitats, e.g. vertebrates like reptiles, amphibians, or even fish, as well as invertebrates, e.g. most insects and some crustaceans. However, most aquatic species are incapable of moving across terrestrial zone due to sedentary lifestyle or poor moving skills or due to intolerance to overland conditions, e.g. desiccation. While passive dispersal of active forms between water habitats seems feasible on meter-wide distances, most freshwater animals are not able to survive long overland journey. Still, many of them are able to disperse passively across hostile terrestrial habitats once they suspend their development. Dormant forms of freshwater species are well suited for overland dispersal (Panov et al. 2004; Panov and Cáceres 2007). They are tolerant to various abiotic and biotic threats (Radzikowski 2013) and may possess various morphological or physiological adaptations facilitating their dispersal.

8.1.3 Routes of Passive Dispersal of Dormant Aquatic Organisms

Organisms producing resting stages disperse passively via different routes, often with multiple processes and vectors involved (Higgins et al. 2003; Incagnone et al. 2015). From the disperser perspective, that is, from the viewpoint of the challenges the animals encounter on route, and thus the mechanisms, or the adaptations, facilitating dispersal via these routes, dormant propagules of freshwater organisms move between habitats, mainly thanks to different processes such as (1) floating on the water surface, (2) sticking to other surfaces, (3) surviving being eaten or (4) floating in the air. The relative importance of these phenomena to effective dispersal and population establishment naturally varies both between organisms and their propagule characteristics, and between habitats. Also, the contribution of the different processes and vectors varies between studies, and is still unresolved (see Bohonak and Jenkins 2003; Cohen and Shurin 2003; Allen 2007; Vanschoenwinkel et al. 2008a; Coughlan et al. 2017).

8.1.3.1 Floating on Water

Water-air interface is the first gateway out of the aquatic habitat. Floating in water – plankton-wise – distributes the propagules within and between interconnected bodies of water, and may aid encountering animate and inanimate vectors, but in itself is not effective in transporting them between the habitats where waterway connections are scarce. Temporarily entering neuston, in its turn, exposes propagules to further potential vectors of dispersal. For aquatic plants, water-mediated spread of seeds, i.e. hydrochory, is recognized as an important dispersal route (Hopfensperger and Baldwin 2009; Pollux et al. 2009), and there are no reasons to expect it to be otherwise for aquatic animals dispersing via dormant forms. As propagules float, thanks to either of the two acting forces, buoyancy (upthrust) or surface tension, water-borne dispersal is facilitated by traits harnessing either of them. Floatability is even used as a proxy for seed dispersal capacity (Kappes et al. 2014; van Leeuwen et al. 2014). For freshwater invertebrates, the dormant stage floatability has also been linked to dispersal capacity, like the increasing proportion of floatoblasts to sessoblasts produced with increasing ramet size of bryozoans (Karlson 1992), the distinction and phenotypic reversals between floaters and sinkers among resting eggs of an anostracan (Pinceel et al. 2013), or the prevalence of floating and sinking ephippia of cladocerans (Pietrzak and Ślusarczyk 2006; Cáceres et al. 2007; Ślusarczyk et al. 2017a). Furthermore, numbers of propagules attached to birds have been recorded to coincide with their availability on the water surface (Brochet et al. 2010a).

8.1.3.2 Sticking to Surfaces

What makes propagules trapped at the water-air phase boundary makes them also stick to other surfaces. Darwin famously noted minuscule creatures actively crawling onto duck feet immersed in water, but in fact, even surface debris, dead or alive, will stick to a resurfacing object provided both have hydrophobic surfaces. For instance, cladoceran ephippia get easily and firmly attached to hair fringes of the abdomen and legs of a heteropteran insect *Notonecta* sp., and thus are carried away from the water surface, with ephippium floating ability being a critical factor here (van de Meutter et al. 2008). Propagules of many aquatic species have been repeatedly found to be externally transported by vertebrates, especially waterbirds, but also mammals (Maguire Jr. 1963; Bilton et al. 2001; Figuerola and Green 2002a). Plant, fungal and protist spores were reported to disperse between aquatic environments in plumage or on feet of ducks, waders and rails (Vivian-Smith and Stiles 1994; Figuerola and Green 2002b; Brochet et al. 2010b; Raulings et al. 2011; Lewis et al. 2014; Reynolds and Cumming 2016). So were invertebrate eggs, including dormant cladocerans (Figuerola and Green 2002b; Brochet et al. 2010a; Reynolds and Cumming 2016). Finally, the omnipresent and extremely migratory species, humans, now provides a multitude of artificial surfaces frequently moved between waterbodies. Tourism, transportation and research, all these activities involve adherence and movement of dormant propagules of aquatic organisms between environments, on equipment from vehicles and vessels to boots (see Waterkeyn et al. 2010; Valls et al. 2016; Bullock et al. 2018), where even a hundred of cladoceran ephippia can adhere to a single fishing line (Jacobs and MacIsaac 2007). Indeed, animal-mediated external dispersal of propagules, i.e. ectozoochory, has been frequently evaluated as an important route, at least on local scales (see Coughlan et al. 2017; van Van Leeuwen 2018, for most recent reviews). Still, a lot, including the impact of factors affecting attachment and detachment of the propagules, like bird preening behaviour or the role of sediment type, e.g. encapsulating mud (Figuerola and Green 2002b), is still understudied (Coughlan et al. 2017).

8.1.3.3 Being Eaten

Even if not directly preyed upon, adherence to various organic substrates will make the dormant stages more easily an unintended prey (Green et al. 2002; Reynolds and Cumming 2016). Wherever on the tree of life, if an aquatic organism produces dormant stages, some of them are likely to survive being eaten and being defecated at another location. Chances of surviving such internal transport might be even larger than chances of surviving external one (Reynolds and Cumming 2015). Comparing to ectozoochory, dispersal involving passage through the gut, i.e. endozoochory, of aquatic invertebrate resting stages and plant seeds may play even a more important role in terms of number and diversity of propagules dispersed between water bodies (Brochet et al. 2010b; Sánchez et al. 2012; Costa et al. 2014; Lewis et al. 2014; Moreno et al. 2019). However, their relative roles may change seasonally with the availability of floating propagules (ingestion and adherence) as opposed to constant availability of sediment propagule bank (ingestion) (Brochet et al. 2010a; Reynolds and Cumming 2015). Viable organisms, from freshwater algae (Proctor 1959) to cladocerans (Brochet et al. 2010a) are found in the lower digestive tracts of migrating birds, and a meta-analysis showed that roughly one in three waterbird droppings collected in the field contains a viable propagule, of which macroinvertebrates are almost as frequent as plant seeds (van Leeuwen et al. 2012).

8.1.3.4 Floating in the Air

As much as for the endozoochory, the presence of dormant forms is considered a prerequisite for effective wind dispersal, i.e. anemochory, of small aquatic metazoans (Panov et al. 2004). Once at the surface, or as soon as the temporary basin dries out, they are exposed to wind. Not only the resistant dormant state gives them good chances for survival outside their aquatic medium, but also, in anhydrobiosis they are lighter, and thus more easily than the hydrated forms taken into and moving in the air (Ricci and Caprioli 2005; Ptatscheck et al. 2018). This is important, as adaptations like ballooning – using silk to fly – have evolved only on land, in spiders, mites and moths (Bell et al. 2005). Recently, nematodes have been found as an exception to the general rule, readily dispersing in the air in active form, their aquatic taxa included (Ptatscheck et al. 2018). Copepods, cladocerans, ostracods and others have been caught moving in the air (Vanschoenwinkel et al. 2008b, 2009), yet, though considered to be an important mechanism of dispersal, the actual rates of airborne transport or the distances thus covered are hardly known (Ptatscheck et al. 2018). Since the early suggestions that winds are the 'most important means of spread' of minute dormant stages of both animals and plants (McAtee 1917), not much has been resolved on how often and how far do aquatic animals in diapause move in reality. A century later, it is proposed that, more specifically, during the dry phase of temporary ponds, wind is the most important dispersal vector of their inhabitants (Incagnone et al. 2015). The floating of the propagules aids to the wind importance (Ślusarczyk et al. 2017a; Sirianni 2017). Not only taxa themselves differ in their 'flying' propensity and its dependence on environmental conditions, but their measured dispersal rates also vary greatly both between studies and methods used (Ptatscheck et al. 2018).

8.2 *Daphnia* as a Model Animal with passively Dispersing Dormant Forms

A good single crustacean taxon to study the mechanisms behind dispersal of its dormant forms will (1) rely only or mostly on the dispersal of dormant forms; (2) disperse via different routes, giving opportunities to study various adaptations; (3) be found in large numbers on at least some of these routes, enabling quantitative sampling for comparative analyses; (4) have wide distributions; and (5) have a record of recent and contemporary invasions, both proxies for effective dispersal. We focus here on the genus *Daphnia*, animals (1) which rarely, if ever, are found dispersing between waterbodies in active form; (2) whose resting eggs are carried over the surface of fresh waterbodies, inside and outside other animals, and in the air; (3) whose resting forms in thousands are periodically found forming near shore carpets covering the surface (Ślusarczyk and Pietrzak 2008; Kaya et al. 2014; Incagnone et al. 2015; Ślusarczyk et al. 2017a), and which together with other cladocerans have been found to be the most abundant group emerging from samples collected from waterbirds (Brochet et al. 2010a); (4) whose species have worldwide distributions (Forro et al. 2008); and (5) which are repeatedly found invading new freshwater habitats of all kinds, natural and artificial, including those experimental (see Sect. 8.2.3). Taking this all together with its being a widely used model in various fields of ecological research, from ecotoxicology to environmental genomics and evolutionary biology, *Daphnia* provides a well-suited model for an aquatic animal dispersing via its dormant forms.

8.2.1 Roles of Dormant Eggs in the life of *Daphnia*

Daphnia dormant eggs together with the encapsulating protective shell form an ephippium, a structure which functionally resembles a plant seed (Pietrzak and Ślusarczyk 2006): a well-protected, by default sexually produced, embryo, ready to endure both temporal challenges of the local environment and those encountered on dispersal routes outside of the native habitat. Thus, these eggs play an important role in both re-establishing local populations and colonizing new habitats.

Daphnia dormant eggs are most often, and primarily, the result of sexual reproduction. When conditions deteriorate, these animals switch from parthenogenetic production of subitaneous eggs to sexual reproduction inseparably linked with dormancy. The development of sexually produced fertilized egg is halted at the early gastrula stage (Zaffagnini 1987). Though some lineages from regions where the growing season is short, like high latitudes and altitudes, skip the sexual part and produce dormant eggs parthenogenetically (Decaestecker et al. 2009), still each genotype newly emerging after recombination first switches on the dormancy developmental program.

Dormant eggs are produced by *Daphnia* in anticipation of regular fluctuations and in response to direct challenges: before summer drying out and winter freezing down of shallow ponds; before winter formation of ice cover on lakes; before periods of intensive predation risk (Ślusarczyk et al. 2006); and under increasing population densities (Berg et al. 2001; Fitzsimmons and Innes 2006). This may occur once a year (Ślusarczyk 2009) after the first or the second summer in life of an individual (Pietrzak et al. 2013) or more often, and at irregular intervals (Mikulski and Grzesiuk 2019, submitted manuscript), according to the actual conditions. At a single sampling up to 80% of the females may be found carrying an ephippium (Ślusarczyk 2009), and in the laboratory up to 100% of the experimental females may produce them under certain conditions (Ślusarczyk 1995; 2004). Individuals hatching from dormancy (females exclusively) may supply overwintering active population of *Daphnia* or re-establish it after seasonal extinction (Cáceres 1998b).

The ephippial eggs in order to hatch need to pass through a refractory period (initial phase of suspended development when embryo remains insensitive to hatching stimuli; Ślusarczyk et al. 2019, submitted manuscript) and receive a still undefined set of environmental triggers including temperature and light (Davison 1969; Radzikowski et al. 2018; Ślusarczyk and Flis 2019).

Their resistance and other features discussed in the next section make them perfect vehicles for overland transport. Though the existence of a general and direct relationship between the ability of producing dormant forms and dispersal efficiency is still controversial (Incagnone et al. 2015), still, we can discuss many of their associated traits as – if not direct dispersal adaptions – traits effectively facilitating their spatial dispersal. Their morphology, as well as maternal life history and behaviour associated with their production and disposal, and life history of dispersing individual itself, as traits aiding the dispersal capacity, are discussed in the next section.

8.2.2 *Adaptations Facilitating Passive Dispersal of Resting Eggs in Daphnia*

Passive dispersal of dormant forms to be effective should (1) coincide in space and time with the occurrence of their vectors and (2) be based on resistance to withstand specific challenges encountered while on travel via different routes. Pinpointing particular traits as being under direct dispersal-driven selection may be difficult, though. Many traits that affect dispersal have other functions (Burgess et al. 2016). Research on plants also shows that long distance dispersal events might be rare and driven by nonstandard mechanisms, and is across adaptations (Higgins et al. 2003). Still, multiple adaptations do exist and work on shorter scales. We focus here on the challenges and the facilitating mechanisms specific to spatial dispersal of these forms, discussing still unverified hypotheses, and giving less attention to mechanisms equally facilitating dispersal both in space and in time.

In the case of *Daphnia*, the time of ephippia production and the place of their deposition by the mother are vital for the dispersing propagule and the dispersal vector coincidence. As for the ephippium itself, its internal and external wall structure providing toughness and aiding its floatation, its sticky appendages aiding both floatation and adherence to other structures, and its size itself are the most important traits.

8.2.2.1 Reaching Water Surface: Oviposition Behaviour

Few vectors of ephippial eggs operate deep in the open water, in the life zone of *Daphnia*, while most of them operate close to the water surface – the dangerous zone for planktonic organisms. Therefore, the resting eggs of *Daphnia* should appear close to the water surface in the right moment to be dispersed between water habitats. We consider two likely options of their transfer to the water surface, these linked to animal properties and not to external forces, like water current movements or air bubbles leaving the sediments, neither of which has been to our knowledge reported to bring *Daphnia* either in active or dormant form to the surface.

Ephippia might be either displaced to the surface due to positive buoyancy, like they are in sessile bryozoans (Karlson 1992), or due to their oviposition by mobile ephippium carrying females (Ślusarczyk and Pietrzak 2008). The first option has been proposed (e.g. by Cáceres et al. 2007), yet not proved so far. Active *Daphnia* are negatively buoyant and must swim constantly, otherwise they sink (Vega and Clausse 1998). Positive buoyancy of ephippia would affect overall buoyancy of ephippia carrying females and would change their swimming behaviour (e.g. hop and sink rate), which, according to our knowledge, has not been reported so far. On the contrary, experimental tests made on three common species of *Daphnia* (*D. magna*, *D. pulex*, *D. longispina*) revealed that all freshly shed ephippia were negatively buoyant, likewise active forms, and sank to the bottom when being shed by females below the water surface (Ślusarczyk and Pietrzak 2008). All ephippia that appeared at the water surface in further tests were left there by ephippial mothers during moulting, despite surface threat imposed on them during the tests (high UV radiation). While all tested ephippial females remained far from the risky water surface for most of the UV exposition period, some of them swam toward the water-air interface a few seconds ahead moulting, where they moulted shedding ephippium and dived down shortly after (Ślusarczyk and Pietrzak 2008).

Daphnia, like most other planktonic organisms, avoid surface zone of aquatic habitats (Dawidowicz and Pijanowska 2018). Close to the water surface, *Daphnia* may be exposed to harmful UV radiation (Alonso et al. 2004), enhanced predation risk of visual predators (Gliwicz 2003), or desiccation if being caught by surface tension and then thrown ashore (personal observation). While ephippial eggs are highly resistant to these surface threats (Radzikowski 2013), active individuals are more vulnerable. To reduce some of these threats, ephippia oviposition at the water surface might occur during night, when both fish predation and UV radiation-associated risks are negligible. Such cyclicity of ephippia deposition at the water surface would be reflected in a higher proportion of ephippial compared to non-ephippial females in subsurface zone during night vs. daytime (Fig. 8.1), or by a higher density of ephippia at the water surface at night and dawn compared to the other periods of a day at a calm weather (Fig. 8.2).

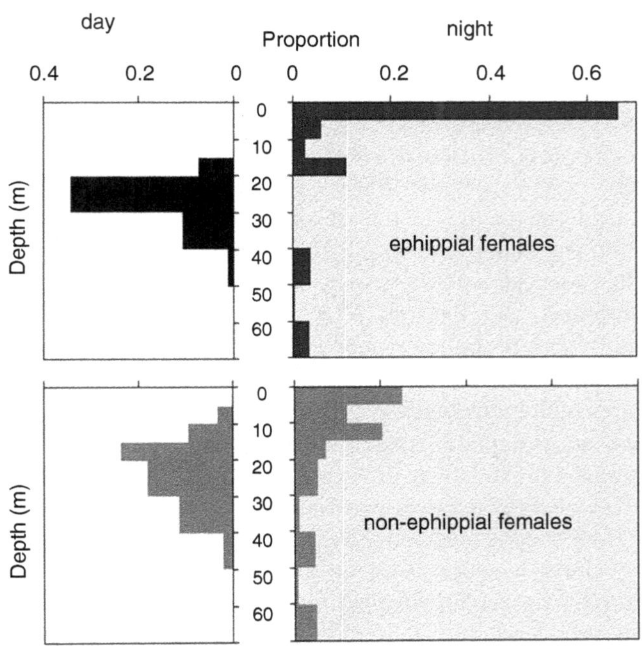

Fig. 8.1 Vertical distribution of *Daphnia pulicaria* females in the deep fishless mountain lake Czarny Staw in the High Tatras, Poland (49°11′18.1″N; 20°04′33.8″E), during the period of ephippia production: top - (black bars) ephippium carrying females; bottom (grey bars) non-ephippial females of the same lineage; during the day (left, white panels) and at night (right, shaded panels). (After Ślusarczyk et al. 2017a, modified)

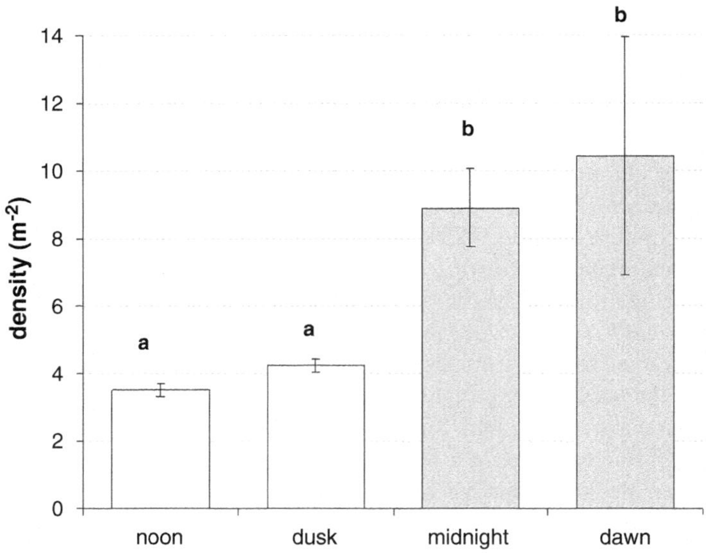

Fig. 8.2 Diurnal changes in ephippia density (mean ± 1SE) of *Daphnia* of the *longispina* group at the water surface in the middle of the small Kociołek Lake (station 5) collected in a single day of an autumn period of ephippia formation. Letters above the bars indicate homogenous groups with similar values according to ANOVA model simplification

8.2.2.2 Floating on Water Surface: External Morphology

Most of the ephippia left by females at the water surface retain negative buoyancy, yet stick to the water plane thanks to weak surface tension, suggesting hydrophobic properties of their outer wall. The role of surface tension was evidenced by sinking of ephippia after water-drop-induced disruption of the water surface or by applying a detergent that reduced surface tension and lowered the proportion of floating ephippia (Ślusarczyk and Pietrzak 2008). The elongated shape (Bormashenko 2016), flattening with protruding edges and relatively small size, all aid in their surface tension-induced floating.

Some ephippia left by *Daphnia* at the water surface gain positive buoyancy due to atmospheric air absorption. Anatomical analysis of the outer structure of the ephippium revealed possible way of air absorption (Bernatowicz et al. 2018). The ephippium of *Daphnia* is made of two concave chitinuous shells that enclose the resting eggs. Each shell is made of double walls (Hiruta and Tochinai 2014) inter-connected by cellular integuments of various shapes (Bernatowicz et al. 2018). In Ctenodaphnia (*D. magna, D. lumholtzi*), the integument has sponge-like structure that hardly absorbs air unless ephippium dries completely. In other Daphnidae (*longispina* or *pulex* group), the wall has a honeycomb structure with hollow cells that readily absorb air after even brief contact with the atmosphere, what makes them positively buoyant. Air probably does not penetrate through external wall,

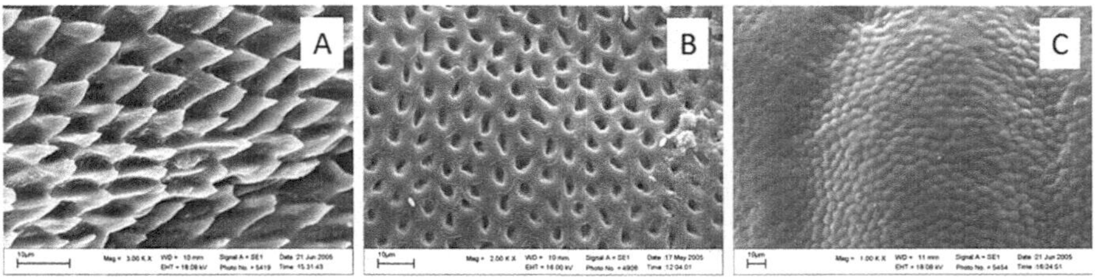

Fig. 8.3 Divergent morphology of the ephippium surface in different groups of *Daphnia* revealed by scanning microscopy. (**a**) *D. magna*, (**b**) *D. pulex*, (**c**) *D. rosea*

which seems impermeable, but most likely enters integuments through a ventral gap then hollows in the internal wall. The surface of the external wall differs considerably between species, which might translate into differences in floating ability: in some species, the wall is dotted with convexities (e.g. *D. rosea*) or with concavities (*D. pulex*, *D. longispina*) or is covered by scaly shaped structures (e.g. *D. magna*) in others (Fig. 8.3).

8.2.2.3 Sticking to Other Surfaces: External Morphology Continued

What makes them hydrophobic and easily caught at the water-air phase boundary makes them also stick to other surfaces. *Daphnia magna*, which hardly ever sheds ephippia at the water surface and is expected to use other mechanisms aiding dispersal, has ephippia that are equipped with tacky filaments, remnants of the moulted carapace which are not seen in other species (Mergeay et al. 2006a), that stay with the ephippium and act as hooks, sticking readily to solid objects (personal observations).

8.2.2.4 Time of Reaching Water Surface: Timing of Life History Events

Ephippia appearance at the water surface was reported not only in laboratory studies but also in the field. Mass occurrence of ephippia at the water surface reaching thousands per square meter in the middle of the lake (Ślusarczyk and Pietrzak 2008) or getting even higher densities on the downwind shore of the waterbodies (Fig. 8.4). It would be tempting to associate seasonal occurrence of ephippia at the water surface with timing of operation of their putative vectors, e.g. with bird migrations period, which seem to us the most effective vectors of ephippial eggs dispersal. However, this would need to be disentangled from the effect of the timing of the environmental deterioration, which might be a stronger selection force than dispersal vector availability.

Fig. 8.4 Floating ephippia accumulating in high densities on a downwind shore within foam or debris

8.2.2.5 Harsh Physical Conditions: Resistance

While the structure of the wall of the ephippia varies greatly between species (Bernatowicz et al. 2018), it invariably gives them toughness and endurance. Also, unlike the subitaneous egg, the dormant embryo invariably has a yet another protective layer: the inelastic and highly impermeable outermost egg shell (chorion). The embryo also contains high concentrations of the heat shock protein Hsp60, and cryoprotective glycerol (Pauwels et al. 2007) and possibly trehalose (Smirnov 2014).

Dormant embryo enclosed in the protective ephippial case of some species of *Daphnia* may withstand prolonged desiccation, extremely high or low temperature, anoxic conditions and passage through the gut of fish or birds (for review, see Radzikowski 2013). For instance, ephippia from northern *Daphnia* from Greenland hatched with success after being keep at −18°C for 18 years (Meijering 2003).

We might expect some differences between resident and dispersal ephippial eggs in their resistance to environmental extremes since different selective forces may shape properties of the two groups. The ephippial eggs of the individuals predestined for passive overland dispersal might be expected to be more resistant to extreme abiotic threats than the resident ones. The latter remain at the bottom of the native habitats where they may never face desiccation nor extreme thermal conditions, which dispersion forms have to cope with. Indeed, ephippial eggs collected at the water surface (dispersing) revealed higher tolerance to elevated temperature and desiccation compared to the resident ones collected at the bottom of some deep lakes in Poland (see Radzikowski 2019 for more details).

Risk of mechanical damage of the resting eggs is relatively low in "soft" aquatic lentic habitats. Unlike, overland dispersing resting eggs are exposed to abrasion and much greater mechanical forces not balanced by water buoyancy or viscosity. Therefore, the ephippial case seems redundant for resident resting eggs residing in deep permanent lakes, more useful in temporarily drying habitats, while vital for those passively dispersing. In this light, the ephippial case could be viewed as a structural adaptation for overland dispersal. Indeed, cladocerans which possess well-developed ephippial case are among the first invaders to newly formed isolated waterbodies (Louette and de Meester 2005). The most advanced ephippial structures may be found in Daphnidae (including Moinidae), while poorly developed cases are found in other Anomopoda (e.g. Bosminidae, Chydoridae, Macrothricidae) and non-existent in other Cladocera: Ctenopoda, Haplopoda or Onychopoda (Fryer 1996).

8.2.2.6 Unknown Hatching Habitat: Sensitivity and Hatching Phenology

The resident and dispersal ephippial eggs might also differ in some other features related to their distinct functions, like hatching phenology or longevity, which has not been probably tested so far. According to Venable and Levin (1985), who compared hatching phenology of dispersal and resident seeds of some terrestrial plants inhabiting unpredictably changing desert habitats, the dispersal seeds equipped with morphological adaptations for spatial dispersal germinated readily and synchronously while the resident seeds germinated more gradually. Some proportion of resident seeds postponed germination until a further favourable occasion as if they hedged against risk of unpredictable environmental deterioration (Venable and Lawlor 1980; Venable and Levin 1985). Diversified revival from the periodic rest (asynchronous germination or hatching) is considered the adaptive mechanism reducing risk of failed decision to resume development at uncertain conditions (Cohen 1966; Seger and Brockmann 1987; Ślusarczyk et al. 2017b).

It is hard to guess which ephippial eggs, either dispersal (shed at the water surface) or resident (released in the water column), should reveal higher longevity since both of them may be exposed to unpredictable environmental fluctuations. The dispersal hitchhikers may be displaced into unpredictable sites before they luckily reach some suitable location. Yet, the fate of the resident ephippia might not be defined either, once most of them may be buried for indefinite time in lake sediments before some external force (burrowing animal, gas bubbles, etc.) bring them back to the sediment surface, only where their development is feasible (Radzikowski et al. 2016). Large quantities of ephippia reside deep in lake sediments (Carvalho and Wolf 1989; Cáceres 1998a; Kerfoot et al. 2004) and may never hatch. For the above reasons, both dispersal and resident ephippial eggs might feature long viability, helping them to survive to a rare occasion for reactivation. While dispersal ephippia might possibly hatch at first favourable occasion (as it might be the last one), the resident should be more careful and reveal bet-hedging way of reactivation to cope with temporal variability of environmental conditions. The uncertainty receives now a new dimension under increasing intensity and scope of anthropogenic hydrological disturbances in the freshwater habitats (Brendonck et al. 2017).

So far we know little about *Daphnia* hatching phenology, length of latency included. We are aware that ephippial eggs of some lake *Daphnia* have very long dormancy periods reaching decades (Cáceres 1998a) or even centuries (Frisch et al. 2014), while those from temporary ponds reveal bet-hedging way of reactivation (Radzikowski 2019) which may prevent coincident resurrection off all resting eggs and diversify the risk of the failed choice in unpredictably changing habitats. We do not know, however, the hatching phenology of ephippial eggs of the lake *Daphnia* nor longevity of ephippial eggs in temporary waters, not mentioning the differences between resident and dispersing ones in that field.

8.2.2.7 More Challenges and Mechanisms

Many more potentially important selective forces can be identified and adaptive mechanisms speculated on. Surviving gut passage demands particular chemical and mechanical resistance. Anemochory and ectozoochory are facilitated by small size of the propagule (van de Meutter et al. 2008), which in case of *Daphnia* is directly linked to mother body size, and female-size-corrected variation in ephippium size has not been shown so far. Lethal effect of ultraviolet light is a vital factor not only for airborne propagules (Maguire Jr. 1963) but also to all exposed for longer periods to direct sunlight. Hydrophobic ephippial case offers enhanced protection against desiccation, and honeycomb-like structure offers enhanced protection against mechanical damage (Bernatowicz et al. 2018) and – when incrusted with black protective pigments – to UV radiation (Shan 1970). Indeed, pigmentation of the shell varies between ephippia (Gerrish and Cáceres 2003), and its association with dispersal versus resident phenotype has neither been shown yet.

8.2.3 Effectiveness of Passive Dispersal

For a long time, it was thought that plankton organisms producing resting stages have high dispersal capacities, enabling them to successfully colonize waterbodies and be widespread in ponds and lakes across large-scale extent (Mayr 1970; Cohen and Shurin 2003). The wide geographical ranges of many freshwater taxa are testimony that passive dispersal by resting stages can be effective for dispersal-gene flow and make many aquatic taxa essentially cosmopolitan and genetically similar.

This assumption was questioned following detailed morphological studies (Frey 1982) and molecular work (Lynch and Spitze 1994; Colbourne and Hebert 1996; De Meester 1996a; Schwenk et al. 1998; Gómez et al. 2000, 2002). Comparison of patterns in gene flow in freshwater taxa with different life cycles support the scenario that high dispersal of organisms favour an effective monopolization of waterbody resources by newly invading species (and genotypes), yielding a strong priority effect (De Meester et al. 2002). This founder effect increases genetic differentiation among zooplankton populations inhabiting nearby ponds and reduces gene flow among populations and waterbodies. Boileau et al. (1992) proposed that the high levels of genetic differentiation in cyclic parthenogenetic species such as *Daphnia* reflect their potential to colonize habitats from a few resting eggs that are resistant to decay for dozens of generations, and their capacity for rapid population growth upon hatching. Such founder effects were illustrated by the persistence of non-indigenous species (*Daphnia exilis* and *D. lumhotzi* in USA; *D. ambigua* and *D. parvula* in

Europe) as resting eggs in sediments and their rapid colonization only promoted by human disturbances (Hairston et al. 1999; Havel et al. 1995; Maier 1996).

De Meester et al. (2002) examined the potential mechanisms causing a discrepancy between high dispersal rates and reduced level of gene flow (the monopolization hypothesis and dispersal-gene flow paradox) in cyclical parthenogens such as *Daphnia*. Founder events upon colonization of waterbodies followed by rapid population growth and local adaptation of daphnids as well as buffering effect of resting eggs banks result in an effective monopolization of resources and a strong priority effect, creating high genetic differentiation among nearby populations (see also Boileau et al. 1992; Jenkins and Buikema 1998; Palsson 2000). Pronounced genetic differentiation among daphnid populations was detected for neutral markers as well as ecological relevant traits (De Meester 1996a, b). Thus, patterns of regional genetic differentiation in daphnid populations may often reflect historical colonization of new habitats via dispersal and priority effect rather than contemporary gene flow. In addition, strong local selection also favours genetic divergence among nearby populations even if dispersal rates are high (Michels et al. 2001; De Meester et al. 2002).

Effectiveness of passive dispersal in freshwater organisms is still debated because it varies among species due to specific differences in functional traits such as body size, resting egg features, resistance to environmental threats during dispersal and adaptation to local habitat (Cohen and Shurin 2003). There is now a substantial amount of evidence that daphnids exhibit significant potential for passive dispersal via long distance transport through wind (Michels et al. 2001; Cáceres and Soluk 2002), direct water flow among connected waterbodies (van de Meutter et al. 2006) or animal vectors such as waterfowl (Proctor and Malone 1965; Figuerola et al. 2003, 2005; Moreno et al. 2019) or aquatic insects (van de Meutter et al. 2008). However, several studies have suggested that aquatic microcrustaceans are relatively efficient dispersers in a variety of landscapes, whereas others have indicated dispersal limitation at large and small spatial scales or under specific circumstances (Juračka et al. 2016; Horváth et al. 2016).

Supporters of effective dispersal have shown that species of cladocerans can disperse effectively over short distance (<1 km) and colonize suitable habitats quickly depending on their ability to tolerate local conditions, such as water acidity and the presence of fish. Louette and De Meester (2005) monitored cladoceran colonization in 25 newly created waterbodies and isolated pools over a wide geographic area. They found high effectiveness of passive dispersal via transport of dormant stages. Daphnids (*D. obtusa*, *Ceriodaphnia* spp., *Simocephalus vetulus* and *Scapholeberis mucronata*) were the most frequently found, representing almost 50% of the colonization events. Experimental studies also gave strong evidence of high dispersal and colonization effectiveness of cladoceran zooplankton, as 60–75% of the regional species pool can be captured in mesocosms after between less than two months and two years depending on the distance to the nearest pond (Cáceres and Soluk 2002; Cohen and Shurin 2003; Lopes et al. 2016). Recently, dispersal limitation was suggested not important in the Arctic young freshwater habitats, where the structure of asexual *Daphnia* populations reflected environmental gradients and differences in clonal ecology (Haileselasie et al. 2016). In another study conducted in Ethiopian reservoirs, these were the founder effects again suggested to structure regional population (Haileselasie et al. 2018).

Effectiveness of passive dispersal of cladocerans has also been evaluated by quantifying the build-up of the dormant egg bank in the sediment during the first six months in newly created egg-free pools over Flanders, Belgium (Vandekerkhove et al. 2005). High densities (60–10000 eggs per m^2) of cladoceran egg banks were observed in the new pools. In parallel, the temporal dormant egg bank on the water surface, reaching densities of hundreds to thousands of ephippia per m^2, has been suggested as another proxy for this high potential (Ślusarczyk et al. 2017a). The high prevalence of floating ephippia in *Daphnia* may make it a more effective disperser than some other cladocerans (Sirianni 2017).

Supporters of dispersal limitation found multiple indications of dispersal limitation affecting the community assembly of microcrustacean communities. On a large-scale survey, spatial patterns in zooplankton community composition were better explained by the geomorphological structure of the landscape than by mere geographic distances between waterbodies (Juračka et al. 2016), suggesting that ridges separating the network of valleys act as dispersal barriers and may channel the dispersal routes of the taxa. Furthermore, dispersal limitation was reflected by a strong positive relationship between species richness and the number of neighbouring waterbodies. On a small-scale extent, although the effect of wind lessen dispersal limitation along its prevailing direction, it persists that dispersal limitation may constraint community assembly in highly mobile organisms even at spatial scale below 5 km (Horváth et al. 2016). Most probably, different processes dominate at different spatial and temporal scales. Heino et al. (2015) suggested dispersal limitation to be increasing with distances between habitats, contrary to the importance of mass effects, and the spatial extents of different processes to be affected by the dispersal ability of organisms. More recently, the need of using novel tools to assess the role of dispersal in shaping communities has been proposed (Heino et al. 2017).

8.2.4 Consequences of Dispersal of Dormant Eggs

Several phenomena associated with the passive dispersal of ephippial eggs may have far reaching non-adaptive implications for these organisms.

8.2.4.1 Ephippia Exploitation

During periods of intense ephippia formation, high quantities of ephippia may accumulate along the downwind shore of the waterbodies (Figs. 8.4 and 8.5). This supra- and meso-littoral aggregation of resting eggs constitutes an underestimated bank of *Daphnia* resting forms in freshwater habitats. The accumulation of the resting eggs of *Daphnia* may be intense enough to exploit it commercially in aquacultures likewise floating eggs of the brine shrimps *Artemia franciscana* and has already been suggested as a source for exploitation for chitin extraction (Kaya et al. 2014).

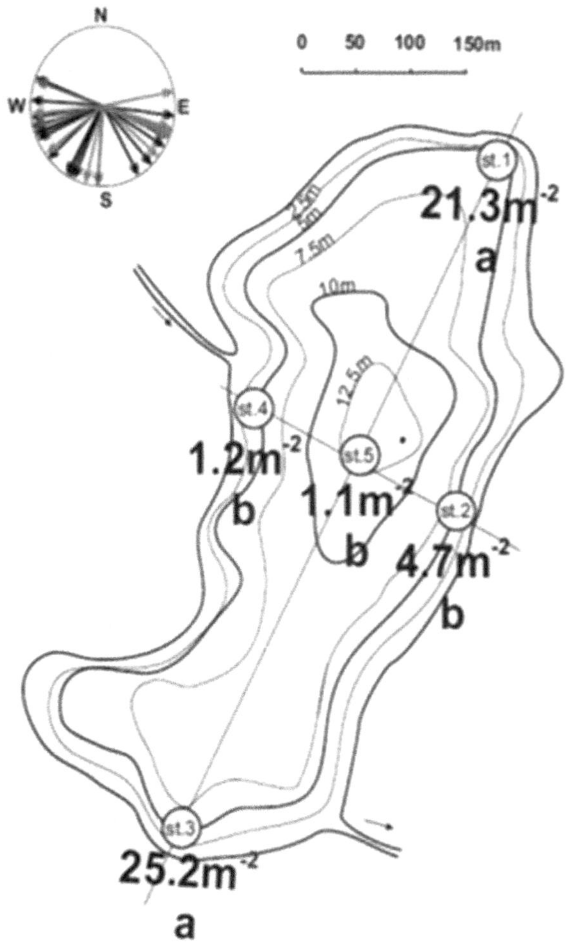

Fig. 8.5 Mean densities of ephippia of *Daphnia* of the *longispina* group at the water surface of the small eutrophic Lake Kociołek (54°03′01″N 22°19′54″E), collected with half-submerged neuston net kept in front of cruising boat at noon at five sampling stations located along prevailing wind directions reported week ahead of sampling period. Letters below the density values indicate homogenous groups with values similar according to ANOVA model simplification

8.2.4.2 Trap More Than Shelter

Ephippia accumulate in deep locations. The resident ephippia which are shed below the water surface, and those oviposited at the water surface that sink with time become unevenly distributed over the lake bottom. The highest densities of ephippia frequently occurs in the deepest zones of lakes (Carvalho and Wolf 1989; Jankowski 2003; Kerfoot et al. 2004) either because of the higher number of females shedding ephippia in cumulated volume of water above those places or, what is more likely, due to their slipway from shallower locations along the bottom slopes. Controversy exists if bottom sediments are shelter for the resting eggs or rather a trap that disable their hatching (Cáceres and Hairston 1998). Dormant forms of aquatic species may reside in lake sediments for long time not deliberately, but due to their burrowing by lake sediments. According to Radzikowski et al. (2016), even a thin sediment layer (less than 2.5 mm) may hamper development of resting eggs of *Daphnia*. This might provoke a significant question on the ultimate function of ephippia deposition at the water surface: is it dispersion between, or rather within aquatic habitats? *Daphnia* could shed ephippia at the water surface not to colonize other waterbodies, but to let them be rafted toward the shallow littoral zone and try to escape the sediment trap of the profundal zone. Higher proportion of ephippia left by Daphnidae at the water surface in shallow lakes with frequently mixed sediments than deep ones with stagnant bottom (Ślusarczyk et al. 2017a) indicate, however, that between-lake dispersal is the more likely selective force behind this phenomenon.

8.2.4.3 Anthropogenic Invasions

Local mass aggregations of ephippia, together with their clinginess to different surface, and with carelessness on the human part, has led to unprecedented dispersal events. Intensified transport of dormant forms by anthropogenic vectors (e.g. by regional and global trades) broke existing intercontinental barriers and facilitated colonization of freshwater habitats by invasive species. This includes encroachment of alien species of *Daphnia* originating from remote locations in a recent time (Havel and Shurin 2004; Panov et al. 2004; Panov and Cáceres 2007), e.g. ongoing invasion of *D. lumholtzi* in North America (Havel et al. 2002) or *D. pulex* in Africa (Mergeay et al. 2006b).

8.2.4.4 Enhanced Biotic Diver sity and Slower Evolutionary changes

The spatial dispersal of dormant forms may enhance biotic diversity on a local (alfa diversity) and regional scale (gamma diversity) due to immigration of novel or recolonization of exterminated genotypes from neighbourhood locations (Chesson 2000) and reduce biotic dissimilarities between habitats within a region (i.e. lower beta diversity). The immigration of dormant forms from neighbourhood locations, like resurrection from the native bank of resting eggs, may slow down competitive exclusion of suboptimal competitors by superior ones and slow down evolutionary changes within genetic pool and species sorting within local habitats (Hairston and de Stasio 1988).

8.3 Conclusions

While the old paradigm of panmictic occurrence of planktonic species due to their easy colonization of aquatic habitats has eroded, the frequent spread of aquatic organisms has not yet been seriously questioned. What stops their successful invasions seems abiotic and biotic filters rather than dispersal barriers (Shurin 2000; De Meester et al. 2002; Incagnone et al. 2015). Frequent overland dispersal may be attributed to dormant forms used not only for temporal but also for spatial dispersal.

Freshwater dormant forms can survive long voyage across hostile terrestrial zone and may hitchhike using various animate or inanimate vectors, especially when aided by behavioural or morphological adaptations. Planktonic crustaceans of the genus *Daphnia* are among the first colonizers of newly founded waterbodies (Louette and de Meester 2005). Ephippial case of resting eggs may boost their spatial dispersal by offering positive buoyancy facilitating their hitchhiking by surface vectors and enhanced protection against mechanical damages or UV radiation. During periods of intense ephippia formation, high quantities of ephippia often accumulate along the downwind shore, where the wind blows them to. The supralittoral and meso-littoral banks of resting eggs may be an underestimated source not only for passive dispersal between waterbodies but possibly also as a refuge preventing their dislocation to the deep trap of profundal sediments.

Daphnia is undoubtedly the best-studied cladoceran species. The question remains, how effective is dispersal of other cladocerans with poorly developed ephippia or no ephippia at all, and what mechanisms of dispersal they utilize instead. An open question remains also, if dormant eggs used for only temporal dispersal differ structurally or physiologically from ones used for spatial dispersal.

Acknowledgments We would like to thank Polish National Science Centre for financial support by two grants (2012/05/B/NZ8/01232 and 2016/23/D/NZ8/01736) that made possible writing this chapter.

References

Allen MR (2007) Measuring and modelling dispersal of adult zooplankton. Oecologia 153:135–143

Alonso C, Rocco V, Barriga JP, Battini MĂN, Zagarese H (2004) Surface avoidance by freshwater zooplankton: Field evidence on the role of ultraviolet radiation. Limnol Oceanogr 49(1):225–232

Bell JR, Bohan DA, Shaw EM, Weyman GS (2005) Ballooning dispersal using silk: world fauna, phylogenies, genetics and models. Bull Entomol Res 95:69–114

Bengtsson BO (1978) Avoiding inbreeding: at what cost? J Theor Biol 73:439–444

Berg LM, Pálsson S, Lascoux M (2001) Fitness and sexual response to population density in *Daphnia pulex*. Freshw Biol 46:667–677

Bernatowicz P, Radzikowski J, Paterczyk B, Bebas P, Ślusarczyk M (2018) Internal structure of *Daphnia* ephippium as an adaptation to dispersion. *Zool Anz* 277:12–22

Bilton DT, Freeland JR, Okamura B (2001) Dispersal in freshwater invertebrates. *Annu Rev Ecol Syst* 32:159–181. https://doi.org/10.1146/annurev.ecolsys.32.081501.114016

Bohonak A, Jenkins D (2003) Ecological and evolutionary significance of dispersal by freshwater invertebrates. Ecol Lett 6(8):783–796

Boileau MG, Hebert PDN, Schwartz SS (1992) Non-equilibrium gene frequency divergence: persistent founder effects in natural populations. J Evol Biol 5:25–39

Boulinier T, McKoy KD, Sorci G (2002) Dispersal and parasitism. In: Clobert J, Danchin E, Dhondt AA, Nichols JD (eds) Dispersal. Oxford University Press, Oxford, pp 169–180

Bormashenko E (2016) Surface Tension Supported Floating of Heavy Objects: Why Elongated Bodies Float Better? J Colloid Interface Sci 463:8–12

Brendonck L, Pinceel T, Ortells R (2017) Dormancy and dispersal as mediators of zooplankton population and community dynamics along a hydrological disturbance gradient in inland temporary pools. Hydrobiologia 796:201–222

Brochet AL, Gauthier-Clerc M, Guillemain M, Fritz H, Waterkeyn A, Baltanás Á, Green AJ (2010a) Field evidence of dispersal of branchiopods, ostracods and bryozoans by teal (*Anas crecca*) in the Camargue (southern France). Hydrobiologia 637:255–261

Brochet AL, Guillemain M, Fritz H, Gauthier-Clerc M, Green AJ (2010b) Plant dispersal by teal (*Anas crecca*) in the Camargue: duck guts are more important than their feet. Freshw Biol 55:1262–1273. https://doi.org/10.1111/j.1365-2427.2009.02350.x

Bullock JM, Bonte D, Pufal G, DaSilva-Carvalho C, Chapman DS, Garcia C, Garcia D, Matthysen E, Delgado MM (2018) Human-Mediated Dispersal and the Rewiring of Spatial Networks. Trends Ecol Evol 33(12):958–970

Buoro M, Carlson SM (2014) Life-history syndromes: Integrating dispersal through space and time. Ecol Lett 17(6):756–767

Burgess SC, Baskett ML, Grosberg RK, Morgan SG, Strathmann RR (2016) When is dispersal for dispersal? Unifying marine and terrestrial perspectives. Biol Rev 91:867–882. https://doi.org/10.1111/brv.12198

Cáceres CE (1998a) Interspecific variation in the abundance, production, and emergence of *Daphnia* diapausing eggs. Ecology 79(5):1699–1710

Cáceres CE (1998b) Seasonal dynamics and interspecific competition in Oneida Lake *Daphnia*. Oecologia 115:233–244

Cáceres CE, Hairston NGJ (1998) Benthic-pelagic coupling in planktonic crustaceans: the role of the benthos. Arch Hydrobiol Spec Issue Adv Limnol 52:163–174

Cáceres C, Christoff A, Boeing W (2007) Variation in ephippial buoyancy in *Daphnia pulicaria*. Freshw Biol 52:313–318

Cáceres CE, Soluk DA (2002) Blowing in the wind: a field test of overland dispersal and colonization by aquatic invertebrates. *Oecologia* 131:402–408

Carvalho GR, Wolf HG (1989) Resting eggs of lake *Daphnia* I. Distribution, abundance and hatching of eggs collected from various depths in lake sediments. Freshw Biol 22:459–470

Chesson P (2000) Mechanisms of maintenance of species diversity. Annu Rev Ecol Syst 31(1):343–366

Cohen D (1966) Optimizing reproduction in a randomly varying environments. J Theoret Biol 12:119–129

Cohen GM, Shurin JB (2003) Scale-dependence and mechanisms of dispersal in freshwater zooplankton. Oikos 103(3):603–617

Colbourne JK, Hebert PDN (1996) The systematics of North American *Daphnia* (Crustacea: Anomapoda): a molecular phylogenetic approach. Philos Trans R Soc Lond B 351:349–360

Costa MJ, Ramos JA, da Silva LP, Timoteo S, Araujo PM, Felgueiras MS, Rosa A, Matos C, Encarnacao P, Tenreiro PQ, Heleno RH (2014) Endozoochory largely outweighs epizoochory in migrating passerines. J Avian Biol 45:59–64. https://doi.org/10.1111/j.1600-048X.2013.00271.x

Coughlan NE, Kelly TC, Davenport J, Jansen MA (2017) Up, up and away: bird-mediated ectozoochorous dispersal between aquatic environments. Freshw Biol 62:631–648. https://doi.org/10.1111/fwb.12894

Dawidowicz P, Pijanowska J (2018) Chapter 9: Diel vertical migration of aquatic crustaceans – adaptive role, underlying mechanisms, and ecosystem consequences. In: Thiel M, Welborn G (eds) The natural history of crustaceans, life histories, vol 5. Oxford University Press, Oxford, pp 231–256

Davison J (1969) Activation of the Ephippial Egg of *Daphnia pulex*. J Gen Physiol 53:562–575

Decaestecker E, De Meester L, Mergeay J (2009) Cyclical Parthenogenesis in *Daphnia*: sexual versus asexual reproduction. In: Lost sex: the evolutionary biology of parthenogenesis. Springer Netherlands, Dordrecht, pp 295–316

De Meester L (1996a) Local genetic differentiation and adaptation in freshwater zooplankton populations: patterns and processes. Ecoscience 3:385–399
De Meester L (1996b) Evolutionary potential and local genetic differentiation in a phenotypically plastic trait of a cyclical parthenogen, *Daphnia*. *Evolution* 50:1293–1298
De Meester L, Gómez A, Okamura B, Schwenk K (2002) The Monopolization Hypothesis and the dispersal–gene flow paradox in aquatic organisms. Acta Oecol 23:121–135
Figuerola J, Green AJ (2002a) Dispersal of aquatic organisms by waterbirds: a review of past research and priorities for future studies. Freshw Biol 47:483–494
Figuerola J, Green AJ (2002b) How frequent is external transport of seeds and invertebrate eggs by waterbirds? A study in Doñana, SW Spain. Arch Hydrobiol 155(4):557–565. https://doi.org/10.1127/archiv-hydrobiol/155/2002/557
Figuerola J, Green AJ, Santamaria L (2003) Passive internal transport of aquatic organisms by waterflow in Doñana, south-west Spain. Glob Ecol Biogeogr 12:427–436
Figuerola J, Green AJ, Michot TC (2005) Invertebrate eggs can fly: Evidence of waterflow-mediated gene flow in aquatic invertebrates. Am Nat 165:274–280
Fitzsimmons JM, Innes DJ (2006) Inter-genotype variation in reproductive response to crowding among *Daphnia pulex*. Hydrobiologia 568(1):187–205
Forro L, Korovchinsky NM, Kotov AA, Petrusek A (2008) Global diversity of cladocerans (Cladocera; Crustacea) in freshwater. Hydrobiologia 595:177–184
Frey DG (1982) Questions concerning cosmopolitanism in Cladocera. Arch Hydrobiol 93:484–502
Frisch D, Morton PK, Chowdhury PR, Culver BW, Colbourne JK, Weider LJ, Jeyasingh PD (2014) A millennial-scale chronicle of evolutionary responses to cultural eutrophication in *Daphnia*. Ecol Lett 17(3):360–368
Fryer G (1996) Diapause, a potent force in the evolution of freshwater crustaceans. Hydrobiologia 320:1–14
Gandon S, Michalakis Y (2002) Multiple causes of the evolution of dispersal. In: Clobert J, Danchin E, Dhondt AA, Nichols JD (eds) Dispersal. Oxford University Press, Oxford, pp 155–168
Gerrish G, Cáceres C (2003) Genetic versus environmental influence on pigment variation in the ephippia of *Daphnia pulicaria*. Freshw Biol 48(11):1971–1982
Gliwicz ZM (2003) Between hazards of starvation and risks of predation: the ecology of offshore animals. International Ecology Institute, Oldendorf/Luhe
Gómez A, Carvalho GR, Lunt DH (2000) Phylogeography and regional endemism of a passive dispersal zooplankter: mitochondrial DNA variation in rotifer resting egg banks. Proc R Soc Lond B 267:2189–2197
Gómez A, Adcock GA, Lunt DH, Carvalho GR (2002) The interplay between colonization history and gene flow in passively dispersing zooplankton: microsatellite analysis of rotifer resting egg banks. J Evol Biol 15:158–171
Green AJ, Figuerola J, Sánchez MI (2002) Implications of waterbird ecology for the dispersal of aquatic organisms. Acta Oecol 23:177–189
Haileselasie TH, Mergeay J, Weider LJ, Sommaruga R, Davidson TA, Meerhoff M, Arndt H, Jürgens K, Jeppesen E, De Meester L (2016) Environment not dispersal limitation drives clonal composition of Arctic *Daphnia* in a recently deglaciated area. Mol Ecol 25:5830–5842. https://doi.org/10.1111/mec.13843
Haileselasie TH, Mergeay J, Vanoverbeke J, Orsini L, De Meester L (2018) Founder effects determine the genetic structure of the water flea *Daphnia* in Ethiopian reservoirs. Limnol Oceanogr 63:915–926. https://doi.org/10.1002/lno.10678
Hairston NGJ, Lampert W, Cáceres CE (1999) Rapid evolution revealed by dormant eggs. Nature 401(30):446
Hairston NGJ, de Stasio BTJ (1988) Rate of evolution slowed by a dormant propagule pool. Nature 336(17):239–242
Hamilton WD, May RM (1977) Dispersal in stable habitats. Nature 269(13):578–581
Havel JE, Mabee MR, Jones JR (1995) Invasion of the exotic *Daphnia lumhotzi* into North American reservoirs. Can J Fish Aquat Sci 52:151–160

Havel JE, Shurin JB (2004) Mechanisms, effects, and scales of dispersal in freshwater zooplankton. Limnol Oceanogr 49:1229–1238
Havel JE, Shurin JB, Jones JR (2002) Estimating dispersal from patterns of spread: spatial and local control of lake invasions. Ecology 83(12):3306–3318
Heino J, Melo AS, Siqueira T, Soininen J, Valanko S, Bini LM (2015) Metacommunity organisation, spatial extent and dispersal in aquatic systems: patterns, processes and prospects. Freshw Biol 60:845–869. https://doi.org/10.1111/fwb.12533
Heino J, Alahuhta J, Ala-Hulkko T, Antikainen H, Bini LM, Bonada N, Datry T, Erős T, Hjort J, Kotavaara O, Melo AS, Soininen J (2017) Integrating dispersal proxies in ecological and environmental research in the freshwater realm. Environ Rev 25:334–349. https://doi.org/10.1139/er-2016-0110
Higgins SI, Nathan R, Cain ML (2003) Are long-distance dispersal events in plants usually caused by nonstandard means of dispersal? Ecology 84:1945–1956. https://doi.org/10.1890/01-0616
Hiruta C, Tochinai S (2014) Formation and structure of the ephippium (resting egg case) in relation to molting and egg laying in the water flea *Daphnia pulex* De Geer (Cladocera: Daphniidae). J Morphol 275:760–767
Holt RD, Barfield M (2002) On the relationship between the ideal free distribution and the evolution of dispersal. In: Clobert J, Danchin E, Dhondt AA, Nichols JD (eds) Dispersal. Oxford University Press, Oxford, pp 83–96
Hopfensperger KN, Baldwin AH (2009) Spatial and temporal dynamics of floating and drift-line seeds at a tidal freshwater marsh on the Potomac River, USA. Plant Ecol 201:677–686
Horváth Z, Vad CF, Ptacnik R (2016) Wind dispersal results in a gradient of dispersal limitation and environmental match among discrete aquatic habitats. Ecography (Cop) 39(8):726–732
Incagnone G, Marrone F, Barone R, Robba L, Naselli-Flores L (2015) How do freshwater organisms cross the "dry ocean"? A review on passive dispersal and colonization processes with a special focus on temporary ponds. Hydrobiologia 750:103–123
Jacobs MJ, MacIsaac HJ (2007) Fouling of fishing line by the waterflea *Cercopagis pengoi*: a mechanism of human-mediated dispersal of zooplankton? Hydrobiologia 583:119–126
Jankowski T (2003). From diapause to sexual reproduction – evolutionary ecology of the daphnia hybrid complex from lake constance. PhD thesis, Universität Konstanz, Konstanz.
Jenkins DG, Buikema AL (1998) Do similar communities develop in similar sites? Ecol Monogr 68:421–443
Juračka PJ, Declerck SAJ, Vondrak D, Beran L, Cerny M, Petrusek A (2016) A naturally heterogeneous landscape can effectively slow down the dispersal of aquatic microcrustaceans. Oecologia 180(3):785–796
Karlson RH (1992) Divergent dispersal strategies in the freshwater bryozoan *Plumatella repens*: ramet size effects on statoblast numbers. Oecologia 89(3):407–411
Kappes H, Tackenberg O, Haase P (2014) Differences in dispersal- and colonization-related traits between taxa from the freshwater and the terrestrial realm. Aquat Ecol 48:73–83
Kaya M, Baran T, Saman I, Ozusaglam MA, Cakmak YS, Menteş A (2014) Physicochemical characterization of chitin and chitosan obtained from resting eggs of *Ceriodaphnia quadrangula* (Branchiopoda: Cladocera: Daphniidae). J Crustac Biol 34:283–288. https://doi.org/10.1163/1937240X-00002221
Kerfoot WC, Wells Budd J, Eadie B, Vanderploeg H, Agy M (2004) Winter storms: Sequential sediment traps record *Daphnia* ephippial production, resuspension, and sediment interactions. Limnol Oceanogr 49(4):1365–1381
Leuven R, van der Velde G, Baijens I, Snijders J, van der Zwart C, Lenders HJR, de Vaate AB (2009) The river Rhine: a global highway for dispersal of aquatic invasive species. Biol Invasions 11(9):1989–2008
Lewis LR, Behling E, Gousse H, Qian E, Elphick CS, Lamarre JF, Bêty J, Liebezeit J, Rozzi R, Goffinet B (2014) First evidence of bryophyte diaspores in the plumage of transequatorial migrant birds. Peer J 2:e424. https://doi.org/10.7717/peerj.424

Lopes PM, Bozelli R, Bini JLM, Santangelo M, Declerck SAJ (2016) Contributions of airborne dispersal and dormant propagule recruitment to the assembly of rotifer and crustacean zooplankton communities in temporary ponds. Freshw Biol 61:658–669

Louette G, de Meester L (2005) High dispersal capacity of cladoceran zooplankton in newly founded communities. Ecology 86(2):353–359

Lynch M, Spitze K (1994) Evolutionary genetics of *Daphnia*. In: Real LA (ed) Ecological genetics. Princeton University Press, Princeton, NJ, pp 109–128

Maguire B Jr (1963) The passive dispersal of small aquatic organisms and their colonization of isolated bodies of water. Ecol Monogr 33:161–185

Maier G (1996) *Daphnia* invasion: population dynamics of *Daphnia* assemblages in two eutrophic lakes with particular reference to the introduced alien *Daphnia ambigua*. J Plankton Res 18:2001–2015

Marszelewski W (2005) Zmiany warunków abiotycznych w jeziorach Polski północno-wschodniej. Wydawnictwo Uniwersytetu Mikołaja Kopernika, Toruń

Mayr E (1970) Populations, species, and evolution. Belknap Press of Harvard University Press, Cambridge

McAtee WL (1917) Showers of Organic Matter. Mon Weather Rev 45:217

Mc Peek MA, Kalisz S (1998) On the joint evolution of dispersal and dormancy in metapopulations. Arch Hydrobiol Spec Issue Adv Limnol 52:33–51

Meijering MPD (2003) The long-lasting resistance of diapausing eggs from Arctic Cladocera at -18°C. Pol Polar Res 24:167–172

Mergeay J, Verschuren D, de Meester L (2006a) *Daphnia* species diversity in Kenya, and a key to the identification of their ephippia. Hydrobiologia 542(1):261–274

Mergeay J, Verschuren D, de Meester L (2006b) Invasion of an asexual American water flea clone throughout Africa and rapid displacement of a native sibling species. Proc R Soc B Biol Sci 273(1603):2839–2844

Michels E, Cottenie K, Neys L, de Meester L (2001) Zooplankton on the move: first results on the quantification of dispersal of zooplankton in a set of interconnected ponds. Hydrobiologia 442:117–126

Mikulski A. & Grzesiuk M. 2019, Sex and sex dependent reproduction strategies in cyclic partenogen - a case study from astatic urban pond. Submitted manuscript.

Moreno E, Perez-MartÃnez C, Conde-Porcuna JM (2019) Dispersal of rotifers and cladocerans by waterbirds: seasonal changes and hatching success. Hydrobiologia 834(1):145–162. https://doi.org/10.1007/s10750-019-3919-6

Palsson S (2000) Microsatellite variation in Daphnia pulex from both side of the Baltic Sea. Mol Ecol 9:1075–1088

Panov VE, Cáceres C (2007) Role of diapause in dispersal of aquatic invertebrates. In: Diapause in aquatic invertebrates theory and human use. Springer Netherlands, Dordrecht, pp 187–195

Panov VE, Krylov PI, Riccardi N (2004) Role of diapause in dispersal and invasion success by aquatic invertebrates. J Limnol 63. (Suppl. 1:56–69

Pauwels K, Stoks R, Verbiest A, de Meester L (2007) Biochemical adaptation for dormancy in subitaneous and dormant eggs of *Daphnia magna*. Hydrobiologia 594:91–96. https://doi.org/10.1007/s10750-007-9091-4

Perrin N, Mazalov V (1999) Dispersal and Inbreeding Avoidance. Am Nat 154(3):282–292

Pietrzak B, Ślusarczyk M (2006) The fate of ephippia - *Daphnia* dispersal in time and space. Pol J Ecol 54:709–714

Pietrzak B, Bednarska A, Markowska M, Rojek M, Szymanska E, Ślusarczyk M (2013) Behavioural and physiological mechanisms behind extreme longevity in *Daphnia*. Hydrobiologia 715:125–134

Pinceel T, Vanschoenwinkel B, Brendonck L (2013) Flexible dispersal dimorphism in zooplankton resting eggs: an example of repeated phenotypic coin flipping? Biol J Linn Soc 110:749–756

Pollux BJA, Verbruggen E, Van Groenendael JM, Ouborg NJ (2009) Intraspecific variation of seed floating ability in *Sparganium emersum* suggests a bimodal dispersal strategy. Aquat Bot 90:199–203

Proctor VW (1959) Dispersal of fresh-water algae by migratory water birds. Science 130:623–624

Proctor VW, Malone CR (1965) Further evidence of the passive dispersal of small aquatic organism via the intestinal tract of birds. Ecology 46:728–729

Ptatscheck C, Milne PC, Traunspurger W (2018) Is stemflow a vector for the transport of small metazoans from tree surfaces down to soil? BMC Ecol 18:43. https://doi.org/10.1186/s12898-018-0198-4

Radzikowski J (2013) Resistance of dormant stages of planktonic invertebrates to adverse environmental conditions. J Plankton Res 35(4):707–723

Radzikowski J (2019). Properties of the resting stages of Daphnidae inhabiting aquatic habitats of varying degrees of variation in environmental conditions. PhD thesis, University of Warsaw, Warszawa.

Radzikowski J, Sikora A, Ślusarczyk M (2016) The effect of lake sediment on the hatching success of *Daphnia* ephippial eggs. J Limnol 75(3):597–605

Radzikowski J, Krupińska K, Ślusarczyk M (2018) Different thermal stimuli initiate hatching of *Daphnia* diapausing eggs originating from lakes and temporary waters. Limnology 19:81–88

Raulings E, Morris K, Thompson R, Mac Nally R (2011) Do birds of a feather disperse plants together? Freshw Biol 56:1390–1402

Reynolds C, Cumming GS (2015) The role of waterbirds in the dispersal of freshwater cladocera and bryozoa in southern Africa. Afr Zool 50:307–311. https://doi.org/10.1080/15627020.2015.1108164

Reynolds C, Cumming GS (2016) Seed dispersal by waterbirds in southern Africa: comparing the roles of ectozoochory and endozoochory. Freshw Biol 61:349–361. https://doi.org/10.1111/fwb.12709

Ricci C, Caprioli M (2005) Anhydrobiosis in bdelloid species, populations and individuals. Integr Comp Biol 45:759–763

Sánchez MI, Hortas F, Figuerola J, Green AJ (2012) Comparing the potential for dispersal via waterbirds of a native and an invasive brine shrimp. Freshw Biol 57:1896–1903. https://doi.org/10.1111/j.1365-2427.2012.02852.x

Schwenk K, Sand A, Boersma M, Brehm M, Mader E, Offerhaus D, Spaak P (1998) Genetic makers, genealogies and biogeographic patterns in the cladoceran. Aquat Ecol 32:37–51

Seger J, Brockmann JH (1987) What is bet-hedging? In: Harvey PH, Partridge L (eds) Oxford surveys in evolutionary biology. Oxford University Press, Oxford, UK, pp 182–211

Sirianni KM (2017) Differential wind dispersal of cladoceran ephippia in a rock pool metacommunity. Aquat Ecol 51:203–218. https://doi.org/10.1007/s10452-016-9611-2

Shan RKC (1970) Influence of light on hatching resting eggs of chydorids (Cladocera). Internationale Revue der gesamten Hydrobiologie und Hydrographie 55(2):295–302

Shurin JB (2000) Dispersal limitation, invasion resistance and the structure of pond zooplankton communities. Ecology 81:3074–3086

Shurin JB, Havel JE (2002) Hydrologic connections and overland dispersal in an exotic freshwater crustacean. Biol Invasions 4:431–439

Ślusarczyk M (1995) Predator-induced diapause in *Daphnia*. Ecology 76(3):1008–1013

Ślusarczyk M (2004) Environmental plasticity of fish avoidance diapause response in *Daphnia magna*. J Limnol 63. (Suppl. 1:70–74

Ślusarczyk M (2009) Extended lifespan traded for diapause in *Daphnia*. Freshw Biol 54(11):2252–2262

Ślusarczyk M, Pinel-Alloul B, Gelinas M (2006) On ultimate reasons for summer diapause of *Daphnia* in a permanent lake. Verh Internat Verein Limnol 29:1440–1442

Ślusarczyk M, Grabowski T, Pietrzak B (2017a) Quantification of floating ephippia in lakes: a step to a better understanding of high dispersal propensity of freshwater plankters. Hydrobiologia 798(1):61–72

Ślusarczyk M, Starzynski J, Bernatowicz P (2017b) How long to rest in unpredictably changing habitats? PLoS ONE 12(4):e0175927

Ślusarczyk M, Pietrzak B (2008) To sink or float: the fate of dormant offspring is determined by maternal behaviour in *Daphnia*. Freshw Biol 53:569–576

Ślusarczyk M, Flis S (2019) Light quantity, not photoperiod terminates diapause in the crustacean *Daphnia*. Limnol Oceanogr 64(1):124–130

Ślusarczyk M, Chlebicki W, Pijanowska J, Radzikowski J (2019) The role of the refractory period in diapause length determination in a freshwater crustacean. Submitted manuscript.

Smirnov N (2014) Physiology of the cladocera. Academic Press, London, p 352

Valls L, Castillo-Escrivà A, Mesquita-Joanes F, Armengol X (2016) Human-mediated dispersal of aquatic invertebrates with waterproof footwear. Ambio 45:99–109. https://doi.org/10.1007/s13280-015-0689-x

Vandekerkhove J, Louette G, Brendonck L, de Meester L (2005) Development of cladoceran egg bank in new and isolated pools. Arch Hydrobiol 162:339–347

Van de Meutter F, Stoks R, de Meester L (2006) Lotic dispersal of lentic macroinvertebrates. Ecography 29:223–230

Van de Meutter F, Stoks R, de Meester L (2008) Size-selective dispersal of *Daphnia* resting eggs by backswimmers (*Notonecta maculata*). Biol Lett 4:494–496

Van Leeuwen CHA, Sarneel JM, van Paassen J, Rip WJ, Bakker ES (2014) Hydrology, shore morphology and species traits affect seed dispersal, germination and community assembly in shoreline plant communities. J Ecol 102:998–1007

Van Leeuwen CHA, Velde G, Groenendael JM, Klaassen M (2012) Gut travellers: internal dispersal of aquatic organisms by waterfowl. J Biogeogr 39:2031–2040. https://doi.org/10.1111/jbi.12004

Van Leeuwen CHA (2018) Internal and external dispersal of plants by animals: an aquatic perspective on alien interference. Front Plant Sci 9:153. https://doi.org/10.3389/fpls.2018.00153

Vanschoenwinkel B, Gielen S, Vandewaerde H, Seaman M, Brendonck L (2008a) Relative importance of different dispersal vectors for small aquatic invertebrates in a rock pool metacommunity. Ecography 31:567–577

Vanschoenwinkel B, Gielen S, Seaman M, Brendonck L (2008b) Any way the wind blows - frequent wind dispersal drives species sorting in ephemeral aquatic communities. Oikos 117:125–134. https://doi.org/10.1111/j.2007.0030-1299.16349.x

Vanschoenwinkel B, Gielen S, Seaman M, Brendonck L (2009) Wind mediated dispersal of freshwater invertebrates in a rock pool metacommunity: differences in dispersal capacities and modes. Hydrobiologia 635:363–372

Vega MP, Clausse A (1998) Hydrodynamic characteristics of *Daphnia middendorffiana*. Int Rev Hydrobiol 83(4):267–277

Venable DL, Lawlor L (1980) Delayed germination and dispersal in desert annuals: escape in space and time. Oecologia 46:272–282

Venable DL, Levin DA (1985) Ecology of achene dimorphism in *Heterotheca latifolia*: I. Achene structure, germination and dispersal. J Ecol 73(1):133–145

Vivian-Smith G, Stiles EW (1994) Dispersal of salt marsh seeds on the feet and feathers of waterfowl. Wetlands 14:313–319

Waterkeyn A, Vanschoenwinkel B, Elsen S, Anton-Pardo M, Grillas P, Brendonck L (2010) Unintentional dispersal of aquatic invertebrates via footwear and motor vehicles in a Mediterranean wetland area. Aquat Conserv Mar Freshwat Ecosyst 20:580–587

Weisser WW (2002) The effects of predation on dispersal. In: Clobert J, Danchin E, Dhondt AA, Nichols JD (eds) Dispersal. Oxford University Press, Oxford, pp 180–191

Zaffagnini F (1987) Reproduction in *Daphnia*. Memorie Dell'Istituto Italiano Di Idrobiologia Dott Marco De Marchi 45:245–284

Chapter 9
Seasonal Diphenism in *Daphnia* from Temperate Environments: Organismal Traits and Molecular Regulation

Kęstutis Arbačiauskas

Abstract In temperate waters, post-diapause and subitaneous offspring in *Daphnia* substantially vary in organismal traits. Post-diapause offspring have elevated metabolic activity. Under rich food, they grow faster, mature at a larger body size and exhibit a greater allocation to progeny resulting in a higher relative fitness. In contrast, subitaneous females are more resistant to starvation and have a higher fitness across limited food concentrations.

These offspring types result from different developmental programs of the same genetic background, representing the phenomenon of phenotypic plasticity. A high-throughput proteomic analysis revealed 176 proteins that were differentially expressed among offspring phenotypes. There were more upregulated proteins with oxyreductase and binding activity in post-diapause offspring, whereas more upregulated proteins with transporter and transferase activity were seen in subitaneous offspring. Over 1.5-fold more of the proteins that were upregulated in post-diapause phenotype are involved in metabolism and biosynthesis. The greatest difference, a fivefold upregulation in post-diapause compared to subitaneous offspring, was recorded for the target of rapamycin-like (TOR) protein. Expression of ribosomal proteins in this offspring phenotype was also increased. These upregulations suggest that the TOR signalling pathway is involved and can be responsible for the regulation of the developmental program underlying post-diapause and subitaneous offspring phenotypes. Gene regulatory patterns observed in post-diapause and subitaneous offspring were in general agreement with observed organismal traits of these *Daphnia* phenotypes.

In temperate waters, *Daphnia* have evolved two alternative seasonal phenotypes matching environmental conditions in which they occur and perform in accordance with predictions of seasonal polyphenism (diphenism in this case). Due to higher metabolic activity, which must lead to increased resource acquisition, post-diapause

K. Arbačiauskas (✉)
Nature Research Centre, Vilnius, Lithuania
e-mail: kestutis.arbaciauskas@gamtc.lt

V. R. Alekseev, B. Pinel-Alloul (eds.), *Dormancy in Aquatic Organisms. Theory, Human Use and Modeling*, Monographiae Biologicae 92,
https://doi.org/10.1007/978-3-030-21213-1_9

offspring are superior to subitaneous offspring under high food conditions, which are expected during early season, but inferior under limiting food environments, which frequently occur later in the season. In seasonal climate, the adjustment of resource acquisition with respect to resource availability may be a general evolutionary trend for multivoltine organisms resulting in a seasonal polyphenism.

Keywords Cladocera · Diapause · Metabolism · Allocation · Fitness · Phenotypic plasticity · Proteomic analysis · TOR (target of rapamycin) · Seasonal polyphenism

9.1 Introduction

Diapause is an adaptation enabling many organisms by programmed arrest of development to survive seasonally or at least in short-term predictable adverse conditions (Hairston Jr. 1998). Advantage of the diapause is tightly associated not only with its timely induction but also with the proper time of diapause termination. These processes are governed by environmental cues usually adjusted by internal control (Tauber et al. 1986; Alekseev 1990; Fox and Mousseau 1998). Obviously, diapause termination should happen in a favourable environment, at least in an environment sufficient for living.

In univoltine organisms, the termination of the diapause within a season as soon as environment become suitable for living seems to be advantageous, whilst in multivoltine organisms, a few emergence strategies are possible. If the environment varies unpredictably, the most adaptive emergence strategy may be to disperse the recruitment over the season in pursuit of spreading the risk of occurring in hostile conditions. If the environment can be predicted as favourable, the synchronization of recruitment might be most adaptive. The predictability of environment during activation further implies that selection can favour specific post-diapause traits which increase recruit fitness in a prospective environment. Therefore, a disparity between post-diapause and subitaneous (immediately/directly developing) offspring can evolve, if they use to inhabit varying environmental conditions.

In this chapter, we analyse seasonal phenotypes of *Daphnia* inhabiting temperate waters. We collate organismal traits of post-diapause and subitaneous offspring phenotypes, provide data on probable molecular regulation of this diphenism and qualify this seasonal phenomenon.

9.2 *Daphnia* Life Cycle

In temperate waters, the cyclic parthenogen *Daphnia* propagates mostly by parthenogenesis, producing subitaneous eggs. With deterioration of environmental conditions, which can be predicted from environmental cues, these animals initiate sexual reproduction and produce diapausing eggs encased in an ephippium (Stross and Hill

1965; Carvalho and Hughes 1983; Pijanowska and Stolpe 1996; Slusarczyk 2001; Koch et al. 2009; Deng et al. 2010), although the timing and magnitude of investment into a dormant stage depend upon local conditions (Walsh 2013). Daphnids as all other cladocerans possess an embryonic diapause. Diapausing eggs start to hatch in the next season but, being resistant to external factors, can remain viable for extended periods, forming a sediment egg bank that facilitates temporal population persistence (Hairston Jr. et al. 1995; De Stasio 2007).

A mass emergence from diapausing eggs in *Daphnia* in temperate lakes and ponds almost exclusively takes place in the early season over a relatively short time (Wolf and Carvalho 1989; Caceres 1998; Hairston Jr. et al. 2000; Gyllström 2004; Rother et al. 2010; Perez-Martınez et al. 2013), although hatching also may happen years later during any seasons and in a strongly altered environment (Frisch et al. 2014). In temperate waters, as suggested by the Plankton Ecology Group (PEG) model, early seasons with abundant nutrients and increasing sunlight and temperature result in massive development of algae suitable for *Daphnia* nutrition – the spring algal bloom (Sommer et al. 1986). Investigations of cues inducing diapause termination suggest temperature is the dominant cue for emergence (reviewed in Gyllström and Hansson 2004), reducing the probability of a mismatch between daphnids exiting diapause and their algal food (Rother et al. 2010). Hence, the majority of post-diapause *Daphnia* offspring hatch into a predictable high-food environment, whereas during the season their subitaneous descendants encounter environments with greatly variable conditions in terms of food quality and availability. These features of living conditions must affect traits of post-diapause and subitaneous offspring in *Daphnia* inhabiting temperate seasonal environments.

9.3 Organismal Traits of Post-diapause and Subitaneous Offspring

Post-diapause offspring hatch from bisexually produced diapausing eggs (although diapausing eggs also can be produced parthenogenetically) in which development is arrested at early embryogenesis and resumes after hatching cues are received, whilst subitaneous offspring hatch from parthenogenetically produced eggs. The two egg types substantially vary in morphology and histology (Zaffagnini 1987). They also differ in biochemical composition and fatty acid profiles (Arbačiauskas 2007, unpublished data). When comparing neonates of differing origin, post-diapause offspring have higher carbon content, which is consistent with the higher content of triglyceride lipids. They also contain larger amounts of nitrogen indicating higher protein levels. Thus, post-diapause and subitaneous offspring clearly vary in their quality and enclosed substances and energetic resources.

The main disparity between offspring origins, however, is associated with their metabolism. The comparison of respiration showed that post-diapause females had respiratory rates about ~11% higher than those of subitaneous females (Fig. 9.1;

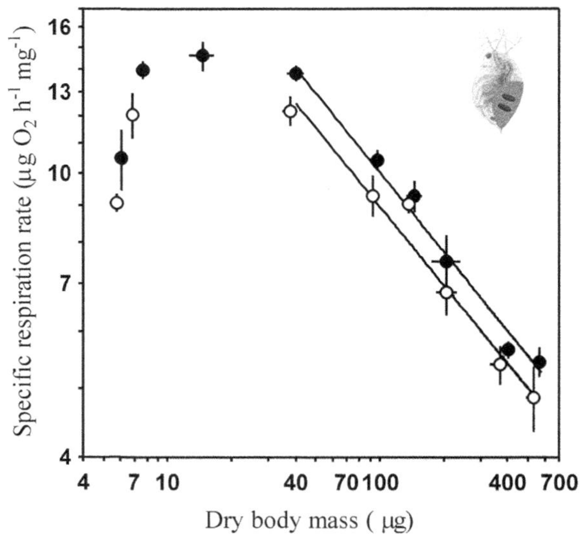

Fig. 9.1 Mass-specific respiration rates (mean ±1SD) in differently sized post-diapause (closed circles) and subitaneous (open circles) offspring of *Daphnia magna*. Note logarithmic scale. (Modified from Arbačiauskas and Lampert 2003)

Arbačiauskas and Lampert 2003). It seems that the metabolic activity is the basic organismal trait distinguishing post-diapause and subitaneous *Daphnia* offspring and affecting their life histories. Higher activity of metabolism in post-diapause females must be related to higher feeding rate, and obviously is adaptive during low water temperatures and abundant food, as is the case in the beginning of the season when hatching from diapausing eggs occurs.

Elevated metabolism, however, can be disadvantageous during severe food shortage. When subjected to starvation, post-diapause neonates as well as adult females exhibit lower survival than subitaneous individuals (Arbačiauskas and Lampert 2003; Arbačiauskas unpublished data).

Under high food concentrations, juvenile post-diapause females grow faster and, consequently, mature at larger body sizes than subitaneous females (Arbačiauskas and Gasiūnaitė 1996; Arbačiauskas 1998). The faster somatic growth of post-diapause offspring, which enabled them to compensate until maturity about twofold smaller initial body size, was observed in *Daphnia magna* also under limited food quantities (Arbačiauskas and Lampert 2003). Thus, when food is not severely limiting, the higher metabolic activity and initial biochemical quality of neonates must be responsible for the fast somatic growth in post-diapause females during juvenile development.

After maturation, at least during early maturity, the somatic growth of post-diapause females under rich food is slower than that in subitaneous females (Arbačiauskas 1998). This pattern is due to the different resource allocation strategy. Under abundant food environments, post-diapause females possess significantly higher reproductive effort during early maturity. When food is limiting, differences in early reproductive effort between offspring origins decrease, and were undetectable when comparing population samples (Arbačiauskas 1998, 2004b).

A greater allocation into reproduction results in increased progeny numbers only under abundant food concentrations. Post-diapause *Daphnia* females produced a first clutch which was from 1.6-fold to more than twice that of subitaneous females (Arbačiauskas 1998; Arbačiauskas and Lampert 2003). Progeny numbers in offspring origins during early maturity were close under limited food availabilities.

The relative fitness of post-diapause offspring in *D. pulex* under rich food was estimated to be ~20% higher than that for subitaneous offspring (Arbačiauskas 1998). The larger early fecundity at high food concentrations resulted in higher fitness of post-diapause females also in *D. magna*; however, a trend for a higher fitness in subitaneous offspring was measured over reduced food concentrations (Arbačiauskas and Lampert 2003). Obtained data clearly suggest that post-diapause offspring in *Daphnia* are superior to subitaneous offspring only under rich food environments, whereas subitaneous offspring perform better when food is limiting.

It seems also that across differing food environments, subitaneous females exhibit a lower than post-diapause female variation in reproductive characteristics, consequently, lower variability of relative fitness in response to feeding conditions (Arbačiauskas 2001). This pattern may be interpreted as an adaptation of subitaneous offspring to unpredictability and lower quality of their environment, i.e. bet-hedging strategy (see Roff 1992).

It is noteworthy that specific post-diapause organismal traits generally are not transferred to the successive subitaneous generations, at least they were not seen when comparing population samples (Arbačiauskas 2004a). The propensity to sexual reproduction of *Daphnia* in post-diapause generation may be reduced in comparison with successive subitaneous generations, as has been detected for rotifers (Gilbert 2002) and aphids (Lees 1960).

9.4 Molecular Regulation of Development of Offspring Phenotypes

Post-diapause and subitaneous offspring types in the same clone must result from different developmental programs of the same genetic background, representing the phenomenon of phenotypic plasticity. To explore the molecular mechanism of this phenotypic plasticity in *D. pulex*, a high-throughput proteomic analysis and expression analysis of several genes were applied. The study revealed 176 proteins that were differentially expressed among offspring phenotypes; 89 proteins were overexpressed in post-diapause offspring and 87 proteins were upregulated in subitaneous offspring (Kaupinis et al. 2017). There were more upregulated proteins with oxyreductase and binding activity in post-diapause offspring, whereas more upregulated proteins with transporter and transferase activity were seen in subitaneous offspring. In comparison with subitaneous phenotype, over 1.5-fold more of the proteins that were upregulated in post-diapause phenotype are involved in metabolism and biosynthesis (Fig. 9.2). Expression levels of several selected genes linked

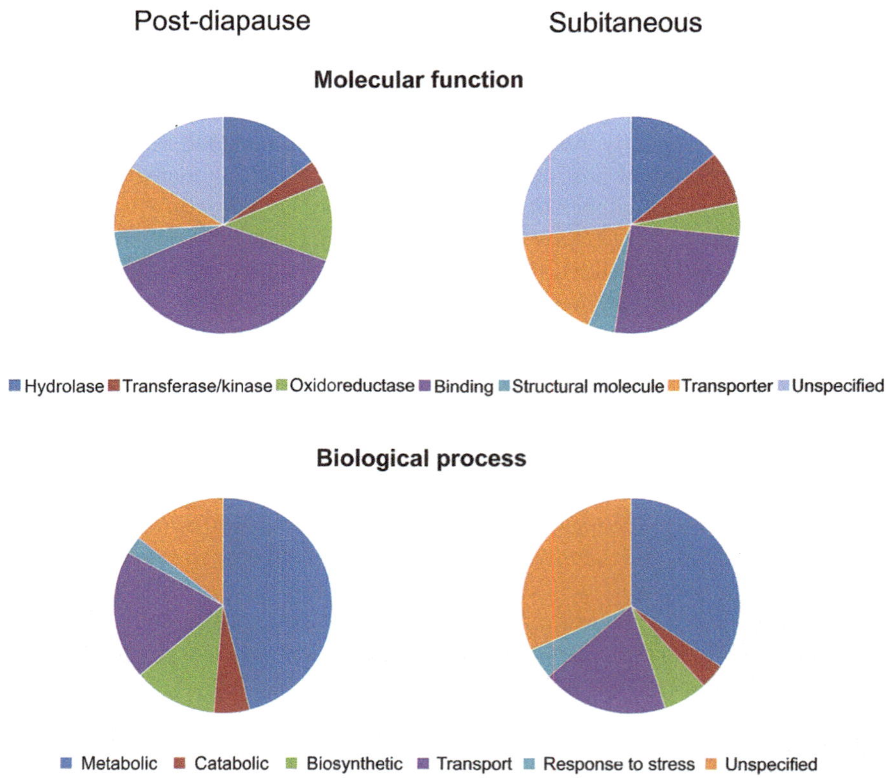

Fig. 9.2 Ontology analysis of upregulated proteins in post-diapause and subitaneous offspring of *Daphnia pulex*. Classification was performed according to the gene ontology terms 'Molecular function' and 'Biological process'. (Modified, for more details see Kaupinis et al. 2017)

to cellular metabolism were also higher in post-diapause females. This pattern perfectly corresponds to previously established higher metabolic rates and faster somatic growth (under rich food) in daphnids exiting diapause. Proteomic data also generally support the previously presumed higher resistance to stress in subitaneous offspring. There were six upregulated proteins related to response to stress, in contrast with only three such proteins upregulated in post-diapause offspring (for details, see Kaupinis et al. 2017).

The greatest difference, a fivefold upregulation in post-diapause compared to subitaneous offspring, was recorded for the target of rapamycin-like (TOR) protein. This nutrient- and energy-sensing kinase is the central component of the conserved eukaryotic signalling pathway that regulates cell and organismal growth (Oldham and Hafen 2003). The TOR signalling pathway is important in the regulation of energy homoeostasis and the control of intracellular metabolic processes in mammals (Andre and Cota 2012); it regulates the synthesis of ribosomal components including transcription and processing of pre-rRNA, expression of ribosomal proteins and synthesis of 5S rRNA (Mayer and Grummt 2006). In association with higher TOR levels, there was a 1.7–1.9-fold overexpression of five ribosomal proteins (Fig. 9.3; Kaupinis et al. 2017). These upregulations of the TOR and ribosomal proteins strongly suggest that the TOR signalling pathway is involved in and can be responsible for the formation of post-diapause and subitaneous phenotypes in *Daphnia*.

The TOR pathway has been reported as having key importance in the regulation of diphenic caste development (queen and worker formation) in the honey bee (Patel et al. 2007), although insulin/insulin-like signalling is considered also to be involved in caste determination (Wolschin et al. 2011; Mutti et al. 2011). TOR-dependent caste determination in the honey bee is conditioned by a rich rather than moderate larval diet (Patel et al. 2007), whilst the involvement of food quality in determining offspring phenotype in *Daphnia* is unlikely, as the phenotype seems to depend upon the egg type from which it originates. Sexually produced diapausing eggs and parthenogenetically produced subitaneous eggs vary substantially in *Daphnia* with respect to biochemical composition. Thus, these eggs types should contain an internal factor switching between developmental programs resulting in post-diapause or subitaneous offspring phenotype.

The highest upregulation (fourfold) in subitaneous offspring was recorded for sideroflexin (Fig. 9.3). This mitochondrial transmembrane protein has been reported to be responsible for skeletal abnormalities and the haematological phenotype of flexed-tail mice and is suggested to be a transporter important in mitochondrial iron metabolism; it may not necessarily transport iron but could mediate the uptake of other metabolites essential for heme synthesis (Fleming et al. 2001). Thus, sideroflexin may be linked to haemoglobins, which also were overexpressed in subitaneous females. Recently, it has been shown that sideroflexin family genes can be important in embryonic development in the clawed frog *Xenopus laevis* (Li et al. 2010). The specific role of upregulation of sideroflexin and haemoglobins in subitaneous *Daphnia* offspring needs further investigation. Haemoglobins are directly related to metabolism via oxygen delivery, and induction of haemoglobins is known to be a response to hypoxia (Zeis et al. 2009; Gerke et al. 2011).

The proteomic study revealed 47 proteins which exhibited at least a 1.5-fold difference (Fig. 9.3). The functions of some of these proteins remain unknown and the roles of many others in relation to fitness traits are not well understood. However, the revealed protein patterns support our findings that metabolic and somatic growth rates are higher in post-diapause offspring, whereas subitaneous offspring are more resistant to stress thus better adapted to unpredictable environments. The most important result, however, is the finding of upregulation of TOR and ribosomal proteins in the post-diapause offspring phenotype, suggesting that the TOR signalling pathway is involved and presumably responsible for the regulation of phenotypic plasticity underlying the determination of post-diapause and subitaneous phenotypes in *Daphnia*.

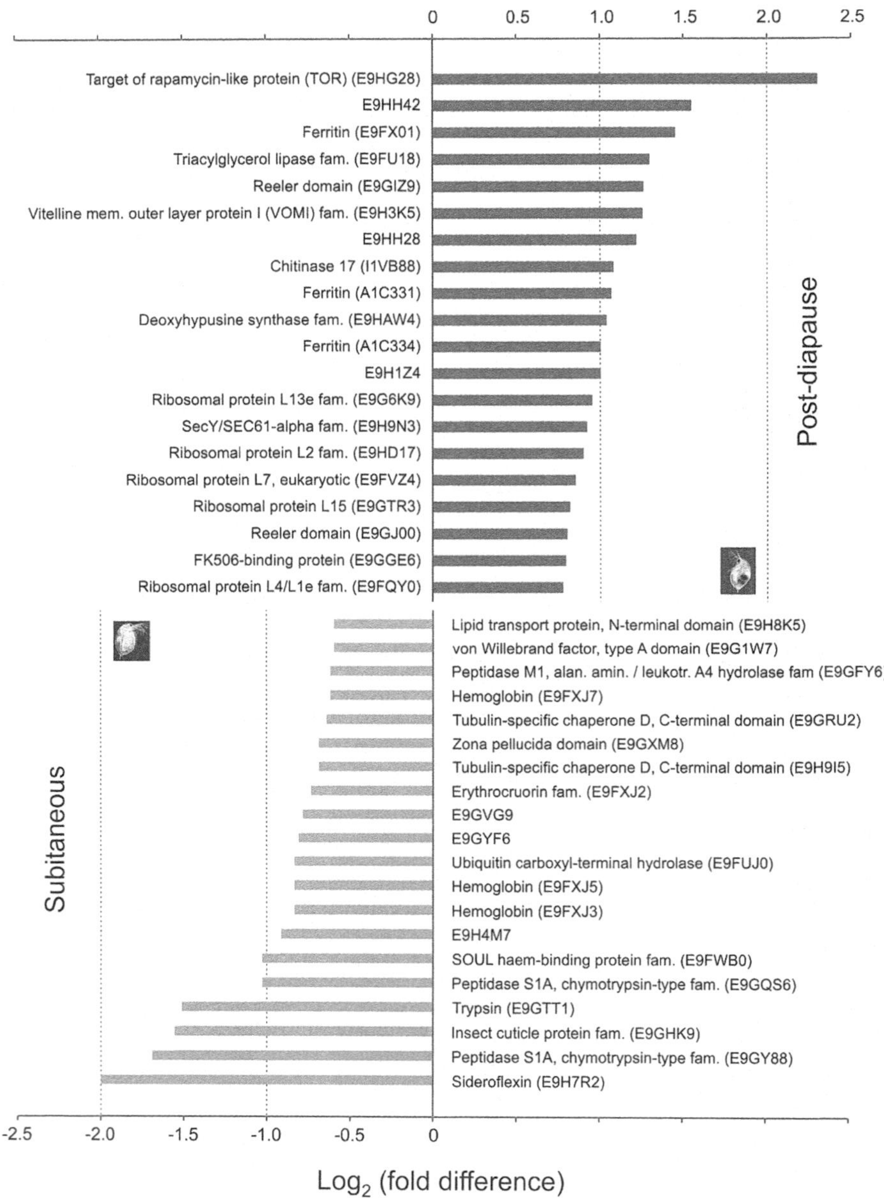

Fig. 9.3 Fold difference (Log_2-transformed) in protein levels between post-diapause and subitaneous offspring of *Daphnia pulex*. Positive values indicate a higher level in the post-diapause phenotype and negative values indicate the opposite. Shown are 20 proteins with largest difference for each phenotype. Protein names (family membership/description) and accession numbers are from the UniProt (Universal Protein Resource database, http://www.uniprot.org/uniprot). (Modified, for more details see Kaupinis et al. 2017)

9.5 Conclusions

Life-history patterns substantially differ between post-diapause and subitaneous offspring in *Daphnia*. Differential gene expression resulting in elevated metabolic activity of post-diapause offspring are primarily responsible for distinction in organismal traits and life-history strategy between offspring phenotypes. During early maturity, post-diapause and subitaneous females possess different resource allocation strategy. The greater allocation to progeny in post-diapause offspring, however, results in increased relative fitness only under rich food environments. On limited nutrition, no advantage over subitaneous females was seen, which even may be superior.

Post-diapause and subitaneous offspring in *Daphnia* are adapted to different environments. Post-diapause offspring are adapted to a favourable environment without food limitation, which is expected during early season when emergence from diapause is observed, whilst subitaneous offspring are adapted to unpredictable and variable food conditions, which are expected later in the season (Arbačiauskas 1998, 2001). This pattern, in turn, implies that selection should favour the synchronous hatching of diapausing eggs in the beginning of the season when the spring algal bloom is to be expected (Arbačiauskas and Lampert 2003).

In temperate environments, the cyclic parthenogen *Daphnia* has evolved two alternative phenotypes matching environmental conditions in which they occur. This seasonal performance of *Daphnia* is consistent with the phenomenon of seasonal polyphenism (specifically diphenism in *Daphnia* case), which predicts the presence of threshold traits affecting life-history patterns and fitness trade-offs across varying environments. The elevated metabolic activity of post-diapause offspring is the threshold trait which determines disparity between offspring phenotypes (see Arbačiauskas 2004b). The transgenerational effect responsible for different quality and, consequently, different life-history patterns of post-diapause offspring in comparison with their mothers has received a definition of a negative maternal effect, which is to be expected when environment for progeny is predictable in long-term, but irrespective of parental environment (see Fox and Mousseau 1998).

Among another animal taxa, the adaptive significance of seasonal polyphenism probably is best examined in butterflies. The seasonal change of wing patterns in butterflies is important for thermoregulation under temperate climate and increases fitness of alternative phenotypes in seasons in which they occur (Brakefield 1996 and references therein). In common with *Daphnia*, seasonal polyphenism of butterflies is associated with metabolic activity, which should condition the enhancement of resource acquisition. The adjustment of resource acquisition with respect to resource availability might be a common evolutionary trend resulting in seasonal polyphenism of multivoltine organisms inhabiting seasonal environments.

Acknowledgments I thank Victor R. Alekseev for the invitation to contribute to this book. Only because of his encouragement and patience this chapter has been written.

References

Alekseev VR (1990) Diapause in crustaceans: ecological and physiological aspects. Nauka Academic Publishers, Moscow, 144 pp (in Russian)

Andre C, Cota D (2012) Coupling nutrient sensing to metabolic homoeostasis: the role of the mammalian target of rapamycin complex 1 pathway. Proc Nutr Soc 71:502–510

Arbačiauskas K (1998) Life-history traits of exephippial and parthenogenetically derived daphnids: indicators of different life-history strategies. Archiv für Hydrobiol Spec Issues Adv Limnol 52:339–358

Arbačiauskas K (2001) Life-history variation related to the first adult instar in daphnids derived from diapausing and subitaneous eggs. Hydrobiologia 442:157–164

Arbačiauskas K (2004a) Life-history characteristics and fitness in descendents of parthenogenetic and ex-ephippio females of *Daphnia magna*. Hydrobiologia 526:211–218

Arbačiauskas K (2004b) Seasonal phenotypes of *Daphnia*: post-diapause and directly developing offspring. J Limnol 63 (Suppl 1):7–15

Arbačiauskas K (2007) Does timing of emergence within a season affect the evolution of post-diapause traits? Post-diapause and directly developing phenotypes of *Daphnia*. In: Alekseev A, De Stasio B, Gilbert JJ (eds) Diapause in Aquatic Invertebrates. Theory and Human Use, Monographiae Biologicae, vol 84. Springer, Dordrecht, pp 167–175

Arbačiauskas K, Gasiūnaitė ZR (1996) Growth and fecundity of *Daphnia* after diapause and their impact on the development of a population. Hydrobiologia 320:209–222

Arbačiauskas K, Lampert W (2003) Seasonal adaptation of ex-ephippio and parthenogenetic offspring of *Daphnia magna*: differences in life-history and physiology. Funct Ecol 17:431–437

Brakefield PM (1996) Seasonal polyphenism in butterflies and natural selection. Trends Ecol Evol 11:275–277

Caceres CE (1998) Interspecific variation in the abundance, production, and emergence of *Daphnia* diapausing eggs. Ecology 79:1699–1710

Carvalho GR, Hughes RN (1983) The effect of food availability, female-density and photoperiod on ephippia production in *Daphnia magna* Straus (Crustacea, Cladocera). Freshw Biol 13:37–46

De Stasio B (2007) Egg bank formation by aquatic invertebrates: a bridge across disciplinary boundaries. In: Alekseev A, De Stasio B, Gilbert JJ (eds) Diapause in Aquatic Invertebrates. Theory and Human Use, Monographiae Biologicae, vol 84. Springer, Dordrecht, pp 121–133

Deng D, Zhang S, Li Y, Meng X, Yang W, Li Y, Li X (2010) Effects of *Microcystis aeruginosa* on population dynamics and sexual reproduction in two *Daphnia* species. J Plankton Res 32:1385–1392

Fleming MD, Campagna DR, Haslett JN, Trenor CC 3rd, Andrews NC (2001) A mutation in a mitochondrial transmembrane protein is responsible for the pleiotropic hematological and skeletal phenotype of flexed-tail (f/f) mice. Genes Dev 15:652–657

Fox CW, Mousseau TA (1998) Maternal effects as adaptations for transgenerational phenotypic plasticity in insects. In: Mousseau TA, Fox CW (eds) Maternal Effects as Adaptations. Oxford University Press, New York, pp 159–177

Frisch D, Morton PK, Chowdhury PR, Culver BW, Colbourne JK, Weider LJ, Jeyasingh PD (2014) A millennial-scale chronicle of evolutionary responses to cultural eutrophication in *Daphnia*. Ecol Lett 17(3):360–368

Gerke P, Börding C, Zeis B, Paul RJ (2011) Adaptive haemoglobin gene control in *Daphnia pulex* at different oxygen and temperature conditions. Comp Biochem Physiol A 159:56–65
Gilbert JJ (2002) Endogenous regulation of environmentally induced sexuality in a rotifer: a multigenerational parental effect induced by fertilization. Freshw Biol 47:1633–1641
Gyllström M (2004) Induction and termination of diapause in a freshwater zooplankton community. Arch Hydrobiol 161:81–97
Gyllström M, Hansson L (2004) Dormancy in freshwater zooplankton: induction, termination and the importance of benthic-pelagic coupling. Aquat Sci 66:274–295
Hairston NG Jr (1998) Time travelers: What's timely in diapause research? Archiv für Hydrobiol Spec Issues Adv Limnol 52:1–15
Hairston NG Jr, van Brunt RA, Kearns CM, Engstrom DR (1995) Age and survivorship of diapausing eggs in a sediment egg bank. Ecology 76:1706–1711
Hairston NG Jr, Hansen A-M, Schaffner WR (2000) The effect of diapause emergence on the seasonal dynamics of zooplankton assemblage. Freshw Biol 22:471–478
Kaupinis A, Aitmanaitė L, Strepetkaitė D, Valius M, Lazutka JR, Arbačiauskas K (2017) Proteomic and gene expression differences between post-diapause and subitaneous offspring phenotypes in the cyclic parthenogen *Daphnia pulex*. Hydrobiologia 798(1):87–103
Koch U, Von Elen E, Straile D (2009) Food quality triggers the reproductive mode in the cyclical parthenogen *Daphnia* (Cladocera). Oecologia 159:317–324
Lees AD (1960) The role of photoperiod and temperature in the determination of parthenogenetic and sexual forms in the aphid *Megoura viciae* Bucton – II. The operation of the "interval timer" in young clones. J Insect Physiol 4:154–175
Li X, Han D, Kam RKT, Guo X, Chen M, Yang Y, Zhao H, Chen Y (2010) Developmental expression of *sideroflexin* family genes in *Xenopus* embryos. Dev Dyn 239:2742–2747
Mayer C, Grummt I (2006) Ribosome biogenesis and cell growth: mTOR coordinates transcription by all three classes of nuclear RNA polymerases. Oncogene 25:6384–6391
Mutti NS, Dolezal AG, Wolschin F, Mutti JS, Gill KS, Amdam GV (2011) IRS and TOR nutrient-signaling pathways act via juvenile hormone to influence honey bee caste fate. J Exp Biol 214:3977–3984
Oldham S, Hafen E (2003) Insulin/IGF and target of rapamycin signaling: a TOR de force in growth control. Trends Cell Biol 13:79–85
Patel A, Fondrk MK, Kaftanoglu O, Emore C, Hunt G, Frederick K, Amdam GV (2007) The making of a queen: TOR pathway is a key player in diphenic caste development. PLoS One 2(6):e509
Perez-Martınez C, Jimenez L, Moreno E, Conde-Porcuna JM (2013) Emergence pattern and hatching cues of *Daphnia pulicaria* (Crustacea, Cladocera) in an alpine lake. Hydrobiologia 707:47–57
Pijanowska J, Stolpe G (1996) Summer diapause in *Daphnia* as a reaction to the presence of fish. J Plankton Res 18:1407–1412
Roff DA (1992) The Evolution of Life Histories. Chapman and Hall, New York, 535 pp
Rother A, Pitsch M, Hülsmann S (2010) The importance of hatching from resting eggs for population dynamics and genetic composition of *Daphnia* in a deep reservoir. Freshw Biol 55:2319–2331
Slusarczyk M (2001) Food threshold for diapause in *Daphnia* under the threat of fish predation. Ecology 82:1089–1096
Sommer U, Gliwicz ZM, Lampert W, Duncan A (1986) The PEG-model of seasonal succession of planktonic events in fresh waters. Arch Hydrobiol 106:433–471
Stross RG, Hill RG (1965) Diapause induction in *Daphnia* requires two stimuli. Science 150:1462–1464
Tauber MJ, Tauber CA, Masaki S (1986) Seasonal Adaptations of Insects. Oxford University Press, New York, 411 pp

Walsh MR (2013) The link between environmental variation and evolutionary shifts in dormancy in zooplankton. Integr Comp Biol 53:713–722
Wolf HG, Carvalho GR (1989) Testing eggs of lake-*Daphnia* II. In situ observations on the hatching of eggs and their contribution to population and community structure. Freshw Biol 22:471–478
Wolschin F, Mutti NS, Amdam GV (2011) Insulin receptor substrate influences female caste development in honeybees. Biol Lett 7:112–115
Zaffagnini F (1987) Reproduction in *Daphnia*. Memorie dell'Istituto Italiano di Idrobiologia 45: 245–284
Zeis B, Lamkemeyer T, Paul RJ, Nunes F, Schwerin S, Koch M, Schutz W, Madlung J, Fladerer C, Pirow R (2009) Acclimatory responses of the *Daphnia pulex* proteome to environmental changes. I. Chronic exposure to hypoxia affects the oxygen transport system and carbohydrate metabolism. BMC Physiol 9:7

Chapter 10
The Role of Chemical Interactions in Embryonic Diapause Induction in Zooplankton

Egor Zadereev and Tatiana S. Lopatina

Abstract Production of resting eggs in zooplankton is controlled by multiple stimuli. In this chapter, we briefly discussed published data that confirm the effect of infochemicals produced by conspecifics, competitors, predators or preys on the production of resting eggs in zooplankton. We found that the effect of conspecific chemicals on the production of resting eggs is the most convincing. Both experimental data and theoretical research demonstrated that this density-dependent reaction often results in a competitive advantage of individuals in the population that follows such a strategy. The data on the effect of chemicals exuded by competitors or predators are controversial. Data on the effect of chemical interaction on the production of resting eggs in natural habitats are almost absent. Most of the studies of chemical interactions are performed with individuals in laboratory experiments with crowded water. Crowded water is water that contains chemicals exuded by the population. Even though this method has the number of drawbacks, it is still widely used in similar studies. There are several studies focused on the identification of the chemical nature of cues responsible for the production of resting eggs in zooplankton. Most probably, chemicals involved are short proteins. However, the exact identification of the chemicals responsible for the production of resting eggs in zooplankton remains an open task. In order to place chemical interactions into a framework of multiple diapause control theory, it is necessary to determine the nature of chemicals involved and to demonstrate population- and ecosystem-level consequences of this phenomenon in natural habitats.

Keywords Resting eggs, Chemical interactions, Zooplankton, Competition, Conspecific chemicals, Kairomones

---

E. Zadereev (✉) · T. S. Lopatina
Institute of Biophysics, Krasnoyarsk Research Center SB RAS, Akademgorodok, Krasnoyarsk, Russia
e-mail: egor@ibp.ru

V. R. Alekseev, B. Pinel-Alloul (eds.), *Dormancy in Aquatic Organisms. Theory, Human Use and Modeling*, Monographiae Biologicae 92,
https://doi.org/10.1007/978-3-030-21213-1_10

10.1 Introduction

Many species of zooplankton are able to enter the embryonic diapause with the embryo encysted as an ephippial (or resting) egg (Alekseev 1990). Usually, resting eggs are produced by animals that change the reproduction mode from parthenogenesis to gametogenesis. In favourable conditions, parthenogenetic females produce female and male offspring, while with the onset of harsh conditions, they switch to gametogenesis and produce ephippial eggs with encysted female embryos. Resting eggs are able to survive such extreme conditions as drying, freezing, salinity and toxic stress (e.g. Radzikowski 2013; Nielsen et al. 2012; Oskina et al. 2018). Resting eggs can survive years in an egg bank at the bottom of the water body (Hairston et al. 1995). Banks of resting eggs support genetic diversity and play an important role in the dynamics of zooplankton communities (Brendonck and De Meester 2003).

The production of resting eggs is controlled by several abiotic and biotic factors including photoperiod, temperature, food availability and infochemicals (e.g. Alekseev and Lampert 2001; Kleiven et al. 1992; Ban 1992). Several studies have demonstrated that the production of diapausing eggs in Rotifera (e.g. Stelzer and Snell 2003), Cladocera (e.g. Slusarczyk 1995; Zadereev and Gubanov 1996) and Copepoda (e.g. Ban and Minoda 1994) could be induced by the effect of chemicals exuded by conspecifics, competitors or predators. However, the role and nature of chemicals involved in the regulation of cyclic reproduction of zooplankters are not resolved.

In this contribution, we will focus on chemical interactions that induce resting eggs production in zooplankton. Within trophic level, chemical interactions include the effect of chemicals produced by conspecifics and chemicals produced by other species at the same trophic level (usually competitors). Chemical interactions between trophic levels include the effect of chemicals produced by either predators (fish or invertebrates) or prey (algae). We will also briefly review methods used to investigate the effect of intra- and interspecific chemical interactions on diapause induction and data on the structure of involved chemicals.

10.2 Intra- and Interspecific Chemical Interactions

Banta and Brown (1929a, b, c) performed classical research on the effect of chemicals exuded by conspecifics on the induction of gametogenesis. They disproved the hypothesis of Weismann (1876) that a fixed, heritable internal cycle causes changes of the mode of reproduction in Cladocera, by maintaining a population of *Moina macrocopa* for 780 parthenogenetic generations under stable conditions. In experiments with water that contained chemicals exuded by conspecifics (crowded water), they came to the following conclusions: (1) male production is induced by the accumulation of metabolic by-products in the medium proportional to the density of the population; (2) metabolic by-products which induce male production are non-volatile, unstable and disappear even after the storage of crowded water; (3) most probably, there is no causal relationship between the accumulation of metabolic by-products and production of resting eggs; and (4) the possible stimulus for the production of resting eggs is food limitation (Banta and Brown 1929a, b, c).

Later, Stross (1987) hypothesized that production of resting eggs requires two stimuli. He concluded that while photoperiod is a regulatory factor that stimulates females to produce resting eggs, density-dependent factor is also very often required to stimulate gametogenesis. He demonstrated that the number of ephippial eggs produced by females in laboratory populations was in direct proportion to population density. This density-related stimulus had not been determined and most probably was not associated only with the depletion of food (Stross 1987).

Recent results confirmed that the detected responses are density-dependent. For example, the rate of mixis in the rotifer *Brachionus plicatilis* increased with the population density used to prepare crowded water (Stelzer and Snell 2003). Zadereev and Gubanov (1996) demonstrated that the water crowded by conspecifics stimulated the change of the reproduction mode in single females of *M.macrocopa*. The proportion of gametogenetic females increased with the density of the population used to prepare crowded water. However, an increase in food concentration reduced the strength of the crowded water effect. The higher the population density used to prepare crowded water, the higher the food concentration needed to prevent gametogenesis induction (Zadereev and Gubanov 1996).

Working with the third major zooplankton order, copepods, Ban and Minoda (1994) demonstrated that females of the calanoid copepod *Eurytemora affinis* that were individually reared in water crowded by conspecifics produced diapausing eggs, and Walton (1985) found crowding induced diapausing eggs production in a diaptomid copepod.

Data on the effect of chemical interactions within a trophic level on diapause induction are contradictory. It was demonstrated that *Daphnia pulex* produced ephippia in water crowded either by conspecifics or by congeners of *D. cuculluta* (Lurling et al. 2003). In the rotifer *B. plicatilis*, water crowded by the brine shrimp *Artemia salina*, as well as by conspecifics, stimulated mixis (Carmona et al. 1993). On the other hand, Gilbert (2003) demonstrated that only metabolic products of conspecifics can induce the production of mictic females in *B. calyciflorus*, while high densities of either congeners or other species, or even a different strain of conspecifics, do not induce mixis (Gilbert 2003).

We also demonstrated that *Moina* species (*M. macrocopa* and *M. brachiata*) produced resting eggs only under the effect of conspecific chemicals. *Moina* species did not change reproductive mode under the effects of either water from crowded cultures of a larger Cladocera species, *D. magna* or water crowded by a population of another *Moina* species (Zadereev and Lopatina 2007). It should be emphasized that *M. macrocopa* is more effective food competitor than *M. brachiata* as it starts to produce resting eggs at lower food concentrations than *M. brachiate* (Zadereev and Lopatina 2007). Thus, neither the weaker competitor reacted to chemicals exuded by a more effective competitor to establish an egg bank nor did the smaller species produced resting eggs under the effect of the exudates of the larger Cladocera species. Theoretical considerations and laboratory study should answer the question whether the production of resting eggs due to chemical interactions between competitors will be beneficial strategy.

Chemical interactions between different trophic levels also induce resting eggs production in zooplankton. They are mediated by kairomones exuded by fish or other predators. Kairomones are chemicals transferring the information between species which impart a benefit to the receiving organism while not being beneficial for the producer of the cue (Pohnert et al. 2007). It was shown that water crowded with fish induced resting eggs production in *Daphnia* (Slusarczyk 1995; Pijanowska and Stolpe 1996). Slusarczyk (1999) also demonstrated that resting eggs are produced only under the effect of infochemicals of fish that coexisted with *Daphnia* used in the experiment. On the other hand, later it was demonstrated that predation cues did not induce diapause in such small bodied cladocerans as *Moina* and *Diaphanosoma* (Santangelo et al. 2010). Before, Hairston Jr. (1998) noted that the effect of fish kairomones on the change of the reproduction mode is not always detected in experiments.

The chemical interactions are also possible between zooplankton and phytoplankton. Laurenmaatta et al. (1997) demonstrated the effect of toxic substances released by the cyanobacterium *Microcystis aeruginosa* on the resting eggs production by *D. pulex*. Meanwhile Burks et al. (2000) found that the chemicals exuded by the macrophyte *Elodea canadensis* reduced the growth rate of *D. magna* but did not change the reproduction mode. In addition, it was demonstrated that the presence of *Microcystis* colonies in the diet was positively associated with the production of ephippia by *D. pulex* but negatively associated with ephippia production for *D. carinata* (Deng et al. 2010).

Most of the results reviewed above were obtained in experiments with individuals treated with crowded water. Crowded water is water which contains chemicals exuded by the group of animals. To produce crowded water, animals that presumably exude infochemicals are cultivated during a fixed period (usually 24 h) in the medium. After the incubation, the medium is filtered to remove suspended particles (e.g. algae cells, bacteria) and used to test its effect on animals of the same or another species.

The protocol to produce crowded water with infochemicals has a number of drawbacks. Interpretations of experiments are usually based on the assumption that the concentration of infochemicals depends on the number of animals used to exude chemicals and the incubation time. However, rates of chemical excretion and decomposition usually remain unknown. The metabolism of animals that exude infochemicals can be "disturbed" due to the elevated population density during the cultivation. Also, the medium produced with this method contains a wide range of chemicals—from nonspecific general metabolites to species-specific signalling chemical cues. In spite of these shortcomings, such method is intensively used in individual experiments.

10.3 Ecological and Evolutionary Implications of Resting Eggs Production Under Infochemical Control

The production of resting eggs under the effect of chemicals from conspecifics can be considered a density-dependent population control factor. Such control has a twofold purpose. First, populations survive in adverse environments because of resting eggs. Theory indicates that resting eggs should be produced before the onset of harsh conditions. This prediction was successfully accounted for in a study of the role and nature of chemicals involved in the production of resting eggs by Hairston and Olds (1987) determining environmental factors stimulating a freshwater copepod *Diaptomus sanguineus* to produce resting eggs. Several experiments on the multifactorial control of resting eggs production demonstrated that metabolic by-products do indeed stimulate females to produce resting eggs before the onset of unfavourable conditions. It was shown that under the effect of metabolic by-products, thresholds for resting eggs production were cued by photoperiod (Stross 1987) and food concentration (Zadereev and Gubanov 1996) increase.

With the onset of favourable environmental conditions, the population is re-established from the bank of resting eggs. Theory indicates that during the reactivation, the population with a higher number of resting eggs will have a competitive advantage over population with smaller egg bank. Specifically, this is the case for temporary ponds, where the egg bank is the main source of the re-establishment of the zooplankton. To ensure the production of the maximal number of resting eggs, threshold values of regulating factors that maximize the production of resting eggs should be selected and stabilized by the evolution. Similar considerations have been developed for some social insects using Pontryagin's maximum principle (Oster and Rocklin 1979). We also demonstrated theoretically that the combined density-dependent control of resting eggs production by density-dependent factors (infochemicals and food concentration) maximizes the number of produced resting eggs in cladoceran *M. macrocopa* (Zadereev et al. 2003). Moreover, it was demonstrated that density-dependent sexual and diapause investments mediate coexistence of facultative sexual species having identical niches (Montero-Pau and Serra 2011).

This effect was later supported by laboratory experiments. The freshwater rotifer *Brachionus calyciflorus* and the cladoceran *M. micrura* were used to test experimentally if diapause could promote coexistence of active individuals under interspecific competition. *B. calyciflorus* exhibited a greater and earlier production of resting eggs compared with *M. micrura*. Costs of earlier diapause investment were translated into reduced probability of *B. calyciflorus* of being outcompeted by *M. micrura* (Aránguiz-Acuña and Ramos-Jiliberto 2014). We also theoretically demonstrated that investment in diapause might promote local and short-term coexistence of competing zooplankton species (Zadereev and Prokopkin 2007). Thus, the experimental data and theoretical research demonstrated that metabolic by-products exuded by conspecifics affect the production of resting eggs to ensure survival of offspring, often resulting in a competitive advantage of individuals in the population that are following such a strategy.

However, direct tests of the effect of chemical interactions on the resting eggs production at population or ecosystem levels are scarce. Slobodkin (1954) manipulated the populations of *D. obtusa* and tried to estimate the role of both density-dependent effects simultaneously (accumulation of metabolic by-products and depletion of food). He demonstrated that the number of males in the population sharply increased after each reproductive effort and was preceded by a reduction of population growth rate. Ephippial eggs appeared later. The author assumed that the increase in male production and the following production of resting eggs were due to food limitation and that there was no experimental proof that metabolic by-products were involved in the regulation of these processes. We performed population experiments with *M.macrocopa* and observed production of resting eggs with food availability favourable for the parthenogenesis of single females. Chemical interactions between animals are the most obvious explanation for these results (Zadereev and Gubanov 2000).

Mitchell and Carvalho (2002) in population experiments with crowded water as the culture medium tried to compare the impact of infochemicals and competition for resources on the population dynamics of *Daphnia magna.* Even though they detected the effect of crowding chemicals on the initial population structure, the timing of population density maxima and subsequent population structure were similar among all treatments. Regarding the timing of diapause induction and numbers of ephippial females, the data were insufficient for analysis.

The serious drawback of using crowded water as the medium for population experiments is the role of maternal effects. With the growth of the population, the metabolic by-products build up in the medium. If conspecific chemicals are an additional cue to assess the danger of starvation in the future, the stable concentration of the chemical in the medium cannot be considered as a cue approximating food limitation. What will serve as the indicator of the growing population and the subsequent depletion of food resources is the increase in concentration of conspecific chemicals. Experiments with individual females demonstrated that animals could adapt to the effect of chemicals that induce resting eggs production. When the offspring of mothers cultured in crowded water were also cultured in crowded water (the constant effect of conspecific chemicals), they produce resting eggs less readily than the offspring of mothers cultured in control medium (Zadereev 2003).

At the ecosystem level, data on the effect of chemical interactions on the production of resting eggs are practically absent. Detailed and frequent data on population biomass and structure, food concentration, temperature and other biotic and abiotic parameters in the water body are needed to assess this effect. For example, Innes (1997) examined the switch from asexual to sexual reproduction in *D. pulex* in two temporary ponds. The dynamics of the number of ephippial females can be partly explained by variation in population density. Similarly, Gerber et al. (2018) documented population dynamics and the occurrence of sexual reproduction in natural populations of *D. magna* across their growing season. They found that the frequency of sexually reproducing females increased with population density and with decreasing population growth rates.

The production of resting eggs under infochemicals is likely to occur in specific habitats and ecological situations, and some ecological and evolutionary considerations were already discussed above. Most probably, such effects can be observed in temporary or other habitats where the onset of unfavourable conditions is unpredictable and the number of resting eggs will significantly determine the competitive advantage of the species.

10.4 The Nature of Infochemicals Stimulating Production of Resting Eggs

The nature of chemicals involved in resting eggs production has not been identified. Theoretical considerations suggest that chemical interactions in aquatic habitats can be mediated by (a) species-specific chemicals affecting particular biological functions ("pheromone-type" communications), (b) function-specific but not species-specific chemicals ("hormone-type" communications); and (c) nonspecific (neither for functions nor for species) chemicals ("toxic-type" interactions) (Zadereev 2002).

Reproductive responses caused by the effect of waterborne chemicals are often explained by chemically induced food limitation. Indeed, feeding rate in zooplankton can be suppressed due to the effect of chemicals from conspecifics and competitors (e.g. Boersma et al. 1999; Helgen 1987; Lurling et al. 2003; Matveev 1993). Published data on the effect of waterborne chemicals on other life history parameters of zooplankton have demonstrated that the observed responses are similar to the responses recorded when animals are food limited (e.g. Burns 1995; Boersma et al. 1999; Lurling et al. 2003). The suppression of feeding rate under the effect of waterborne chemicals released by different species most probably is an example of "toxic-type" interactions. However, none of the researchers directly measured assimilation/fitness losses due to the suppressed feeding rate under the effect of waterborne chemicals to find the relationship between the decreased feeding rate and the production of resting eggs.

Studies of the chemical interactions in plankton have been growing recently because of broader opportunities to identify chemical cues (Pohnert et al. 2007). Such chemical cues can contain information on states of populations and under optimal abiotic conditions can influence not only individual characteristics but also population dynamics of zooplankton (Van Donk 2007). In spite of the numerous studies, data on the chemical nature of the reactants are scant. Just several studies were focused on the chemical nature of substances involved in the resting eggs production in zooplankton.

Snell et al. (2006) used size exclusion, enzymatic degradation, protease protection assays, selective binding to anion ion exchange and C3 reversed-phase HPLC columns and the sequence of 17 N-terminal amino acids to study the chemical signal, which induces the production of resting eggs in rotifer *B. plicatilis*. The 39 kDa protein was identified as an excellent candidate for the rotifer mixis induction signal. Recently, it was discussed that rotifers excrete a variety of proteins accumulating in the medium until they reach a threshold concentration which is probably the key trigger of the mictic reproduction (Snell 2017). These proteins have been concentrated, and partial amino acid sequences have been obtained for four proteins, 42, 39 and 23 kDa by N-terminal sequencing and 22 kDa by mass spectrometry. Immunoprecipitation of the 22 kDa protein from rotifer-conditioned medium greatly diminishes its ability to induce mictic reproduction. This makes the 22 kDa protein the prime candidate for the mixis-inducing pheromone (Snell 2017).

We demonstrated that chemicals stimulating resting eggs production in *M. macrocopa* are non-volatile compounds. Also, the ability to induce production of resting eggs was retained after the crowded water was boiled. We concluded that most likely this chemical is a short peptide (Lopatina and Zadereev 2015). We also effectively bounded and concentrated chemicals which induce resting eggs production in *M.macrocopa* with C_{18} solid-phase cartridge (Zadereev and Lopatina 2015). This funding supports our assumption that chemicals involved are short peptides and open the perspective to identify their structure.

Pohnert et al. (2007) discussed the main difficulties in the isolation and identification of chemical substances: (a) the extremely low effective concentrations of signalling substances in the medium; (b) the effect of the target substances can be the result of exposure to several chemicals; and (c) the lack of methods for determining the target components. In general, substances such as dissolved gases, hydrocarbons, peptides and proteins are considered as signals involved in chemical interactions in aquatic ecosystems (Rittschof and Cohen 2004; Pohnert et al. 2007). Peptides are active signals in the aquatic systems, being very soluble in water, with a short existence, due to the rapid consumption by microbes. Many peptides have a high degree of activity in small amounts (less than picomole) (Rittschof and Cohen 2004). Zooplankton also excretes organic sulphates that can act as chemical stimuli. For example, 8-methyl sulphate, which was originally isolated from water crowded with *D. pulex* and *D. magna*, induced morphological changes in green algae (Yasumoto et al. 2006). Similarly, *D. pulex* kairomones causing changes in the morphology of the freshwater alga *Scenedesmus gutwinskii* were characterised as 2,6-dimethylsulfate and 6-methyl-ethyl-sulfate (Uchida et al. 2008).

There are active research focused on chemical ecology in marine ecosystems (e.g. Sieg et al. 2011; Schwartz et al. 2016) with several advances in identification of chemical-mediated interactions in these habitats. For example, polyunsaturated aldehydes have been implicated in a range of functions, including intraspecific signalling and programmed cell death, anti-grazing defences, allelopathy and bacteria–phytoplankton interactions. Several recent reviews concentrated on the chemical ecology of marine copepods (Heuschele and Selander 2014; Kiørboe 2013). These reviews illustrated the importance of chemical interactions for ecology of copepods, but also mentioned the lack of identified infochemicals to confirm the function of sensory structures. Thus, despite the large number of studies, which demonstrated the effect of infochemicals on various life history traits (including production of resting eggs), data on the chemical nature of these substances, are scarce.

10.5 Conclusion

This chapter was focused on the role of chemical cues in resting eggs production in zooplankton. We briefly discussed the experiments that confirm the effect of info-chemicals produced by conspecifics, competitors, predators and prey on the production of resting eggs. Summarizing the progress in this area, we found that the effect of conspecific chemicals on diapause induction is the most convincing. These reactions are density-dependent. There are good theoretical explanations of such density-dependent responses. We found that the data on the effect of chemicals exuded by competitors and predators are controversial. Most probably, the adaptive value of such reactions and history of the populations should be carefully considered before any future experimental research.

In spite of the numerous studies on chemical-mediated interactions in aquatic organisms, the nature of active components is poorly known. The identification of the chemicals responsible for the production of resting eggs in zooplankton remains an open task. It is necessary to theoretically explain and experimentally determine the nature of chemicals involved and to investigate population- and ecosystem-level consequences of this phenomenon in order to place chemical interactions into a framework of multiple diapause control theory.

References

Alekseev V, Lampert W (2001) Maternal control of resting-egg production in *Daphnia*. Nature 414:899–901

Alekseev VR (1990) Diapause in crustaceans: ecological - physiological aspects. Nauka, Moscow. (in Russian, English summary

Aránguiz-Acuña A, Ramos-Jiliberto R (2014) Diapause may promote coexistence of zooplankton competitors. Journal of Plankton Research 36:978–988

Ban S (1992) Effects of photoperiod, temperature, and population density on induction of diapause egg production in *Eurytemora affinis* (Copepoda: Calanoida) in Lake Ohnuma, Hokkaido, Japan. Journal of Crustacean Biology 12:361–367

Ban S, Minoda T (1994) Induction of diapause egg-production in *Eurytemora affinis* by their own metabolites. Hydrobiologia 293:185–189

Banta AM, Brown LA (1929a) Control of sex in Cladocera. 1. Crowding the mothers as a means of controlling male production. Physiological Zoology 2:80–92

Banta AM, Brown LA (1929b) Control of sex in Cladocera. 2. The unstable nature of the excretory products involved in male production. Physiological Zoology 2:93–98

Banta AM, Brown LA (1929c) Control of sex in Cladocera. IV. Relation between the rate of the mother's development and the sex of her young. Physiological Zoology 2:302–308

Boersma M, De Meester L, Spaak P (1999) Environmental stress and local adaptation in *Daphnia magna*. Limnology and Oceanography 44:393–402

Brendonck L, De Meester L (2003) Egg banks in freshwater zooplankton: evolutionary and ecological archives in the sediment. Hydrobiologia 491:65–84

Burks RL, Jeppesen E, Lodge DM (2000) Macrophyte and fish chemicals suppress Daphnia growth and alter life-history traits. Oikos 88:139–147

Burns CW (1995) Effects of crowding and different food levels on growth and reproductive investment of *Daphnia*. Oecologia 101:234–244

Carmona MJ, Serra M, Miracle MR (1993) Relationships between mixis in *Brachionus plicatilis* and preconditioning of culture-medium by crowding. Hydrobiologia 255:145–152

Deng D, Zhang S, Li Y, Meng X, Yang W, Li Y, Li X (2010) Effects of *Microcystis aeruginosa* on population dynamics and sexual reproduction in two Daphnia species. Journal of Plankton Research *32*:1385–1392

Gerber N, Kokko H, Ebert D, Booksmythe I (2018) *Daphnia* invest in sexual reproduction when its relative costs are reduced. Proc. R. Soc. B 285. https://doi.org/10.1098/rspb.2017.2176

Gilbert JJ (2003) Environmental and endogenous control of sexuality in a rotifer life cycle: developmental and population biology. Evolution & Development 5:19–24

Hairston NG Jr (1998) Time travelers: What's timely in diapause research? Archiv für Hydrobiologie 52:1–15

Hairston NG, Olds EJ (1987) Population differences in the timing of diapause: a test of hypotheses. Oecologia 71:339–344

Hairston NG, Vanbrunt RA, Kearns CM, Engstrom DR (1995) Age and survivorship of diapausing eggs in a sediment egg bank. Ecology 76:1706–1711

Helgen JC (1987) Feeding rate inhibition in crowded *Daphnia pulex*. Hydrobiologia 154:113–119

Heuschele J, Selander E (2014) The chemical ecology of copepods. Journal of Plankton Research 36:895–913

Innes DJ (1997) Sexual reproduction of *Daphnia pulex* in a temporary habitat. Oecologia 111:53–60

Kiørboe T (2013) Attack or attacked: the sensory and fluid mechanical constraints of copepods' predator-prey interactions. Integrative and Comparative Biology 53:821–831

Kleiven OT, Larsson P, Hobæk A (1992) Sexual reproduction in Daphnia magna requires three stimuli. Oikos 65:197–206

Laurenmaatta C, Hietala J, Walls M (1997) Responses of *Daphnia pulex* population to toxic cyanobacteria. Freshwater Biology 37:635–647

Lopatina T, Zadereev E (2015) Assessment of the volatility and thermal stability of substances that stimulate female *Moina macrocopa* (Cladocera) to produce diapausing eggs. Russian Journal of Ecology 46:103–108

Lurling M, Roozen F, Van Donk E, Goser B (2003) Response of *Daphnia* to substances released from crowded congeners and conspecifics. Journal of Plankton Research 25:967–978

Matveev V (1993) Investigation of allelopathic effects of *Daphnia*. Freshwater Biology 29:99–105

Mitchell SE, Carvalho GR (2002) Comparative demographic impacts of 'info-chemicals' and exploitative competition: an empirical test using *Daphnia magna*. Freshwater biology 47:459–471

Montero-Pau J, Serra M (2011) Life-cycle switching and coexistence of species with no niche differentiation. PLoS ONE 6:e20314. https://doi.org/10.1371/journal.pone.0020314

Nielsen DL, Smith D, Petrie R (2012) Resting egg banks can facilitate recovery of zooplankton communities after extended exposure to saline conditions. Freshwater Biology 57:1306–1314

Oskina N, Lopatina T, Anishchenko O, Zadereev E (2018) High Resistance of Resting Eggs of Cladoceran *Moina macrocopa* to the Effect of Heavy Metals. Bulletin of Environmental Contamination and Toxicology. https://doi.org/10.1007/s00128-018-2473-7

Oster GF, Rocklin SM (1979) Optimization models in evolutionary biology. In: Levin SA (ed) Lectures on mathematics in the life sciences, vol 11, pp 21–88

Pijanowska J, Stolpe G (1996) Summer diapause in *Daphnia* as a reaction to the presence of fish. Journal of Plankton Research 18:1407–1412

Pohnert G, Steinke M, Tollrian R (2007) Chemical cues, defence metabolites and the shaping of pelagic interspecific interactions. Trends in Ecology and Evolution 22:198–204

Radzikowski J (2013) Resistance of dormant stages of planktonic invertebrates to adverse environmental conditions. J Plankton Res 35:707–723

Rittschof D, Cohen JH (2004) Crustacean peptide and peptide-like pheromones and kairomones. Peptides 25:1503–1516

Santangelo JM, Bozelli RL, Esteves FDA, Tollrian R (2010) Predation cues do not affect the induction and termination of diapause in small-bodied cladocerans. Freshwater Biology 55:1577–1586

Schwartz ER, Poulin RX, Mojib N, Kubanek J (2016) Chemical ecology of marine plankton. Natural Product Reports 33:843–860

Sieg RD, Poulson-Ellestad KL, Kubanek J (2011) Chemical ecology of marine plankton. Natural Product Reports 28:388–399

Slobodkin LB (1954) Population dynamics in *Daphia obtusa* Kurz. Ecol. Monographs 24:69–88

Slusarczyk M (1995) Predator-induced diapause in *Daphnia*. Ecology 76:1008–1013

Slusarczyk M (1999) Predator-induced diapause in *Daphnia magna* may require two chemical cues. Oecologia 119:159–165

Snell TW (2017) Analysis of proteins in conditioned medium that trigger monogonont rotifer mictic reproduction. Hydrobiologia 796:245–253

Snell TW, Kubanek J, Carter W, Payne AB, Kim J, Hicks MK, Stelzer C-P (2006) A protein signal triggers sexual reproduction in *Brachionus plicatilis* (Rotifera). Marine Biology 149:763–773

Stelzer CP, Snell TW (2003) Induction of sexual reproduction in *Brachionus plicatilis* (Monogononta, Rotifera) by a density-dependent chemical cue. Limnology and Oceanography 48:939–943

Stross RG (1987) Photoperiodism and phased growth in *Daphnia* populations: coactions in perspective. Mem. Ist. ital. idrobiol. "Dott.M.Marchi" 45:413–437

Uchida H, Zweigenbaum JA, Kusumi T, Ooi T (2008) Time-of-flight LC/MS identification and confirmation of kairomone in *Daphnia magna* cultured medium. Bulletin of the Chemical Society of Japan 81:298–300

Van Donk E (2007) Chemical information transfer in freshwater plankton. Ecological Informatics 2:112–120

Walton WE (1985) Factors regulating the reproductive phenology of *Onychodiaptomus bergei*. Limnology and Oceanography 30:167–179

Weismann, A. (1876). *Beitrage zur Naturgeschichte der Daphnioden*. Leipzig

Yasumoto K, Nishigami A, Kasai F, Kusumi T, Ooi T (2006) Isolation and absolute configuration determination of aliphatic sulfates as the *Daphnia* kairomones inducing morphological defence of a phytoplankton. Chemical and Pharmaceutical Bulletin 54:271–274

Zadereev E, Lopatina T (2007) The induction of diapause in Moina by species-specific chemical cues. Aquatic ecology 41:255–261

Zadereev E, Lopatina T (2015) Extraction of info chemicals inducing the production of resting eggs in cladocerans. Doklady Biochemistry and Biophysics 461:127–130

Zadereev ES (2002) Chemical interactions between planktonic crustaceans [in Russian, with an English summary]. Zhurnal Obshchei Biologii 63:159–167

Zadereev ES, Gubanov VG (1996) The role of population density in gamogenesis induction in *Moina macrocopa* (Cladocera: Crustacea) [in Russian, with an English summary]. Zhurnal Obshchei Biologii 57:360–367

Zadereev ES, Prokopkin IG (2007) Role of diapause in coexistence of zooplankton species: Model analysis. Doklady Biological Sciences 412:73–75

Zadereev ES, Prokopkin IG, Gubanov VG, Gubanov MV (2003) Individual-based model of the reproduction cycle of *Moina macrocopa* (Crustacea: Cladocera). Ecological Modelling 162:15–31

Zadereev YS (2003) Maternal effects, conspecific chemical cues, and switching from parthenogenesis to gametogenesis in the cladoceran *Moina macrocopa*. Aquatic ecology 37:251–255

Zadereev YS, Gubanov VG (2000) Density dependent regulation of the reproduction cycle of Moina macrocopa (Cladocera). Crustacean Issues 12:535–542

Chapter 11
Aquatic Invertebrate Dormancy and Medicine

Mei-Hiu Tseng, Jiang-Shiou Hwang, and Victor R. Alekseev

Abstract Aquatic organism dormancy has interesting medical applications, which include dormancy in parasite species, dormancy in intermediate hosts of parasites, and several diseases that seem to present a kind of dormancy. Among these, the most remarkable is the state of dormancy in some types of cancer, one of the most serious human diseases. Holmes, in the middle of the twentieth century, recognized, as one stage of cancer, a long period of delay in carcinogenesis, similar to dormancy. Later on, Makrushin described an "evolutionary hypothesis" of carcinogenesis and recognized this disease as an old genetically related program—a kind of archaic program that is normally dependent on "sleeping" genes. His hypothesis did not explain several important peculiarities of cancerous tumors. The most important one is fast dissemination of the tumor and creation of a large number of metastases. In this chapter, we suggest another hypothesis of carcinogenesis, also related to sleeping genes, that is responsible for the metabolic rate found in dormant cells and postdiapause cells. A possible hormonal mechanism underlying this hypothesis is discussed.

Keywords Carcinogenesis · Evolutionary origin · Medical aspects of dormancy · Metabolic theory

M.-H. Tseng
Medical College, National Taiwan University, Taipei, Taiwan

J.-S. Hwang
Institute of Marine Biology, National Taiwan Ocean University, Keelung, Taiwan

V. R. Alekseev (✉)
Zoological Institute of Russian Academy of Sciences, St. Petersburg, Russia
e-mail: alekseev@zin.ru

V. R. Alekseev, B. Pinel-Alloul (eds.), *Dormancy in Aquatic Organisms. Theory, Human Use and Modeling*, Monographiae Biologicae 92,
https://doi.org/10.1007/978-3-030-21213-1_11

11.1 Introduction

Dormancy is a profound and ancient adaptation found in a wide spectrum of plants and animals in all habitats (Alekseev 1990). One of the most fascinating and profound adaptations related to dormancy is diapause (Danilevsky 1961). In diapause the switch between active and dormant states is driven by a complicated and multilevel hormonal mechanism, which usually includes a photoperiodic pacemaker and several other signals coming from the environment. Temperature, food limitation, and some other stress factors of different origins are known as external drivers of dormancy induction (Alekseev et al. 2007; Nelson et al. 2010). In recent years, dormancy studies on a wide variety of topics have demonstrated that it may be possible to develop diapause switch mechanisms to create novel applications in biotechnology, science, and medicine (Alekseev et al. 2006, 2007; Alekseev 2010).

Several aspects of aquatic organism dormancy can also be interesting in medicine. For example, dormancy in mosquito larvae can be important for distribution of several tropical diseases in a polar direction from the equator (Vinogradova 2007). Many cyclopoid copepods are intermediate hosts for parasitic organisms such as protists and worms. The peculiarities of dormancy in such cases become crucial factors for parasitic diseases in humans and in human-cultivated organisms (Evseeva 1996). It might be useful to consider dormancy and its genetic mechanisms as a model of some medical situations related to activation/suppression processes that are also observed in some human diseases (Table 11.1). Diapause hormonal and biochemical mechanisms may also parallel mechanisms of lethargic dream syndrome, some posttraumatic syndromes, some categories of disability, carcinogenesis, and aging (Craigie et al. 1951; Makrushin and Lyanguzova 2006).

The most fascinating hypothesis, however, could involve the possible impact of this dormancy theory on our understanding of carcinogenesis in humans.

11.2 Makrushin's Evolutionary Hypothesis of Carcinogenesis

Makrushin (2004) was the first person to propose that destructive and proliferative phenomena in dormancy in some lower invertebrates are similar to known destructive processes in cancer in vertebrates. He speculated that there is a parallel between these particular pathological processes in vertebrates and seasonal body reduction associated with formation of the resting stage known to occur in primitive metazoans (*Spongia*, Hydrazoa, etc.) According to his hypothesis, cancer cells appear as a result of regressive development of tissue cells caused by stochastic activation of an ancient program (a sleeping gene) for dormancy induction.

Table 1.1 Physiological states of organisms based on metabolic level and on the cause of the arrestation of vital functions (after Alekseev et al. 2007 with changes)

<table>
<tr><th>Active Life</th><th>Boundary States</th><th colspan="6">Dormancy or Hypobiosis</th></tr>
<tr><td rowspan="3">Normal metabolism, development, growth, and breeding</td><td rowspan="3">Short-time metabolism declining</td><td colspan="6">Suppressed metabolism, with long-time cessation of development, growth, and breeding</td></tr>
<tr><td colspan="3">Quiescence</td><td colspan="3">Diapause</td></tr>
<tr><td colspan="3">(hibernation and/or aestivation) driven by unfavorable/favorable conditions for life</td><td colspan="3">driven by hormones, induced by both signal and vital factors</td></tr>
<tr><td rowspan="3"></td><td rowspan="3">Sleep, physiological stress, thermal shock, some illness and medical/artificial coma in humans</td><td colspan="3">Latent life</td><td colspan="3">Suppressed life</td></tr>
<tr><td rowspan="2">Cryptobiosis</td><td rowspan="2">Anabiosis</td><td rowspan="2">Abiosis</td><td>Superpause</td><td>Mesopause</td><td>Oligopause</td></tr>
<tr><td>More than 1 year</td><td>3–12 months</td><td>Less than 3 months</td></tr>
</table>

When the idea of dormancy in cancer cells was first published, the hypothesis was not taken seriously (Craigie et al. 1951). There was very little critique among medical scientists and biologists, and the hypothesis was by no means considered testable. But, to everyone's surprise, data obtained from molecular genetic studies of dormancy in the nematode *Caenorhabditis elegans* supported this idea of a common genetic basis for diapause and aging (Gerisch and Antebi 2004).

Recent studies have suggested that in *C. elegans*, gene expression changes choose between the third larval diapause and reproductive development. A cascade of genes and their gene products mediates this choice between diapause and active development in *C. elegans*. These include *daf-9* (a cytochrome P450 gene related to steroidogenic hydroxylases) and *daf-12* (a nuclear receptor gene encoding for lipophilic hormones that control the physiological status of the organism) (Gerisch and Antebi 2004).

A simple model is that the *daf*-9 gene produces a hormone regulating *daf-12*. Its gene product bypasses diapause, promotes reproductive development, and perhaps also shortens the life-span. This hormone might be a sterol (Gerisch et al. 2001). Expressed in potential endocrine matter, *daf-9* appears to control developmental decisions for the entire organism. Recent findings implicate *daf*-9 as a central point of developmental control, producing hormonal signals that regulate the life history of *C. elegans* (Gerisch and Antebi 2004).

The choice between development and dormancy can be regulated by several substances. Insulin is such a substance. Other signals that control the diapause program throughout the body are special peptides (e.g., transforming growth factor β (TGFβ) and serotonergic signaling) (Finch and Ruvkun 2001). Insulin and TGFβ peptides are synthesized in response to environmental stimuli, mainly from sensory neurons. Alternatively, a complex of the *daf-3* and *daf-5* genes may effect a shift to diapause in adverse environments. TGFβ inactivates *daf-3* and *daf-5*, thus allowing reductive development (Gerisch et al. 2001). There is evidence that both insulin and TGFβ receptors convey signals through downstream secondary endocrines and that *daf-2* regulates diapause and the life-span through systemic signals (Apfeld and Kenyon 1998; Wolkow et al. 2000). The mechanism of verification, based on experiments with mutants, is in other publications (Alekseev 2010; Alekseev et al. 2006).

The genes *daf-9* and *daf-12* linked with diapause have now been found in a very wide range of organisms, from yeasts to vertebrates. The genetic program coding for dormancy can either be actively used by the organism or simply kept "in reserve" by the genome without being expressed. This discovery by molecular biologists comes back to Makrushin's hypothesis of the evolutionary origin of cancer in vertebrates that appears as activation of sleeping dormancy genes. Makrushin also postulated that the final aim of a dormant tumor, which is almost never realized, is creation of a kind of dormant embryo similar to one produced by primitive Coelenterata organisms in harsh environmental conditions. This statement is not well supported by medical observations.

11.3 Hypothesis of Metabolic Mechanism Breaking in Cancer Cells

A new version of the hypothesis of carcinogenesis based on dormancy gene expression has been suggested to overcome some contradictions in Makrushin's theory (Alekseev et al. 2006). At the cancer stage of a dormant tumor, there is a sudden shift from a quiet state to a state of fast division and dissemination of new fast-growing "seeds" called metastases. Cells in metastases have a higher metabolic rate than normal cells, so the metastases grow very fast and finally kill the organism within a short time.

We propose that a similar mechanism responsible for diapause, for normal metabolism, and also for more active metabolism can be observed in diapausing organisms after termination of dormancy. Further, this mechanism results from activation of a single gene (or a group of genes?) and is hormone driven. Normal metabolism is the standard state of cells in an organism, and it is regulated by a balance between the hormone activator and inhibitors of metabolism. Both balance and misbalance, as normal adaptations to environmental demands, have, in fact, been demonstrated in the molting and life cycles of some decapods (Aiken 1969; Alekseev 1998).

In this way, carcinogenesis can thus appear as the result of an abnormal hormonal misbalance in some cells. The mechanism behind this could be triggered in a cell as a result of gene damage, whether from chemicals, radioactive or viral agents, or other causes. In this hypothetical cell, some normally sleeping genes responsible for metabolism would suddenly become activated, and the organism would lose control of metabolism and division in these cells. In a normal, healthy organism, the immune system is able to recognize and kill these out-of-control cells. In our supposed scenario, however, such a cell would not be killed by the immune system and would start intensively reproducing itself like a postdiapause embryo, developing into a carcinogenic tumor. Among the possible explanations for a scenario so negative for the organism could be many reasons for immune system weakness: stress, other illness, age, etc. This explains the so-called poly-reason nature of carcinogenesis.

In our hypothesis, in contrast to Craigie's and Makrushin's hypotheses, cells in cancer tumors are similar to postdiapausing embryonic cells driven by an out-of-control hormone activator. They are characterized by higher metabolic activity and division, as shown by a reactivated *Daphnia* embryo (Arbacauskas and Lampert 2003).

Our hypothesis, if substantiated through testing, will provide clinical scientists with dormancy-derived hormones (many of them are known!) as a new instrument for cancer research and treatment. With this comes the promise of new medicines or types of treatment aimed at normalizing the balance in hormonal mechanisms of cell metabolism.

11.4 Conclusion

Diapause is a very profound and ancient adaptation found in many classes of aquatic animals and plants. It appears likely that there is a monophyletic origin for this adaptation, as there are similarities in the molecular basis and genetics of diapause mechanisms among organisms. Research on diapause in aquatic organisms opens up many new scientific directions and technologically important applications. When medical conditions originate from a misbalance in the activation/suppression process, diapause may offer a useful model for innovative treatment approaches. A new hypothesis of the carcinogenesis process states that tumor cells parallel postdiapause embryonic cells.

Acknowledgements We acknowledge financial support from Taiwan (NSC 94-231 l-B-019-009) and Russia (05-04-90588-HHC_a, 17-04-00027) in a bilateral collaboration. The study was partly supported by the Russian Academy of Science (topics 65.4 and 65.5).

References

Aiken DE (1969) Photoperiod, endocrinology and the crustacean molt cycle. Science 164:149–155

Alekseev VR (1990) Diapause in crustaceans: ecological and physiological aspects. Nauka Academic, Moscow. [in Russian]

Alekseev VR (1998) Biochemical and physiological characteristics of crustaceans in diapause: the internal mechanism of reactivation. Arch Hydrobiol 52:463–476

Alekseev VR (2010) Physiological and molecular biological mechanisms underlying diapause in aquatic invertebrates. Russ J Dev Biol 41(61):1608–3326. [in Russian]

Alekseev V, Hwang JS, Tseng MT (2006) A review of diapause in aquatic organisms: what's known and what's next in research and application? J Mar Res Nat Taiwan Ocean Univ 14(2):3–28

Alekseev VR, de Stasio BT, Gilbert JJ (eds) (2007) Diapause in aquatic invertebrates: theory and human use, Monographiae Biologicae 84. Springer, Dordrecht. 257 p

Apfeld J, Kenyon C (1998) Cell nonautonomy of *C. elegans* daf-2 function in the regulation of diapause and lifespan. Cell 95:199–210

Arbacauskas K, Lampert W (2003) Seasonal adaptation of ex-ephippio and parthenogenetic offspring of *Daphnia magna*: differences in life-history and physiology. Funct Ecol 17:431–437

Craigie J, Lind PE, Hayward ME, Begg AM (1951) Preliminary observation on a "dormant" state of sarcoma cells with special references to resistance to freezing and drying. J Pathol Bacteriol 63:177–178

Danilevsky AS (1961) Photoperiodism and seasonal development of insects. Leningrad University Press, Leningrad. [in Russian]

Evseeva N (1996) Diapause of copepods as an element for stabilizing the parasite system of some fish helminths. Hydrobiologia 320:229–233

Finch CE, Ruvkun G (2001) The genetics of aging. Annu Rev Genomics Hum Genet 2:435–462

Gerisch B, Antebi A (2004) Hormonal signals produced by DAF-9/cytochrome P450 regulate *C. elegans* dauer diapause in response to environmental cues. Development 131:1765–1776

Gerisch B, Weitzel C, Kober-Eisermann C, Rottiers V, Antebi A (2001) A hormonal signaling pathway influencing *C. elegans* metabolism, reproductive development, and life span. Dev Cell 1:841–851

Makrushin AV (2004) Evolutionary predecessors of oncogenesis and old age involution. Adv Gerontol 13:32–43
Makrushin AV, Lyanguzova IV (2006) The shell of invertebrates and plant propagules: selective permeability and barriers. J Obshch Biol 67:115–122. (in Russian)
Nelson RJ, Denlinger DL, Somers DE (2010) Photoperiodism: the biological calendar. Oxford University Press, New York, p 596
Vinogradova EB (2007) Diapause in aquatic insects, with an emphasis on mosquitoes. In: Alekseev VR et al (eds) Diapause in aquatic invertebrates. Springer, New York, pp 83–113
Wolkow CA, Kimura KD, Lee MS, Ruvkun G (2000) Regulation of *C. elegans* life-span by insulin-like signaling in the nervous system. Science 290:147–150

Part III
Dormancy, Biological Information and Modeling

Chapter 12
Biological Information and Dormancy in Aquatic Population Modeling

Victor R. Alekseev

Abstract The importance of biological information in life activity of organisms and superorganisms at the overlying levels is not disputed by anyone, but it remains an area with scant investigation. There are several reasons, starting with the blurriness of the very concept of information and ending with methodological difficulties in its quantitative data tallying and measurement. Diapause is one of the most studied adaptations based on the detection of information signals about the state of the environment and its upcoming seasonal reconstructions and seems to be one of the most convenient bioinformational object for studying at this stage of development of this branch of science. The very significant progress of object-oriented modeling in recent decades provides a methodological basis for such studies. Different aspects of Shannon's theory of information in biological systems are also discussed.

Keywords Shannon's theory of information · Use of biological information · Effectiveness of biological functions · Shannon index, Informative links · Evaluation of biological information · Biological informative system

12.1 A Brief Review of the History of Bioinformation Research

The need to study the flow of information, along with the flow of energy and matter, was proclaimed at the beginning of the modern phase of the study of aquatic ecosystems (Khailov 1966). Major environmental programs have called for the study of theses flows but in most specific content remain empty, leaving biologists without firm foundations to follow-up research.

What is the reason for such long-lingering futility at one of the most important directions in the environment, assessing the state of organisms and its ability at reconstruction by neighboring activity? And what is the actual locus of information

V. R. Alekseev (✉)
Zoological Institute of Russian Academy of Sciences, St. Petersburg, Russia
e-mail: alekseev@zin.ru

V. R. Alekseev, B. Pinel-Alloul (eds.), *Dormancy in Aquatic Organisms. Theory, Human Use and Modeling*, Monographiae Biologicae 92,
https://doi.org/10.1007/978-3-030-21213-1_12

links in superorganismic biosystems? In this paper, we propose one option for answering these questions.

The peak of interest in problems related to the assessment of the role of information in biological suborganismic systems occurred in Russia and abroad in the 1960s and 1970s (Shmalhausen 1968; Seravin 1973; Naumov 1973; Setrov 1975, etc.). By the beginning of the 1980s, the amount of work related to this topic decreased significantly, mainly because of limitations of starting theoretical assumptions. Turning to the origins of the bioinformation "boom," one should point to cybernetics (Wiener 1961), the systemic approach (von Bertalanffy 1969) and, of course, Shannon's mathematical theory of communication of optimal coding (1948). If the first two theories related to information would largely apply to descriptive constructs, pointing at the significance and the place of information links in the management of complex systems, Shannon's constructs allowed to quantify the quality of information for biology.

The famous Shannon index served as a quantitative basis for science—unprecedented in the breadth of application and the alleged effect of an explosive search for information links in highly divergent areas of science: from physics to linguistics. The Shannon index is extremely engaging in its simplicity for assessing information:

$$H = -\sum_{t=1}^{t=n} P_i * \log_2 P_i$$

where P_i is the probability of an event.

In addition to that, Shannon proposed several theorems describing the patterns of information transfer via communication channels from source to receiver; Shannon also considered the mathematical Boltzmann's H-theorem for thermodynamics and adapted it to the needs of information research. Among the most important contributions of Shannon, we find also strict descriptions of communication channels, sources of interference, and reliability criteria for communication.

However, having disregarded the meaning of the message being transmitted in the quantitative measure (H), Shannon posed a challenge to his apologists, one that has not been solved until now, as it appeared to be insoluble in principle. A wide recognition of that fact did not come immediately to the fore, including in the field of biology.

It should be said that the field biology is largely conservative in relation to constructs and models, which is also said of or even more so in applied sciences. Nevertheless, an indisputable impetus for biologists to search for and find meaningful information links in biosystems at various levels happened in 1953 with the decoding of the structure of the biological DNA code by Watson and Crick. This example, now part of students' textbooks, of the leading role of information in the implementation of such a determining function of the organism as it is reproduction, made an extremely strong impression and gave rise to numerous followers in different biological disciplines. However, to the great disappointment of researchers on the role of information in biology, the situation with the genome was the exception rather than the rule. As the organization of biological systems becomes more complex, the probabilistic (balancing) principle of organizing information links comes to replace the rigid rules of the information bank (DNA). This is due to the fact that the internal environment of the organism, as well as its structure, turned out to be a more rigid system than the one that develops in superorganismic structures. Thus, it is very difficult to program and implement all possible variants of the relationship of the organism with the environment (as opposed to the stages of embryonic development, for instance). It became apparent that the question of the role of information in the biological systems at the superorganism level is decided differently than in the underlying systems.

Moreover, even among the followers of Shannon, voices began to be heard more and more loudly that although the theory of information is so broad that it can be applied to any subject, it is nevertheless advisable to apply it in cases where the decisive factor is the stream of communication rather than the meaning of the message being transmitted (Ichas 1960).

In other words, Shannon's information theory is not applicable to assess the role of information in biological systems, but only to study and evaluate the effectiveness of its transmission in coded form. In this case, this theory has a methodological rather than a conceptual significance. A serious analysis of this aspect was made by L. Seravin (1973) and M. Setrov (1975).

12.2 Universal Language and Information Flows in the Ecosystem

Attempts to apply the theory of optimal coding for studying the role of information in biological systems, including complex ones such as biocenoses, did not stop. The idea of an ecological code—one for the ecosystem—was put forward and actively discussed (see Levich 1983 for review). It was assumed that there is a language (ecocode) that can be deciphered, in which information is exchanged by all participants in the ecosystem. The study of the ecocode was supposed to be similar to mastering language. The scheme of understanding the ecocode by the researcher, in principle internally consistent, was constructed with the involvement of the logic of formal linguistics.

Among the texts of the ecological code were included such indicators as abundance and species composition of planktonic organisms. Mandelbrot (1972) proposed to equate communities with texts, types with words, individuals of one species with word forms, etc. Levich (1983), using his version of analogies, equated individuals with letters, one-time communities with words, and succession of communities with text. All these analogies gave their authors formal grounds to draw as mathematical patterns of numbers obtained in the analysis of textual materials, in particular Zipf's law (Zipf 1949):

$$n(i) \sim \frac{1}{i}$$

where i is the rank of the word in the ranked series and $n(i)$ is the occurrence of a word of rank i.

Despite the apparent dubiousness of such associations for the biologist, certain grounds for using regularities like Zipf's law (1949) have proven useful. Even before Zipf's law was published, Willis (1922) used a rank distribution for the analysis of the fauna of Ceylon and derived quite convincing material concerning the distribution of species in isolates. It is quite obvious that these studies are grounded on general statistics laws underlying the distribution of a wide range of objects that obey the law of large numbers in probability theory. It could be said that authors overlooked the notions of information significance of the rank-size distribution, and this would also be true for the inhabitants of a biocenosis. There is almost no doubt that organisms are completely deprived of both the ability and the biological equipment for analyzing the species structure of their ecosystem, as well as using information that could be obtained through them. But the fact that the ecocode should be understood not only by the researcher but also by the biota's organisms follows from the definition of ecocode and how it is perceived through collected or derived data.

Thus, we have come to another fairly common misconception regarding the evaluation of the role of information in biological systems inferred on the basis of probabilistic regularities of indices and indicators and their identification (including the usage of Shannon index). We also tend to draw conclusions from the analysis of indices purportedly to obtain biological information, but rather with the category of problems of studying the role of information in biological systems. In the best case, such studies allow us to build a mathematical apparatus developed in, and for, other fields of science detecting all kinds of diversity and heterogeneity in biological systems. These techniques are likely to be ultimately unsuitable to accommodate the role of biological information as regards to the structures and functions of biological systems.

The question arises: Which way is more productive? To answer this, it is necessary to determine what exactly gives information to the individual of the population and the biocenosis. First, the advantage in performing a particular function that will life expectancy, i.e., again. The question of the usefulness of information for the organism is deemed as an important one because, only in this case, it is possible to develop appropriate adaptations fixed either in a series of generations (e.g., be it instincts or biorhythms) or during the lifespan of an individual (conditioned reflexes, to mention one). So what is the measure of usefulness of information can be used for the organism? L. Seravin (1973) proposed two elementary indicators in this capacity: the real gain $C_p = P_2 - P_1$ and the relative gain $C_o = P_2/P_1$, where P_2 and P_1 are the probabilities of the event, respectively, with and without the information. Enduring systems, including biological ones, are designed for maximum efficiency; therefore, the value of information increases for unlikely events and decreases for probable ones.

Thus, the main reason (and therefore the criterion for researchers) for information links can be considered limited or uneven spatial-temporal distribution of any of the factors (resources) affecting the success of the vital functions. Previously, a number of authors have addressed some striking examples of the participation of information in the implementation of such functions (Seravin 1973; Zelikman 1977; Levich 1980, 1983) which enables us to infer an approximate structure in information links of the organism as an element of the population and an element of the superorganismic system (Fig. 12.1). Almost all the communication channels known for biological objects can act as carriers of information in providing each of the functions in question: chemical (such as metabolites, attractants, etc.), optical (such as visual image, photoperiod), and mechanical vibrations of the environment, electromagnetic waves (Zelikman 1977). Interaction between individuals of one population can be considered as a sort of multinodal information network covering the habitat of the population and increasing the probability of survival of any of the members of the network. In the diagram, this is indicated by the selection of the plane of intrapopulation interactions.

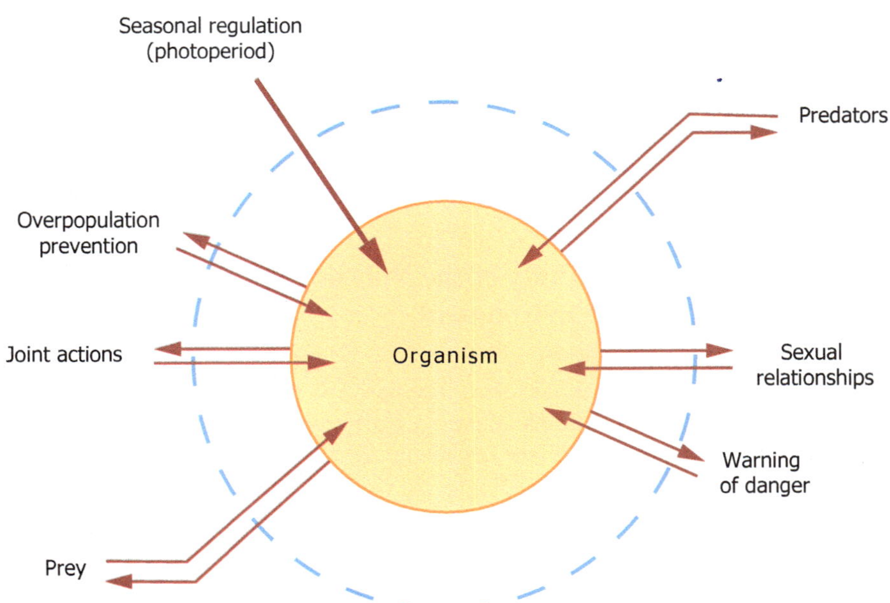

Fig. 12.1 Schematic diagram of vertical and horizontal information links of the organism. Dashed line shows the plane of intrapopulation links

The relationship between the higher and lower levels should be considered as an element of the "pyramid of information links," which in some ways resembles a trophic system. Since one type food feeding and control by the same predator is very rare in nature, vertical information interaction is carried out in a multidimensional space. In our opinion, vertical and horizontal information communications in many respects give concrete outlines and certain stability to superorganismal structures.

Returning to the question of the biological language of ecosystems (i.e., Levich's ecological code), it should be said that it was not possible to detect it due to the fact that the original premises did not correspond to reality. In real-world ecosystems, instead of just one universal language, the "Babylonian confusion" of extremely simple dialect functions, which is understandable to the circle of organisms, is limited by information links. Each organism in search of food and trying to evade a predator is constantly forced, as a first task, to collect information about these factors in order to increase the effectiveness of actions. During periods of reproduction or collaborative interactions (hunting, collective protection in a pack, etc.), individuals send and perceive information that is understandable only to those to whom it is addressed, that is, it does not go beyond members of the same population co-inhabiting in the same biotope. Exceptions are rare and refer to species with similar niches (Alekseev 1990).

Of the few languages that can claim to be universal because of direct connection with such a generally necessary adaptation, as the seasonal adjustment of communities, the photoperiod and temperature can be considered very simple codes. The perception of specific photoperiod values in the binary system (day/night) is largely species-specific. Nevertheless, the photoperiod and partly the temperature, passing and acting approximately unambiguously, causing the completion of the active part of the life cycle and leading to the formation of various kinds of biological dormancy among ecosystem members, can claim the role of flows of biologically significant information penetrating through a dense multidimensional multi-member information network of organisms of the whole ecosystem. In this sense, the factors of transition to biological dormancy can be considered as flows of bioinformation.

12.3 Features of the Study of the Role of Bioinformation in the Aquatic Environment

Specific study of information links should be based on the structure of the information system (Fig. 12.2). Problems arising in the study of information links can be grouped in accordance with the elements of the information system: (a)

Fig. 12.2 Schematic diagram of the information system. (Modified from Shannon 1948)

identification of the information source and destination; (b) establishment of the type of communication channel (chemical, optical, etc.); (c) identification of the code (chemical formula, frequency of oscillations, photoperiod); (d) evaluation of the upper and lower thresholds of signal perception; (e) determination of the range and noise immunity of the communication system; (f) clarification of the form of response to the signal; and (g) study of the mechanism of the transmitter and receiver of signals.

The works of the S. Shvarts school, on amphibian larvae (Schwarz et al. 1976), and the Leningrad school of entomology (Danilevsky 1961), on terrestrial insects, can serve as good research examples of this kind of information links, even though they reached there from their own path each from others. In addition, some environmental studies of energy and substance flows, in particular those studying the efficacy of predator nutrition, do not always include an explicit evaluation of information parameters (Ivlev 1961; Krylov 1989). Such indicators, which are of interest from the point of view of studying information, may include the radius of the predator reaction (assessment of the ability of nontactile detection), the proportion of successful predator attacks, and the assessment of the ability of the victim to detect and avoid the predator by analyzing chemosignals—as exemplified by kairomones (Lampert 2011).

A special place in the study of the role of biological information belongs to studies of seasonal adaptations and, foremost among them, the diapause or biological dormancy. This evolutionary adaptation comprises a complex mechanism of detection and analysis of heterogeneous information (including photoperiod, temperature, chemosignals), and simultaneously it materializes this information in the form of biological dormancy. Furthermore, the reactivation and subsequent induction of diapause coincide with the active part of the life cycle of organisms, and that affects the development of other species, primarily foraging elements and predators of diapause-using organisms. In other words, it optimizes the seasonal cycle of population development, combining the whole ecosystem into a single whole. Hence, diapause has a biorhythmical function, which, perhaps, is even more important than the protective function (Alekseev 1990).

Vertical information links in biological systems appear primarily in the experimental work of a physiological and biochemical kind, and it is usually well endowed from the methodological side—this is confirmed by results already obtained (Schwarz et al. 1976; Danilevsky 1961; Lampert 2011). The greatest difficulties for the formation of a full-fledged theory about the role and place of information in biological systems at the superorganism level lie in the need to perform an extremely large amount of work comparable only with the decoding of the human genetic code. Significant advances are achieved through model studies on the role of information in the functioning of both populations and more complex biosystems at different trophic levels (Alekseev and Fiks 1989; Alekseev and Umnov 2002; Alekseev and Kazantseva 2007). Research in this direction will shed light on the problem of a more accurate model description of the relationship of the species to the environment and assess the role of bioinformation not only qualitatively but also quantitatively (Alekseev and Kazantseva 2015).

Acknowledgments I appreciate very much the kind help in improving my English and productive discussion with Professor Eduardo Vincente (Valencia University, Spain). This section was prepared with the financial support of RFFI Grant 17-04-00027. The study was partly supported by Russian Academy of Science, topics 65.4 and 65.5.

References

Alekseev VR (1990) Diapause in crustaceans: ecological and physiological aspects. Nauka Academic Publishers, Moscow, 144 pp (in Russian)
Alekseev VR, Fiks BV (1989) The population dynamics model of the planktonic Cyclopid Mesocyclops leuckarti (Claus), based on Photoperiodism. Transof Zool Inst USSR Acad sci 205:96–107 (in Russian)
Alekseev VR, Kazantseva TI (2007) The usage of individual-oriented model in studies on the role of the maternal effect in the reproductive switch in Cladocera. Zhurnal obshchei biologii 68(3):242–251
Alekseev VR, Kazantseva TI (2015) The role of chemoreception in forming of Daphnia longispina sustainable population (the simulation experiments). Zhurnal obshchei biologii 76(5):377–389
Alekseev VR, Umnov AA (2002) The role of trophic and signal factors in diapause induction and seasonal dynamics in a Daphnia longispina population: a simulation model. Verh Int Verein Limnol 27:3927–3932
Danilevsky AS (1961) Photoperiodism and seasonal development of insects. Leningrad University Press (in Russian)
Ichas M (1960) Protein text. Theory of information in biology, Moscow, 399 p (in Russian)
Ivlev VS (1961) Experimental ecology of the feeding of fish. Yale University Press, New York, p 302
Khailov KM (1966) The problem of links and organization in life systems' evolution. Russ J Philos:4
Krylov PI (1989) Freshwater predative zooplankton feeding. Hydrobiology, Moscow, 146 p (in Russian)
Lampert W (2011) Daphnia: development of a model organism in ecology and evolution. Excell Ecol 21:1–275
Levich AP (1980) Structure of ecological communities. MGU Publication, pp 3–181 (in Russian)
Levich AP (1983) Biological system languages. Human and biosphere 8:68–77 (in Russian)
Mandelbrot B (1972) Theory of information and psycholinguistic. Mathematical methods in social sciences
Naumov NP (1973) Signal biological fields and their significance to animals. J Commun Biol 74:P808–P817 (in Russian, English summary)
Schwarz SS, Piaskolova OA, Dobrinskaya AA, Runkova GG (1976) Group effect in aquatic animals population and chemical ecology. Nauka Academic Publishers, 151 p (in Russian)
Seravin LN (1973) Theory of information from a point of view of biologist. Nauka Academic Publishers, 160 p (in Russian)
Setrov MI (1975) Informative processes in biological systems. Nauka Academic Publishers, 155 p (in Russian)
Shannon CE (1948) A mathematical theory of communication. Bell Syst Tech J 27:379–423, 623–656
Shmalhausen II (1968) Cybernetic questions in biology. Nauka Academic Publishers, 224 p

von Bertalanffy L (1969) General system theory. George Braziller, New York, p 136
Wiener N (1961) Cybernetics or control and communication in the animal and the machine. M.I.T. Press, New York, 212p
Willis JC (1922) Age and area, Cambridge. https://doi.org/10.5962/bhl.title.30741
Zelikman EA (1977) Non-trophic regulatory interactions among marine invertebrates. Ocean Biol 2 (in Russian)
Zipf GK (1949) Human behavior and the principle of least effort, Cambridge

Chapter 13
Dormancy and Individual-Based Modeling of *Daphnia* Population Dynamics

Tamara I. Kazantseva and Victor R. Alekseev

Abstract A modification of the previously designed individual-based simulation model of population dynamics of a mass pelagic species, *Daphnia longispina*, is presented. This model includes several important patterns related with dormancy. The model describes a population as an ordered list of individuals with specific sets of traits. For each individual, its energy traits are calculated, and the probabilities of different outcomes are assigned or calculated for various life situations. This variant significantly differs from the previous chemosignal blocks that modify *Daphnia* behavior. After receiving a chemical signal (predator kairomones or sex pheromones), adult *Daphnia* of the model population can avoid predation with a certain probability by diel vertical migrations and can increase the efficiency of sexual reproduction through the active search for a sexual partner. Simulation experiments in which the ability to detect different chemical signals was excluded, sequentially or in combination, were used to investigate the role of chemoreception in the population dynamics and the seasonal adaptations of planktonic crustaceans (dormancy, based on photoperiod accepting along with food ability with maternal transmission of environmental information in a chain of generation). For the first time, a quantitative estimate for the role of such information has been obtained as the ratio between the annual production levels of model populations that possess or lack the ability to perceive chemical signals of predators and mating partners. According to the present version of the model, chemoreception increased the annual production of the model population by 1.5 times. Thus, biological information is a significant factor enabling individual hydrobiont populations to fine-tune their adaptations to the particular environmental conditions of a given aquatic habitat.

T. I. Kazantseva (✉)
Zoological Institute of the Russian Academy of Science, University Emb., St. Petersburg, Russia
e-mail: tamara.kazantseva@zin.ru

V. R. Alekseev
Zoological Institute of Russian Academy of Sciences, St. Petersburg, Russia
e-mail: alekseev@zin.ru

V. R. Alekseev, B. Pinel-Alloul (eds.), *Dormancy in Aquatic Organisms. Theory, Human Use and Modeling*, Monographiae Biologicae 92,
https://doi.org/10.1007/978-3-030-21213-1_13

Keywords Chemosignals · Individual-based modeling · Vertical migration · Photoperiod accepting · Effectiveness of mating · Quantative role of biological information

13.1 Introduction

Biorhytmic function of dormancy serves to organize life cycle of a population to fit the most optimal way to environmental demands, such as seasonality in temperature, food presence, and predator absence. Being dependent on time and local factors, the life cycle is based on several environmental ques, such as photoperiod and so named vital determining functions (temperature, oxygen, water presence, predator presence, etc). Coming in dormant start as well as reactivation aquatic organisms of different trophic levels should be well corresponding to each other, so filtrators must start active breedings when algae have blumed as well as predators have appeared when their prey become dense.

This function of dormancy is so far based on accepting and realization of biological information. No doubts that as a life cycle organizer dormancy presents an interest to population life modeling. Such a model created to describe a daphnia population dynamic is described hereafter.

The model in question was developed in order to demonstrate that biological information chains can really play an important role in seasonal adaptation of a *Daphnia* population to environment demands and can affect its population dynamics. The model parameters describing the environmental conditions (water temperature, photoperiod, fish, and invertebrate predation) and the life cycle of *Daphnia* (maximal duration of development, egg caloric weight, critical photoperiod responce), as well as the only control parameter—the annual dynamics of real population size—were set according to field observation data and were not changed in the course of model testing. Our hypotheses as to how different signal types can influence the population were based on our own observations or on the data from available publications, most of which were purely phenomenological.

In particular, this concerns the intensity of fish signaling compounds and their perception by *Daphnia*. It is known that the signal about the presence of a predator has a complex nature (Lampert 2011). It comprises secretions of fish themselves and signaling compounds from the bodies of prey. In our own experiments, we analyzed how *Daphnia* kept in a limnotower reacted to an inflow of water from a different aquarium containing fish. If fish were fed with mosquito larvae or another fodder, *Daphnia* did not perform any vertical migrations. However, as soon as the fish fodder in the aquarium was changed for the same species of *Daphnia*, the *Daphnia* in the limnotower began diel vertical migrations. Moreover, the intensity of migrations (their depth or the density of aggregation in the aphotic zone) was directly correlated to the amount of *Daphnia* consumed by the fish. In response to an imitation of strong pressure, nearly all *Daphnia* concentrated at the bottom. These observations formed the conceptual basis of the assumed dependencies describing the interaction between *Daphnia* and fish. The exact functions describing this and other hypotheses, as well as the quantitative values of the involved parameters, were set and adjusted in such a way as to achieve the best possible agreement between the dynamics of the model and the real population dynamic. We were able to achieve this purpose with an appropriate set of parameters, which indicates that our model was adequate to the real object. The fact that our basic assumptions enabled us to simulate in general the real dynamics of the population size shows that they do not contradict the reality and therefore the model can be employed for simulation experiments. The performed simulations in fact analyzed the model's sensitivity to the biological information perceived by individuals. As a result, we obtained quantitative estimates of the effects gained by the use of chemoreception in a fine adjustment of life cycle of organism to the local combination of environmental factors. Or in other words—to a microevolution process in the species population.

13.2 The Model Description

To test the hypotheses concerning the role of chemoreception in the development and behavior of *Daphnia* in freshwater habitats, as well as to evaluate quantitatively the efficiency of chemical signals as bioinformation agents, we modified the previously developed individual-based simulation model of a *Daphnia* population (Alekseev and Umnov 1996, 2002; Alekseev and Kazantseva 2015).

The model population was composed of individuals (adults and eggs) characterized by a certain set of traits; each individual was assigned its proper values of these traits. The main qualitative trait of an individual as a population element was its type. An individual belonged to one of the four types: male, female, summer resting eggs, or winter resting eggs. Each individual was also characterized by its weight and age, which was calculated as the sum of degree-days accumulated by the given moment of time. Parthenogenetic eggs, depending on their genesis, were divided into three groups denoted as weak, medium, and strong. Eggs of each group were characterized by their so-called growth potential, which reflected the trophic conditions under which the mother (1) developed herself as an embryo, (2) reached sexual maturity, and (3) laid this particular egg. The egg growth potential determined the curve of the maximal possible weight growth of an individual hatched from the given egg, i.e., the individual's growth potential, and also affected the individual's fertility and its probability of changing its reproduction type (gamogenetic shift). Individuals that hatched from resting eggs had the highest potential, which was equal to 4 (Fig. 13.1).

Thus, the model described the nongenetic, mother-to-offspring transmission of information about the environmental trophic conditions (the maternal effect). Depending on the environmental factors, which change over the course of the year (water temperature, photoperiod, and available food amounts), on the current moment of time, and on the initial potential, a model individual hatched as a male

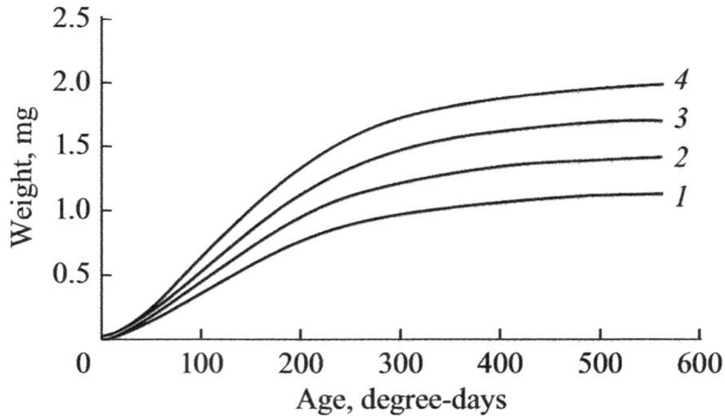

Fig. 13.1 Possible curves of age-related growth depending on the individual's potential (1–4)

or a female, grew, reached sexual maturity, produced eggs of different quality, and died from different causes. Each individual was characterized by its own traits and its own fate; all of the individuals together constituted a population. All events occurring during an individual's life were determined stochastically. As a result of the addition of newly hatched individuals and the removal of dead ones, the size, age, and quality composition of the population changed with time. This determined the annual dynamics both of the population as a whole and of its particular subgroups defined by sex, age, or any other trait.

For each individual, its energy traits were calculated, and the probabilities of different outcomes in various life situations were either assigned or also calculated according to specified rules. The rate of death from natural causes (the so-called background death rate) was set as a constant coefficient and was the same for the whole population. If the actual growth rate of an individual's weight was lower than that potentially possible by more than 45%, the probability of its death increased. An adult that did not die from natural causes could be consumed by a predator with the set constant probability of 0.02.

The model included the following environmental factors as function of time: water surface temperature (°C), photoperiod (h), and the available amount of food (mg/L). The model step was 1 day, and the working volume was 1 L.

The prototype of the model was a well-studied population of *Daphnia longispina* O.F. Müller, 1785 of the nursery Lake Krasnen'koe, Pskov oblast (lake area, 1.2 ha, average epilimnion depth, 3.1 m, maximal hypolimnion depth, 6.5 m; there is a stable thermocline with an oxygen-deficient zone), according to data from 1986 (Alekseev 1987). The model was adjusted to achieve the best possible agreement between the annual dynamics of the model population size, i.e., the total number of males and females of all ages, and the dynamics of the real population.

The initial version of the model (Alekseev and Umnov 2002) and its subsequent modifications (Alekseev and Kazantseva 2007; Kazantseva and Alekseev 2007) confirmed the possibility of the maternal effect in *Daphnia* (Bernardo 1996; Mousseau and Fox 1998). This effect implies that information about the alimentary and light (photoperiod) conditions existing at the moment a parthenogenetic egg was laid can be transmitted in three generations of females (Alekseev and Lampert, 2001).

To evaluate the role of chemoreception in the seasonal dynamics of *Daphnia* populations, the previous model was considerably modified. Two types of responses were to be investigated:

1. Individuals' reactions to information about the presence of predators (primarily fish); the reactions involve vertical migrations from the surface layer to the hypolimnion after a certain level of fish pressure is achieved (Alekseev and Lampert 2003, 2004).
2. Pheromone-dependent active search for mature females by adult males; the pheromone effects increase with rising water temperatures (Fries 1964; Alekseev and Khozyaikin 2009).

Thus, the present variant of the model describes the formation of annual population dynamics in *Daphnia* that respond to information signals of different natures (Fig. 13.2).

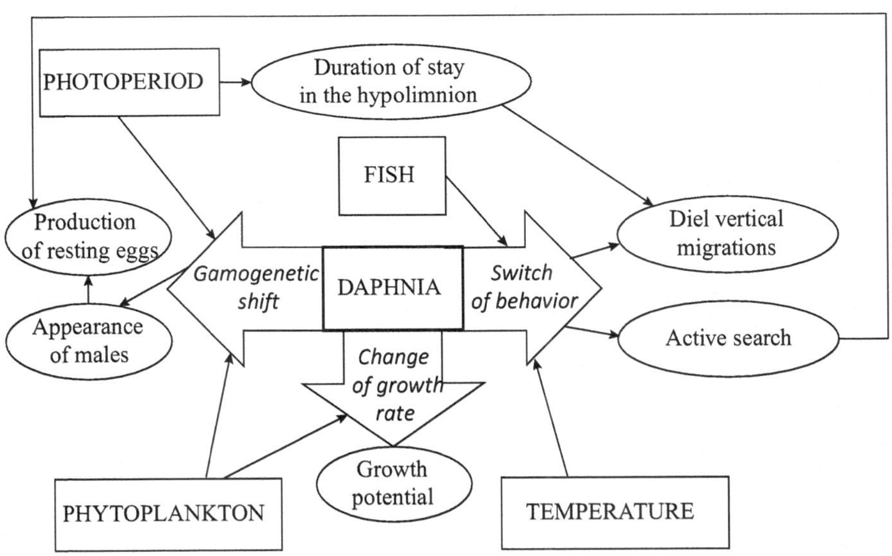

Fig. 13.2 A basic block diagram of biological information chains in the model population. It combines the classical trophic chain "fish–*Daphnia*–phytoplankton" with the effects of biological information signalling (food shortage, female pheromones, fish kairomones) and abiotic environmental factors (temperature, photoperiod) that determine the choice of alternative strategies growth, reproduction, and behavior in *Daphnia*

13.3 Experiments with the Model to Test the Role of Biological Information

The model described above was employed to obtain quantitative estimates for the role of particular components of environmental information in the seasonal adaptation of *Daphnia* populations. Thus, the model describing the normal population behavior was modified by subsequent exclusion of the perception of particular signals.

13.3.1 Experiment A: Assessing the Role of Information About the Presence of Fish

In this simulation, the ability of adult *Daphnia* to perceive the presence of fish and, accordingly, to perform downward migrations was excluded. Under more favorable feeding and temperature conditions, the growth rate and parthenogenetic fertility of mature females increased. The probability for an adult individual to be consumed by a fish was still determined with formula (1) and did not exceed 0.04. However, egg-carrying females were larger than other individuals and were more easily noticed by fish. As a result, their predation by fish increased dramatically in comparison to other individuals, and their specific death rate was increased. The model described this as follows:

p_down = 0;
yield_fish (♀ov) = yield_fish (standard) + S.

If the specific death rate of egg-carrying (S) had increased by up to 0.38, the model population remained sustainable in the multiyear dynamics. But the peak population size in the second part of the season dropped significantly in comparison to the standard (N) variant (Fig. 13.3b). Population created less dormant eggs in fall time, when in Standard variant it was not so. However, with a further increase of S, the population lost stability and was then extinguished as early as in the first simulation years. The critical value was S = 0.41: the population could either persist for a fairly long time, or perish very quickly (Fig. 13.4). Such a situation can occur in shallow water bodies (lacking the hypolimnion) with insignificant fish pressure. It was also described in field observations, when strong coregonid fish cultivated in a forest lake without the hypolimnion completely eliminated *Daphnia cristata* population at the end of summer before females could produce overwintering dormant eggs (Alekseev 1987; Alekseev et al. 1989).

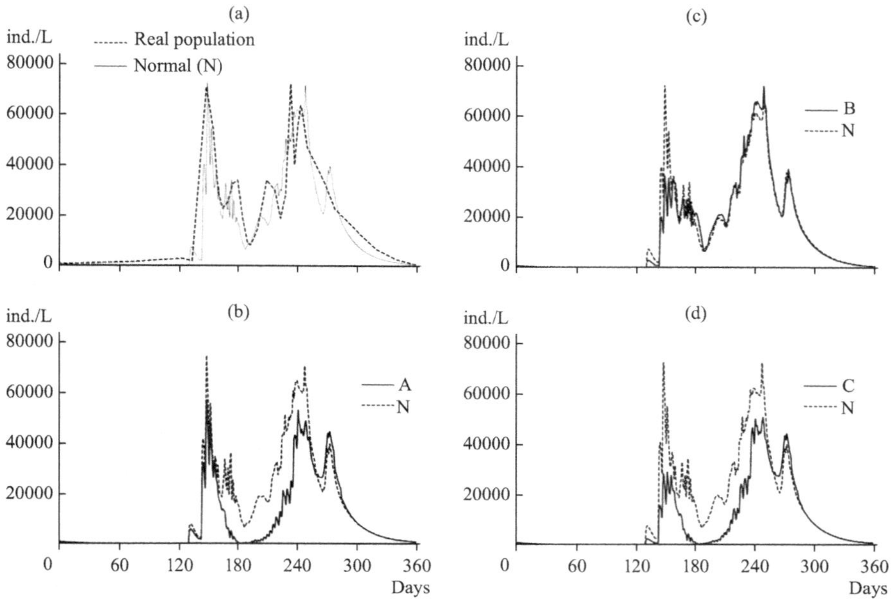

Fig. 13.3 Comparison of population size dynamics: (**a**) the real population and the model variant Standard (N); (**b**) the model population variants Standard (N) and A(1) for S = 0.35; (**c**) the model population variants Standard (N) and B; (**d**) the model population variants Standard (N) and C for S = 0.35

Fig. 13.4 Variants of unstable multiyear dynamics of the model population size in experiment A(1) at S

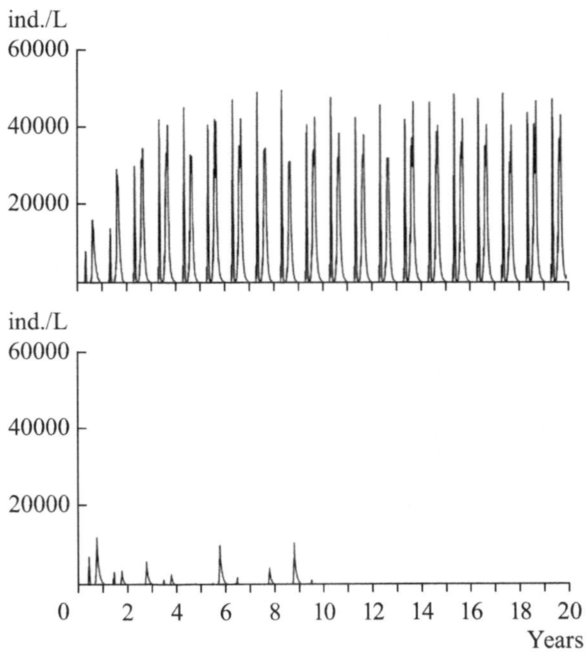

13.3.2 Experiment B: Assessing the Role of Information About the Presence of a Mature Female

In the next simulation, individuals were able to react to the presence of predators, but the ability of males to perceive the presence of mature females was excluded. Under such conditions, the probability of an encounter between mature male and female individuals was at any temperature equal to the ratio between their numbers (Fig. 13.5a). As a result, the efficiency of sexual reproduction and creation of dormant eggs was decreased, and the stock of summer and winter resting eggs diminished (Fig. 13.5b). Consequently, the size of the subsequent spring generation, which depends on the number of overwintered ephippia containing dormant eggs was significantly lower than under normal conditions for sexual contacts in the *Daphnia* population model (Fig. 13.3c). In the mid-summer, during the laying of summer resting eggs, fewer females switched to gamogenesis than normally, and the remaining parthenogenetic females were able to restore the population size during the fall peak. The fall generation develops at lower temperatures and with decreasing photoperiod. By the moment when the temperature falls below 12 °C, most overwintering eggs have already been laid, but the stock of dormant eggs was less than in normal model version. The results of Experiment B also indicate the necessity to take into consideration the competition for food in aquatic crustacean communities. The simulation showed than the sufficiency of gamogenesis in the summer–fall period and the number of overwintering dormant eggs produced determine the size of the spring generation of *Daphnia*. At the same time, the outcome of interspecies competition for common food resources (small size algae blooming in spring) is largely determined by the size of the spring generation of competing

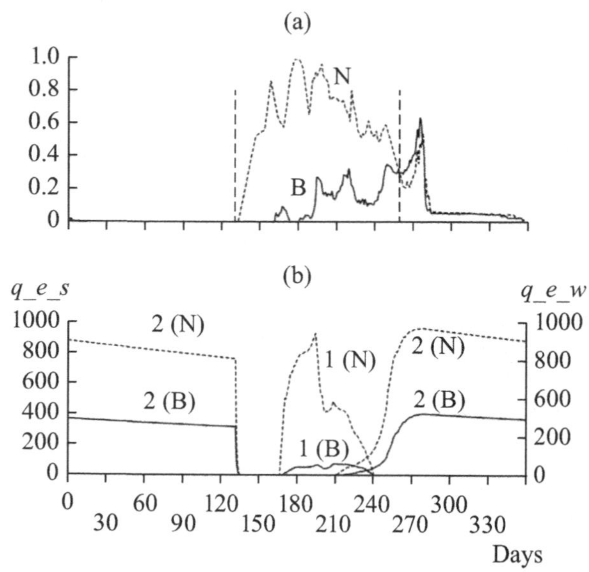

Fig. 13.5 Simulation experiment B: (**a**) probability of an encounter between a male and a female in variant B and in the standard variant (N). Vertical lines mark the period with water temperatures of 12–22°C; (**b**) the number of summer (q_e_s, 1) and winter (q_e_w, 2) resting eggs in Variant B and in the standard variant (N), ind./L

populations of filtators. Therefore, the accumulating delay in the production of winter resting eggs is bound to significantly diminish the competitive abilities of a species that lacks chemoreception in comparison to other species that possess such an ability or are otherwise more advanced in bisexual breeding and dormant egg production in the critical periods of seasonal development. As a natural consequence, the first species will be gradually eliminated from the community and replaced with competing species with similar trophical and dynamical characteristics. These considerations suggest how individual-based models could possible be modified further in order to simulate competitive interactions between close species.

13.3.3 *Experiment C: Assessing the Role of Chemoreception in the Annual Dynamics of a* Daphnia *Population*

In this simulation, chemoreception was excluded completely by combining variants A and B. The moderate value of fish pressure on egg-carrying females (S = 0.35) was selected so that a stationary population could persist. Under these conditions, the model population retained the same character of annual dynamics, but the population size diminished for both the spring and the fall generations (Fig. 13.3d). Thus, in the absence of chemoreception, *Daphnia* population failed to make full use of the available resources and to reach the maximal possible size as well as storage of overwintering dormant eggs. As a result, competitive ability in such population was declined in all seasons.

13.4 *Daphnia* Population Strategy to Survive in Extremal Environmental Conditions

The unsteady-state situation discussed above, in which the population of *Daphnia* does not react to the presence of fish and to a risk being almost extinguished (Fig. 13.4). It depends on the whole set of assumed conditions, in particular, on the character and the magnitude of pressure from invertebrate predatorical fish pressure on adult *Daphnia*, to achieve sustainability of the model population by changing the predation of *Chaoborus* larvae on juvenile *Daphnia*. To find out the answer to this question, the model was tested under the conditions of variant A with S = 0.41 via two modifications of chaoborus pressure as follows.

1. The *Chaoborus* pressure on juvenile *Daphnia* was decreased from May 10 to June 23, i.e., during the development of the first *Daphnia* generation, when fish pressure on adult individuals was absent or weak.
2. The *Chaoborus* pressure was decreased during the development of the second generation from July 7 to August 19. In both cases, the pressure was decreased by 20% of the initial value for the period of 6 weeks. It was found that even such

a moderate decrease in invertebrate pressure on juvenile *Daphnia* was enough to allow the model population to persist steadily, although at a lower quantitative level, despite the increased predation of egg-carrying females. If invertebrate pressure on juvenile *Daphnia* in the spring had no significant effect on the population dynamics.

The fact that the initial generations are highly sensitive to negative effects due to predation or insufficient food supply is well known (Meester et al. 2002). The situation simulated in Experiment A (adult *Daphnia* do not respond to fish pressure by migrating to deeper water levels) may well occur in shallow water bodies that do not have a hypolimnion. According to our observations, moderate fish pressure in a lake without hypolimnion resulted in a considerable decrease of the size of the *Daphnia* population, whereas strong fish pressure brought about complete disappearance of *Daphnia* (Alekseev et al. 1987). The critical value of the daily death rate of egg-carrying *Daphnia* obtained both from field research data (Alekseev et al. 1989) and in simulation experiments, S = 0.41, was not random. It corresponds to the situation in which a significant share of the population is able to produce only one generation of offspring, which are able to avoid fish pressure due to their small size and thus manage to maintain a sufficient population size. An investigation of the model showed that this value was related to the duration of embryonic development of egg carried by a female. If the term of the embryonic development diminished by 15% in comparison to the assumed "standard" value, the critical S value increased, and vice versa. In our opinion, this fact suggests that the duration of embryonic development can be employed to predict the resistance to fish pressure in crustaceans of similar ecological traits and also makes possible the use of optimal combinations of population traits in the modeling of interspecies competition.

13.5 Quantitative Evaluation of the Role of Chemical Signaling in the Development of a *Daphnia* Population

To evaluate quantitatively the role of biological information, it is necessary to compare the efficiency of a given functional activity in the presence and absence of the signals in question. Professor Lev Seravin (1973) proposed to describe the utility of information with two quantitative parameters (for details, see above the overview on biological information):

1. Cn, an actual gain, defined as the difference between the probabilities of performing a vital function in the presence and in the absence of given information (P2 and P1, respectively):

$$Cn = P2 - P1$$

2. Cr, relative gain, defined as the ratio of the same probabilities: Cr = P2/P1.

The Cn value shows by what amount the use of the information increases the probability of the given event, while Cr shows how many folders this probability increases due to the information using.

This seemingly simple design is, however, hardly applicable when considering chemical signaling in aquatic environments, mainly because the chemical nature of signaling agents is insufficiently understood, their active concentrations are extremely low, and equipment for their detection does not exist. The difficulties associated with direct quantitative measurements of the role of information signals in the adaptation of real populations can be surmounted to a certain extent with computer simulations. To evaluate quantitatively the contribution of each of the above factors on the functioning of the model population, we analyzed the total annual population production, including both somatic and generative production as a control parameter. Based on these results, we calculated the values of Cr, which shows the relative efficiency of using chemoreception in different simulated situations. By making the full use of chemical signaling (in the Standard variant), the model population achieved a 1.28-fold gain in annual production in comparison to the variant lacking information on the presence of fish (experiment A at S = 0.35), a 1.06-fold gain in comparison to the absence of information on mating partners (experiment B), as well as a 1.56-fold gain in comparison to a complete lack of chemoreception (experiment C). We showed that absence of chances to avoid predation usually led to much more dramatical results such as complete extinction of prey population in lakes without hypolimnion.

13.6 Discussion and Conclusion

By comparing the results of simulations performed for different models of chemoreception involvement in the annual dynamics of the population size, we reached certain conclusions concerning the role of active predator avoidance and search for a mating partner. Obviously, the size of the spring generation diminished sequentially from variant N to variant A, then to variant B, and to variant C, according to the decrease in the number of overwintering eggs in dormancy state. The spring generation peak was the lowest in variants B and C, where males were unable to perceive female pheromones. The fall generation size differed between two groups of situations: in variants N and B, *Daphnia* could avoid predators, while there was no avoidance in variants A and C (Fig. 13.3). Thus, the size of the summer–fall generation depends on the active predator avoidance.

On the basis of above discussed model that comprises the role of biological signal information we conclude:

1. Transmission of biological information in the chain of generation (maternal effect) forms major peculiarities of *Daphnia* annual population dynamics in particular trophic conditions.

2. The maternal effects also play an important role in switching from parthenogenesis to the sexual reproduction in *Daphnia longispinal* population. Photoperiod (day length declining) is more important factor in changes of *Daphnia* type of breeding from parthenogenesis to bisexual reproduction than food limitation.
3. The possibility of accumulation of maternal effects in long series of generations apparently increased the resistance of *Daphnia* population to interannual fluctuations of environmental parameters.
4. The size of the spring generation of the model *Daphnia* population was determined by the number of overwintered resting eggs and was increased in the presence of an active search for mature females mediated by chemoreception.
5. The size of the summer–fall generation of the model population was nearly independent of the initial population size but was instead determined by the environmental conditions present throughout the season, primarily by predation. Chemoreception helps adult individuals to avoid predation and thus serves to maintain high population numbers.
6. In the absence of chemoreception, an increase of predator pressure exceeding a certain critical threshold, which depends on the duration of prey's embryonic development, leads to instability of the model population and to a decrease in its size, up to complete extinction.
7. The model provided a quantitative estimate for the contribution of chemoreception to the population dynamics of *Daphnia*. In our model, chemoreception allowed the *Daphnia* population to increase its annual production by up to 1.5 times. In the absence of chemoreception, the population failed to make full use of the habitat resources and to reach the maximal size.

Acknowledgments This chapter was written with the partial support from RFBR (grant 17-04-00027). The study was partly supported by Russian Academy of Science, topics 65.4 and 65.5.

References

Alekseev VR (1987) Influence of intensive measures on development of zooplankton and seasonal cycles of crustacean in the lake whitefish farms. Sb Nauchn Tr Gos Nauchno-Issled Inst Ozern Rechn Rybn Khoz 264:20–32

Alekseev VR, Kazantseva TI (2007) The usage of individual-oriented model in studies on the role of the maternal effect in the reproductive switch in Cladocera. Zh Obshch Biol 68(3):242–251

Alekseev VR, Kazantseva TI (2015) The role of chemoreception in forming of Daphnia longispina sustainable population (the simulation experiments). Zh Obshchei Biol 76(5):377–389

Alekseev VR, Khozyaikin AA (2009) Transformation of seasonal cycles in Cladocera depending on temperature. Tr Zool Inst Ross Akad Nauk 313(1):10–22

Alekseev V, Lampert W (2001) Maternal control of resting-egg production in Daphnia. Nature 414:899–901

Alekseev V, Lampert W (2003) Study on the role of biological information in local seasonal signal formation in Daphnia. Tr Zool Inst Ross Akad Nauk 299:15–24

Alekseev V, Lampert W (2004) Maternal effects of photoperiod and food level on life history characteristics of the cladoceran Daphnia pulicaria Forbes. Hydrobiologia 526:225–230

Alekseev VR, Umnov AA (1996) Integration of production-energetic, population-dynamics, and bioinformational approaches for analysis of life strategies of Daphnia: simulation model, Mater. VII s"ezda Vses. gidrobiologicheskogo obshchestva. Proc VII Congr All-Union Hydrobiol Soc Kazan 1:38–41

Alekseev VR, Umnov AA (2002) The role of trophic and signal factors in diapause induction and seasonal dynamics in a Daphnia longispina population: a simulation model, Verh. Int Verein Limnol 27:3927–3932

Alekseev VR, Lavrent'eva GM, Teslya AY (1987) Specific energy flow in the lake whitefish farms. Sb Nauchn Tr–Gos Nauchno-Issled Inst Ozern Rechn Rybn Khoz 264:124–137

Alekseev VR, Gurinovich AM, Teslya AYa (1989) Dynamics of the dry matter content and fat in the organisms of peled juveniles at the different development of food resources in farming pools, Mater. 3-go Vses. simp. po troficheskim svyazyam i produktivnosti vodnykh soobshchestv. In: Proceedings of the third symposium on trophic relationships and productivity of water communities, Chita, pp 18–19

Bernardo J (1996) Maternal effects in animal ecology. Am Zool 36:83–105

Fries G (1964) Über die Einwirkung der Tagesperiodik und der Temperature auf den Generationswechsel, die Weibchengrösse und die Eier von Daphnia magna Straus. Z Morphol Ökol Tiere 53:475–516

Kazantseva T, Alekseev V (2007) Use of a mathematical model to study the role of maternal effects in population dynamics and diapause formation in Daphnia. Fund Appl Limnol 169(4):293–306

Lampert W (2011) Daphnia: development of a model organism in ecology and evolution. Excell Ecol 21:1–275

Meester L, de Gome A, Okamura B, Schwenk K (2002) The monopolization hypothesis and the dispersal-gene flow paradox in aquatic organisms. Acta Oecol 23(3):121–135

Mousseau TA, Fox CW (1998) Maternal effects as adaptations. Oxford University Press, New York, pp 159–177

Seravin LN (1973) Teoriya informatsiya s tochki zreniya biologa (The theory of information from the biological point of view). Leningrad Gos. Univ, Leningrad

Chapter 14
The Role of Biotic and Abiotic Interactions in Summer Diapause in Cyclopoids: Conceptual Model and Field Validation in Southern Quebec Boreal Lakes

Bernadette Pinel-Alloul and Victor R. Alekseev

Abstract Cyclopoid copepods are a dominant component of freshwater zooplankton and the Cyclopidae is the most diverse family. Dormancy is a fundamental process of cyclopoid life history in permanent lakes and temporary ponds. In boreal lakes, cyclopoid diapause has been documented mainly in Northern Europe but more rarely in Northern America. We present the first assessment on summer diapausing cyclopoids (SDC) in 22 boreal lakes in southern Québec, which vary in their morphometry, tropic status, and predation pressure by fish or invertebrates. We developed a conceptual model to test the hypothesis that diapausing patterns of SDC in boreal lakes are a complex response depending primarily on morphometric and trophic features and secondly on the intensity of hypolimnetic anoxia and predation by fish or invertebrates. Using morphometric indices, lakes were classified in groups varying by the importance of wind-mixing, the strength of thermal stratification during summer, the potential of meromixis, and the risk of predation by fish or chaoborids. Three cyclopoid species (*Diacyclops thomasi*, *Cyclops scutifer*, *Mesocyclops edax*) dominated in all groups of lakes, while *Diacyclops nanus* and *Acanthocyclops vernalis* were found in small abundance and only in few lakes. The highest abundances of SDC (density and biomass) were found in sediments of thermally stratified mesotrophic lakes with intermediate wind-mixing index (Ko) and low potential for meromixis (Pm). Strong positive correlations were found between SDC and meiobenthos abundances in all groups of lakes. High biomass of *Chaoborus* larvae or presence of large populations of fish negatively affected SDC abundances in meiobenthos. Morphometric features, wind-mixing and thermal

B. Pinel-Alloul (✉)
GRIL, Département de sciences biologiques, Université de Montréal,
Montréal, QC, Canada
e-mail: bernadette.pinel-alloul@umontreal.ca

V. R. Alekseev
Zoological Institute of Russian Academy of Sciences, St. Petersburg, Russia

V. R. Alekseev, B. Pinel-Alloul (eds.), *Dormancy in Aquatic Organisms. Theory, Human Use and Modeling*, Monographiae Biologicae 92,
https://doi.org/10.1007/978-3-030-21213-1_14

stratification, as well as the abundance of *Chaoborus* and fish predators are the key factors governing species diversity and abundance of summer diapausing cyclopoids in boreal lakes.

Keywords Cyclopoid copepods · Summer diapausing patterns · Wind-mixing index · *Chaoborus* and fish predators · Boreal lakes

14.1 Introduction

Cyclopoid copepods are a dominant group of freshwater crustacean zooplankton in the north-temperate region of North America. The Cyclopidae represents the most diverse family comprising 114 species and contributing to 33% of the total species richness (Boxshall and Defaye 2008). Freshwater cyclopoids are known to have complex life cycles including winter or summer larval diapause, usually in deep waters and sediments of lakes (Dahms 1995; Alekseev and Starobogatov 1996; Williams-Howze 1997; Alekseev et al. 2004). Diapause in cyclopoids has been observed more often in permanent lakes and temporary ponds of northern temperate regions than in tropical regions (Dahms 1995; Alekseev 2007). Diapause patterns in cyclopoids vary highly among species and habitats in relation to taxa and ontogenetic stages, latitudinal gradients in photoperiod and temperature, and other environmental factors such as food condition and predation from fish and invertebrates (Watson and Smallman 1971; Hairston 1987; Dahms 1995; Hansen 1996).

Dormancy in cyclopoid copepods is considered a response to changes in abiotic and biotic conditions (Elgmork 1967, 1981, 2004; Naess et al. 1993; Hansen 1996; Williams-Howze 1997; Medland and Taylor 2001; Alekseev et al. 2004, 2007). In general, larval diapause is induced during summer for cold-water cyclopoids (*Cyclops* and *Diacyclops*) (Santer and Lampert 1995; Santer and Boldt 1998; Hansen and Hairston 1998; Santer and Hansen 2006) and during winter for warm-water cyclopoids (*Mesocyclops* and *Thermocyclops*) (Naess et al. 1993; Dahms 1995). Winter diapause as a response to seasonal decline in temperature and photoperiod has been the most investigated (Elgmork 1996; Kobari and Ban 1998, Alekseev 1998). Summer diapause in cold-water cyclopoids has remained largely unexplored because the seasonal disappearance of cyclopoids from summer plankton community was wrongly assumed to be due to production of resting eggs as seen in calanoid copepods and cladocerans (Alekseev et al. 2007). The discovery of important concentrations of cyclopoid copepodites IV–V in the deep waters and sediments of lakes during summer, and their reappearance at spring in the plankton community, has led to a better understanding of the benthic/pelagic coupling between diapausing and active stages of copepods (Alekseev 2007). This is a singular phenomenon in the life cycle of cold-water cyclopoids and other freshwater zooplankton (Gyllström and Hansson 2004).

In general, cold-water cyclopoids (*Cyclops scutifer*, *Diacyclops thomasi*) disrupt their ontogenetic development at the copepodite stage IV in early summer when water temperature rises to 12–15 °C (Hansen and Hairston 1998). Diapause, however, may also occur at the copepodite stage V (*Diacyclops nanus*, Lake Magog, this study) and in fertilized females (*Cyclops strenuus*, Naess and Nilssen 1991). The dormant stages sink and accumulate in deep waters and surface sediments sometimes in higher abundance than the active stages in the water column (Elgmork 1981, 2004). In autumn, wind-mixing and water turnover redistribute the dormant copepodites which can survive during winter by feeding on resting stages of rotifers and crustaceans, as well as algae, protozoa, and bacteria, as observed in laboratory (Krylov et al. 1996). Diapausing cyclopoid stages restart active growth and complete sexual maturation at 2–3 °C in March–April at the end of the ice-covered period. They then reproduce intensively in the open water during May and June. Recruitment of the active stages in the pelagic zone lasts 20–30 days until copepodites reach stage IV and restart summer diapause. Through this benthic-pelagic coupling, the dormant reservoir of copepods can have considerable impact on the ecological and evolutionary dynamics of crustacean zooplankton and trophic interactions in planktonic food webs (Elgmork 1980; Brendonck et al. 1998; Hairston et al. 2000; Brendonck and De Meester 2003; Alekseev et al. 2007). In north-temperate lakes, the population dynamics of spring cyclopoids are highly dependent on the abundance and emergence of diapausing stages. Studies should therefore consider both the active and dormant stages of the population in order to fully understand their life history strategies. However, studies addressing the dormant stages are rare in comparison to the large number of studies addressing the active stages (Gyllström and Hansson 2004). Increased knowledge of summer diapausing patterns in cyclopoids and of their environmental controls is needed to improve our ability to understand population dynamics of crustacean zooplankton and the functioning of plankton food webs in lake ecosystems.

Diapause is a powerful hormone-driven strategy (Alekseev 1991, 1998; Alekseev et al. 2004, 2007) analogous to insect diapause (Elgmork and Nilssen 1978). It not only allows survival during harsh environmental conditions but also synchronizes growth and reproduction with periods of favourable environmental conditions, rich in food resources and devoid of high competition and predation pressure (Santer and Lampert 1995; Brendonck and De Meester 2003; Gyllström and Hansson 2004; Santer and Hansen 2006; Radzikowski 2013). Diapause patterns in cyclopoids are driven by both abiotic and biotic factors, which vary between freshwater habitats. Changes in photoperiod and temperature are the most important abiotic factors for inducing diapause in zooplankton, as they are proximal cues signalling seasonal variation in food resources or predation (Williams-Howze 1997; Gyllström and Hansson 2004; Alekseev 2007). Winter diapausing is considered to be a common strategy to overcome periods of limited food resources, as well as low temperature and oxygen conditions in Northern European lakes (Elgmork 1980, Elgmork et al. 1990; Naess et al. 1993). On the other hand, summer diapause in cyclopoids is seen as a long-term decline in metabolic processes providing encysted or passive dormant stages with the capability to survive in cold and oxygen-depleted deep waters of thermally stratified lakes and avoid food resource limitation and fish predation (George 1973; Gliwicz and Rowan 1984; Wyngaard 1988; Santer and Lampert 1995; Hansen 1996; Santer and Boldt 1998; Kobari and Ban 1998; Santer and

Hansen 2006; Alekseev et al. 2007). However, diapausing cyclopoids may suffer from predation by invertebrates, such as chaoborids, that migrate and survive in cold and anoxic deepwater zones. Assessing the relative influence of abiotic and biotic factors on summer diapause of freshwater cyclopoids is a major challenge (Gyllström and Hansson 2004) because variation in these factors depends on morphometry and trophic status of lakes, as well as hypolimnetic anoxia and predation pressure.

In Northern European boreal lakes, the life cycles of cyclopoids have been well studied for the past several decades (Halvorsen and Elgmork 1976; Naess et al. 1993; Maier 1994; Elgmork 1981, 1996; Rivier 1996; Williams-Howze 1997; Frisch 2002; Gyllström and Hansson 2004; Santer and Hansen 2006; Seebens et al. 2009). Fewer studies have investigated the life cycles of cyclopoids in Northern Canada (Boers and Carter 1978; Paquette and Pinel-Alloul 1982; Elgmork 2004). Several studies examined summer diapause patterns of large body-sized *Cyclops* species in temperate lakes of Europe (George 1973; Halvorsen and Elgmork 1976; Gliwicz and Rowan 1984; Santer and Lampert 1995; Hansen 1996). In contrast, since the first mention of a "resting cyst" of *Diacyclops thomasi* in the Great Lakes by Birge and Juday (1908), only four studies have described summer diapause and the emergence of cyclopoids in North American lakes. Hansen and Hairston (1998) and Hairston et al. (2000) studied diapause and the emergence patterns of cyclopoids (*Diacyclops thomasi*, *Mesocyclops edax*, *Acanthocyclops vernalis*) in response to food limitation in Lake Oneida (NY, USA). Elgmork (2004) investigated life cycle patterns in *Cyclops scutifer* in eight lakes along a latitudinal gradient and observed both direct and delayed developments; he did not, however, find diapausing cyclopoids in lake sediments. Medland and Taylor (2001) described diapause emergence patterns in cyclopoid species in a temporary pond in South Carolina (USA) and inferred five different strategies for *Acanthocyclops*, *Microcyclops*, *Diacyclops*, *Macrocyclops*, and *Tropocyclops* according to seasonal hydroperiods.

This chapter provides the first documentation of dormancy patterns of cyclopoid copepods in boreal lakes of Quebec (Canada) in relation to lake environmental features. We developed a conceptual model as a framework to evaluate the role of abiotic (wind-mixing, thermal stratification, trophic state) and biotic (fish and invertebrate predation) processes in controlling diapausing patterns in cyclopoids in boreal lakes in southern Quebec. Modelling and field validation revealed that dormancy in cyclopoid copepods is a complex response to spatial heterogeneity in lacustrine environments.

14.2 Dormancy Patterns of Cyclopoids in Boreal Lakes of Southern Quebec

We studied summer diapause in cyclopoids in 22 lakes located in the Eastern Townships and Laurentian regions of southern Quebec (45.24–45.86°N; 71.54–74.48°W). These lakes are of glacial origin and cover the typical range of morphometry and trophy of Eastern Canadian boreal lakes (Keller and Conlon 1994). They also represent the spatial heterogeneity of zooplankton community structure typically encountered in southern Québec lakes (Masson et al. 2004). In most of the lakes, we measured temperature and oxygen depth profiles with 0.5–1 m intervals using an YSI sonde and water transparency with a Secchi disk. Water samples were taken in lakes with hypolimnetic anoxia in order to measure dissolved oxygen concentration using Winkler's method.

Lake and catchment morphometry and water quality conditions were highly variable among lakes (Table 14.1). Catchment area (DA) ranged from less than 1 km^2 to almost 2000 km^2 and lake area (LA) and volume (LV) from 1 to 1439 ha and 0.04 to 87 km^3, respectively. Mean depth (Zmean) varied from 1.8 to 12.9 m and maximum depth (Zmax) from 3.2 to 30.5 m. Trophic status of lakes ranged from oligotrophic to meso-eutrophic; total phosphorus (TP) concentrations varied 10-folds from 3 to 34 $\mu g.L^{-1}$ and chlorophyll a biomass (Chl.) 20-folds from 0.3 to 7.6 $\mu g.L^{-1}$. Secchi water transparency (SD) varied from 1.2 to 6.5 m. Water temperature in the hypolimnion (T, hypo) increased from 4 to 6 °C in strongly stratified lakes, to around 8–12 °C in moderately stratified lakes, and up to 12–21 °C in weakly strati-

Table 14.1 Limnological features of the studied lakes in each group according to the wind-mixing index (A, Ko > 20; B, 1 < Ko < 10; C, Ko < 1)

Group	Lake	LA, ha	LV, km3	DA, km2	Zmean, m	Zmax, m	Ko, m	Pm	TP, µg/L	Chl., µg/L	SD, m	T, hypo °C	DO hypo mg/L
A	Achigan	537	69,5	88,7	12,9	26,8	20,0	0,6	5,6	0,5	6	8	NA
A	Brome	1439	84,7	172,9	5,9	12,2	118,0	0,2	26,6	4,7	1,8	19,8	8
A	Écho	176	4,3	10,2	2,5	9,1	19,3	0,2	13,5	1,6	3	21,5	NA
A	Lovering	494	43,5	41,3	8,8	22,9	21,6	0,5	10,2	1,6	3	11,8	NA
A	Magog	1095	87,4	1939,2	8	19,2	57,0	0,3	24,3	3,1	3,9	14	NA
A	Waterloo	147	4,9	32,5	3,4	4,9	30,0	0,1	33,9	7,6	1,2	20	12
B	Beaulac	87	4,6	4,2	5,4	9,3	9,4	0,3	8,1	0,6	6	12,5	NA
B	Connelly	121	9,5	22,1	7,9	20	6,1	0,6	10,4	1,4	NA	6,5	NA
B	D'Argent	100	4,5	64	4,5	15,6	6,4	0,5	16,2	1,4	2	9,3	4
B	Desmarais	24	0,6	1,6	2,5	7	3,4	0,3	4,5	2,2	1,5	10,5	1
B	Marois	99	6,6	6	6,7	22,7	4,4	0,7	5,8	0,7	4,5	6,5	1
B	Ouimet	59	3,1	5,8	5,2	18,3	3,2	0,7	4,3	0,7	5	6	2
B	Patrick	151	16,3	61,7	10,8	30,5	5,0	0,9	10,1	0,6	NA	5,2	NA
B	Cristal	29	2,8	1,9	9,5	18,5	1,6	0,8	3,1	0,3	6,5	4,3	1
B	Croche	19	0,8	1,07	7,5	11,5	1,7	0,6	3,6	1,7	5	6	1
B	Cromwell	11	0,3	10,1	3,1	10	1,1	0,5	12,3	2,0	2	6	1
B	Desjardins	6	0,1	0,2	2,1	4,6	1,3	0,3	14,5	2,2	2,5	18	NA
C	Geai	1	0,3	0,2	3,5	8,2	0,1	0,8	8,0	2,3	2,6	5	1
C	Pin Rouge	15	0,8	6,6	5,1	15,2	1,0	0,8	11,9	2,0	2,3	5	0,5
C	Tracy	8	0,7	0,1	8,9	24,4	0,3	1,5	4,7	0,4	6	5,5	0,5
C	Triton	2	0,04	0,2	1,9	3,2	0,6	0,3	6,3	1,0	3,2	18	NA
C	Brassard	4	0,1	0,1	1,8	6,1	0,7	0,4	5,3	6,1	NA	6	1,5

fied or mixed lakes. Oxygen concentrations in deep waters (DO, hypo) also varied according to thermal stratification regime and ranged from maxima > 10 mg/L^{-1} in well-mixed shallow lakes to minima < 2 mg/L^{-1} in deep and well-stratified lakes.

To assess the abundance of active and dormant stages of cyclopoids, we sampled crustacean zooplankton in the water column and the sediments of the 22 study lakes several times during October 1998 and July and August 2000. The active stages of cyclopoids in the water column were collected at the deepest site of the lakes using a cantilevering plankton net (square opening, 0.04 m^2; mesh size, 53 μm) (Filion et al. 1993) hauled either from 1 m above the sediments to the lake surface in the shallow and well-mixed lakes and across the epi-, meta-, and hypolimnion zones in the stratified lakes. Zooplankton organisms were first narcotized with soda water and then preserved in a sugar solution of 4% formaldehyde (Prepas 1978). The non-encysted and encysted stages of diapausing copepods were collected in the meiobenthos of sediments using Sorokin's sediment core sampler tube (surface, 10 cm^2; length, 40 cm). To select diapausing cyclopoid stages, we collected the meiobenthos in the upper 5 cm liquid part of the sediment core only, which contains over 95% of diapausing cyclopoid copepods (see Table 14.2), based on studies in Northern European lakes (Nilssen and Elgmork 1977). Three sample cores were collected as replicates from each lake, pooled in a single sample and washed in a handle net (100 μm). Meiobenthos samples were conserved in a 2% formaldehyde solution and analysed 1–2 days after sampling. In the laboratory, adult and copepodite stages of cyclopoid species, as well as other crustaceans (calanoids and cladocerans), and *Chaoborus* larvae collected in the water column samples, were identified and counted under a stereomicroscope Leica Wild M3B. In the meiobenthos, we counted all organisms with body length from 0.3 to 3 mm, including diapausing cyclopoid stages, ostracods, worms (mainly oligochaetes), insect larvae (mainly chironomids), and small bivalve molluscs. Biomass (wet weights) of crustaceans and meiobenthos were estimated from body lengths using Balushkina and Winberg (1979) equations. Chaoborid larvae were weighted in a microbalance with 0.01 mg resolution to estimate their biomass.

14.3 Conceptual Model of Environmental Control of Summer Diapause of Cyclopoids

We developed a conceptual model (Fig. 14.1) representing the relative influence of morphometry and trophic state, as well as fish and invertebrate predation on summer diapausing cyclopoids. We tested the hypothesis that spatial patterns in the composition and abundance of summer diapausing cyclopoids (SDC) in boreal lakes are a complex response depending primarily on morphometric and trophic features of lakes and secondly on the intensity of hypolimnetic anoxia and predation pressure from fish or invertebrates. Based on our conceptual model (Fig. 14.1), wind-exposed large and relatively deep lakes would not present strong thermal

Table 14.2 Abundance (D, density, and B, biomass) of meiobenthos (MB) and diapausing cyclopoid copepods (DCyc.), biomass of zooplankton (ZB) and active cyclopoid copepods (ZCycB), and abundance of *Chaoborus* (DCha, BCha) in the studied lakes. Dominance patterns in diapausing cyclopoids (dominant or codominant species in bold)

Group	Lakes	MB mg/m²	MD ind./m²	DCycB mg/m²	DCycD ind./m²	ZB mg/m³	ZCycB mg/m³	Dcha ind./m²	BCha, mg/m²	SDC species dominance
A	Achigan	230	3980	24	330	544	28	0	0	***M. edax***
A	Brome	4650	37200	447	15100	346	62	0	0	***M. edax, D. thomasi***
A	Echo	2400	15800	45	2000	510	33	0	0	***D. thomasi***, *A. vernalis*
A	Lovering	900	3250	22	1250	450	123	100	24	***D. thomasi***
A	Magog	1840	44000	493	34500	1314	182	0	0	***D. nanus***, *D. thomasi*
A	Waterloo	759	18300	280	8500	1263	112	24	3	***M. edax***, *A. vernalis, D. thomasi*
B	Beaulac	5866	297800	5087	287200	1060	197	0	0	***D. thomasi***, *A. vernalis*
B	Connelly	2100	101750	1720	96800	496	80	48	100	***D. thomasi***, *A. vernalis*
B	D'Argent	2847	102850	1474	79235	1109	164	88	41	***D. thomasi***, *M. edax*
B	Desmarais	710	10625	178	7100	1440	91	1749	4713	***D. thomasi***
B	Marois	3150	9150	14	450	815	87	24	27	***D. thomasi, C. scutifer***
B	Ouimet	2209	95025	1447	74250	683	89	0	0	***D. thomasi***, *C. scutifer, A. vernalis*
B	Patrick	692	2700	319	17400	390	43	25	107	***D. thomasi***, *A. vernalis*
B	Cristal	336	14900	289	13500	826	78	0	0	***D. thomasi, C. scutifer***
B	Croche	5021	144000	4474	132000	681	28	408	2092	***C. scutifer***

(continued)

Table 14.2 (continued)

Group	Lakes	MB mg/m^2	MD ind./m^2	DCycB mg/m^2	DCycD ind./m^2	ZB mg/m^3	ZCycB mg/m^3	Dcha ind./m^2	BCha, mg/m^2	SDC species dominance
B	Cromwell	46	1810	16	820	716	123	245	1263	***D. thomasi***, *C. scutifer*
B	Desjardins	329	2600	0	0	603	69	0	0	
C	Geai	6	300	5	200	157	10	425	7425	***C. scutifer, D. thomasi***
C	Pin Rouge	371	10040	281	7333	821	98	64	107	***C. scutifer***, *D. thomasi*
C	Tracy	6846	1032	122	6800	608	10	1175	8061	***D. thomasi***, *A. vernalis*
C	Triton	105	1500	0	0	285	90	0	0	
C	Brassard	96	1180	8	333	357	289	1050	20615	***D. thomasi***

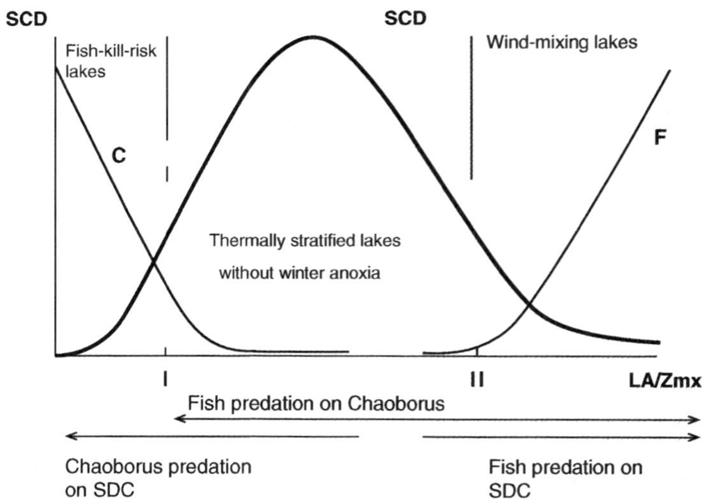

Fig. 14.1 A hypothetic model relating diapausing patterns of cyclopoid copepods (SDC) to morphometric features (Ko = wind-mixing index), and fish (F) and *Chaoborus* (C) predation in boreal lakes. Ko = LA/Zmx – I: Ko < 1; I–II: 1 < Ko < 20; II: Ko > 20

stratification and anoxia in deep waters during summer. Consequently, fish (F) can inhabit deep waters and easily prey on both invertebrate predators and summer diapausing cyclopoids (SDC) in deep waters. In this situation, summer diapause has no adaptive value, and the abundance of SDC should be low. In contrast, in wind-protected, small, and relatively deep lakes with strong thermal stratification and some potential for meromixis, SDC may find a refuge against fish predation in deep anoxic waters and sediments. However, they may suffer important predation pressure by voracious invertebrate predators such as chaoborids (C), which can stay in deep waters during the daytime, avoiding fish predation. For these lakes, SDC abundance should also be low. Lakes with intermediate morphometry combining permanent thermal stratification during summer with lower risk of anoxia in deep waters, and moderate predation by coexisting fish and small chaoborids, should be the safest habitat for summer diapausing cyclopoids. SDC abundances should therefore be the highest in these lakes but might also depend on lake trophic status, measured by changes in total phosphorus and chlorophyll a concentrations. In more productive lakes where cyclopoids account for the majority of crustacean zooplankton and where fish predation is limited by hypolimnetic anoxia, we expect to observe higher abundances of SDC. In less productive lakes with lower risk of anoxia in deep waters and higher predation both by fish and invertebrates, SDC abundances should be lower. According to this conceptual model, optimal survival conditions for dormant cyclopoids during summer are created in permanently stratified lakes with low anoxia in deep waters but might depend on fish-chaoborid predatory interactions.

To estimate the effects of wind-mixing, summer thermal stratification, and potential meromixis on diapausing patterns of cyclopoids, we calculated two indices based on morphometric variables:

(1) *The wind-mixing index (Ko) estimated by the ratio of lake area (ha) to maximum depth (m), Ko = LA/Zmax*. Lakes with Ko > 20 are well-mixed during summer and do not stratify. In contrast, lakes with Ko < 1 do not mix during summer, are strongly dimictic, and may have some potential for meromixis. Lakes with intermediate values of Ko (1–20) are generally dimictic with permanent thermal stratification during summer.
(2) *The index of potential meromixis (Pm) estimated by the ratio of maximum depth (m) to lake area*$^{0.25}$ *(m*^{2}*), Pm = Zmax/LA*$^{0.25}$. In meromictic lakes, Pm > 1, summer thermal stratification is very strong, and deep waters are generally anoxic (Berger 1971; Nürnberg 1988, 1995). Lakes with Pm < 1 have no potential for meromixis.

Based on the morphometric index (Ko), the study lakes were assigned to three different groups (A, Ko > 20; B, 1 < Ko < 20; C, Ko < 1) which differ by the strength of wind-mixing or thermal stratification during summer, the potential for meromixis, and the risk of predation by fish and/or chaoborids (see groups of lakes in Table 14.1). Among-groups differences in morphometric and trophic features are illustrated in Fig. 14.2.

The A group is composed of six large (mean LA = 648 ha) and relatively deep (mean Zmax = 16 m) lakes subjected to high wind-mixing (Ko > 20; mean Ko = 44) without any potential for meromixis (Pm < 1; mean Pm = 0.31). Summer thermal

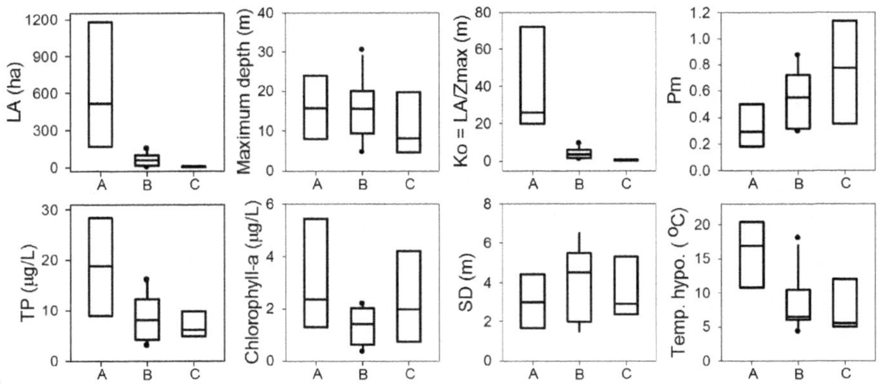

Fig. 14.2 Morphometric and trophic conditions in the different groups of lakes based on the wind-mixing index Ko. LA: lake area (ha); Ko = wind-mixing index (m); Pm = potential of meromixis; TP: total phosphorus (µg/L); SD Secchi transparency (m) (Group A: Ko > 20; Group B: 1 < Ko < 20; Group C: Ko < 1). The boundaries of the box are the 25th and 75th percentiles; the line within the box is the median, and the whiskers above and below the box are outliers.

stratification is generally weak; deep waters are warm (T hypo > 8 °C, mean T hypo = 16 °C) and well-oxygenated (8–12 mg/L^{-1}). These lakes are generally meso-eutrophic (mean TP = 19 µg/L^{-1}, mean Chl. = 3.2 µg/L^{-1}, except for Lake Achigan) and moderately transparent (Secchi depth = 3 m). They include the most productive lakes (TP > 22 µg.L^{-1}) in the Eastern Townships region (Brome, Magog, Waterloo). These lakes are all inhabited by fish (Masson et al. 2004) and largely free of chaoborids, except for Lovering and Waterloo where they are present in low numbers. Benthivorous fish are the main predators of dormant cyclopoid stages in these lakes.

The B group includes 11 dimictic lakes of moderate area (mean LA = 64 ha) and relatively deep (mean Zmax = 15 m), with an intermediate wind-mixing index (Ko, 1–10; mean Ko = 4). They have permanent thermal stratification during summer and no potential for meromixis (mean Pm = 0.56). They are moderately transparent (Secchi depth = 4 m) and oligo-mesotrophic with mean concentrations of total phosphorus (TP) and chlorophyll *a* (Chl.) of 8 and 1.3 µg.L^{-1}, respectively. Hypolimnetic waters are cooler (T hypo 4–8 °C) and depleted in oxygen (< 4 mg/L^{-1}), except in a very shallow lake (Desjardins). Most of the lakes in the B group are in the Laurentian region and contain planktivorous fish (Masson et al. 2004) as well as small-sized chaoborids. In these lakes, large-sized chaoborids are eliminated by fish, which cannot also suppress dormant cyclopoid stages in deep anoxic waters.

The C group includes five small lakes of the Laurentian region (mean LA = 6 ha) of moderate depth (mean Z max = 11 m) with a very low wind-mixing index (Ko < 1; mean Ko = 0.54) and a moderate potential for meromixis (mean Pm = 0.87). Two very small and relatively deep lakes in this group (Geai and Tracy) show strong thermal stratification and exhibit a higher potential for meromixis (Pm = 0.82 and 1.45). The C group lakes are oligo-mesotrophic with low concentrations of TP and Chl. (means, 7 and 2 µg.L^{-1}, respectively). Lakes Geai, Tracy, and Brassard are either fishless or with low numbers of fish (Masson et al. 2004) and inhabited by substantial populations of chaoborids, including the large-sized *C. americanus* (Masson and Pinel-Alloul 1998). All deep and strongly stratified lakes of the C group are cold (T hypo < 6 °C) and anoxic (DO < 1.5 mg/L^{-1}) in the hypolimnion. One exception is a wind-protected shallow lake (Triton) that due to lack of summer stratification is warm, well-oxygenated, rich in small planktivorous fish, and free of chaoborids.

To determine the most important abiotic and biotic environmental gradients in the study lakes, we performed a principal component analysis (PCA) using the statistical software R (R Development Core Team 2016, Austria) following Legendre and Legendre (2012). PCA analyses were done with the FactoMineR library (Husson et al. 2011). PCA analysis explained 67% of the total variation in limnological features and distinguished the three groups of lakes along the two first axes of the ordination (Fig. 14.3). Axis 1 (43%) reflected the abiotic gradients: (i) higher wind-mixing (Ko), lake's size (LA), and trophy (TP) in shallow meso-eutrophic lakes on the positive side and (ii) higher transparency (SD) and potential for meromixis (Pm) in deep oligotrophic and dystrophic lakes on the negative side. Axis 1 separated group A from groups B and C, distinguishing the largest and most productive lakes in the Eastern Township region (Magog, Brome, Waterloo) from the deepest (Achigan, Patrick), clearest (Beaulac, Cristal, Ouimet), and potentially meromictic (Tracy) lakes of the Laurentian region. Axis 2 (24%) reflected the biotic gradient of predation by the chaoborids (Dcha, Bcha) and discriminated lakes with planktivorous fish and no or few chaoborids (positive side: most of the lakes in the group A and some lakes in the group B) from lakes with a dominance of invertebrate predators (*Chaoborus*) (negative side: Brassard, Desmarais, Geai, Tracy, Cromwell from groups B and C).

14.4 Spatial Variation in Composition and Abundance of Diapausing Cyclopoids

All freshwater cyclopoid taxa known to have dormancy belong to the Cyclopidae family, which are represented mainly as non-encysted and encysted copepodites (Dahms 1995). In Quebec lakes, diapausing stages were mainly produced by three species (Table 14.2): *Diacyclops thomasi* that dominated in 59% of samples and occurred in all lakes groups; *Cyclops scutifer* that dominated in 29% of samples mainly in groups B and C lakes; and *Mesocyclops edax* that dominated in 9% of samples but only in lakes from group A. The rare species *Diacyclops nanus* accounted for 88% of diapausing cyclopoids (at copepodite stage V) in Lake Magog. *Acanthocyclops vernalis* was also found in low abundance in some lakes.

Diapausing cyclopoids were found in abundance in the meiobenthos during both summer and fall (before water turnover) in most of the lakes, except for two small and shallow lakes (Desjardins and Triton). The abundance of meiobenthos and summer diapausing cyclopoids (SDC) were highly variable among lakes and groups (Table 14.2, Fig. 14.4). Meiobenthos density and biomass ranged, respectively,

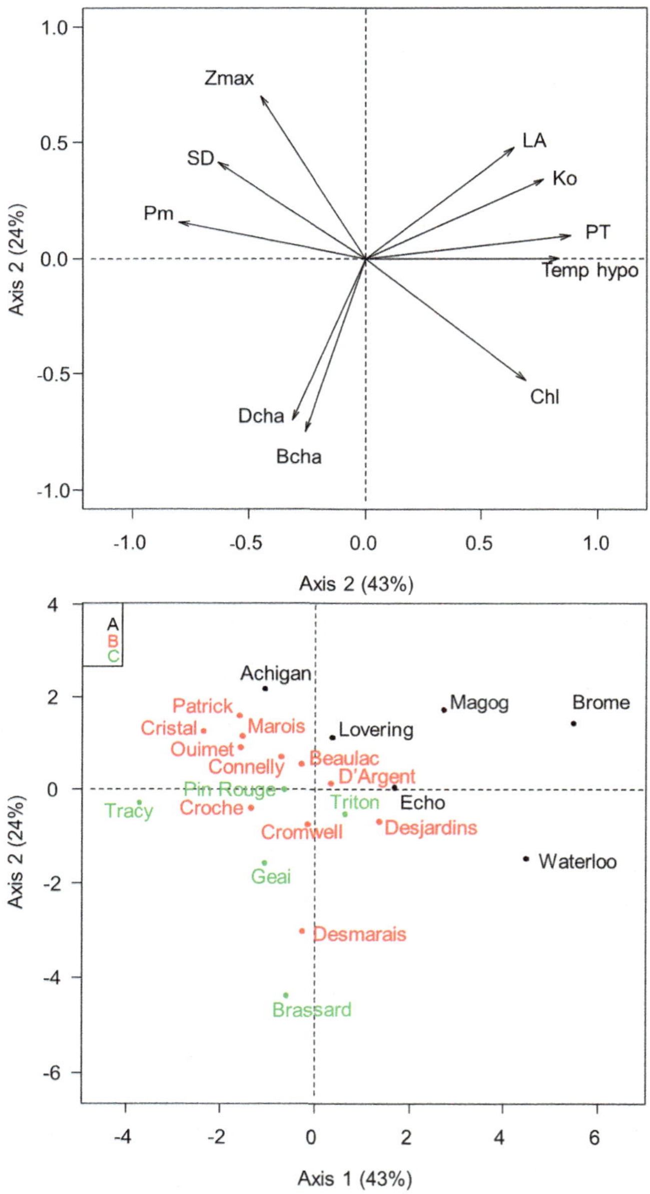

Fig. 14.3 PCA analysis of the environmental variables (upper graph) and ordination of the studied lakes in each group (bottom graph) based on limnological features. Group A: black; Group B: red; Group C: green

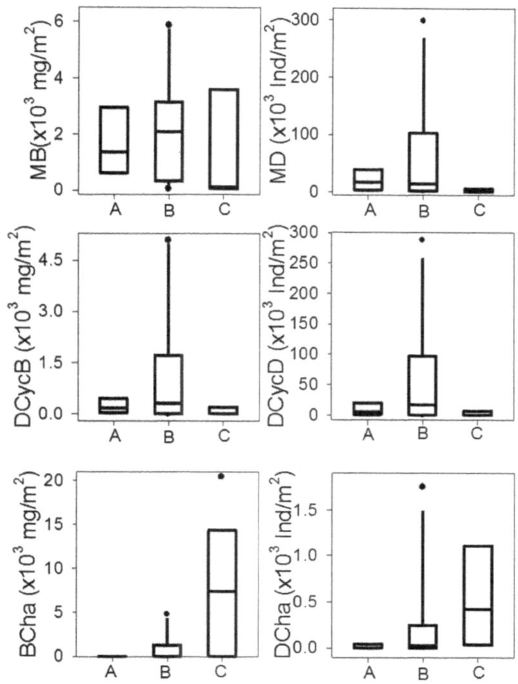

Fig. 14.4 Abundance (biomass and density) of meiobenthos (MB, MD), diapausing cyclopoid copepods (DCycB, DCycD), and *Chaoborus* (Bcha, Dcha) in the different groups of lakes based on the wind-mixing index Ko. (Group A: Ko > 20; Group B: 1 < Ko < 20; Group C: Ko < 1) The boundaries of the box are the 25th and 75th percentiles; the line within the box is the median, and the whiskers above and below the box are outliers.

from 300 to 297800 ind./m^2 and 6 to 5866 mg/m^2 among lakes. Comparatively, summer diapausing cyclopoids (SDC) ranged from 200 to 287200 ind./m^2 in density and 5–5087 mg/m^2 in biomass (not considering Triton and Desjardins). For both between and within each group of lakes, there was a strong relationship ($R2 = 0.9$; $p < 0.001$) between SDC and meiobenthos density, indicating that summer diapausing cyclopoids were an important component of the meiobenthos in most lakes. The relationship between SDC and meiobenthos biomass was weaker ($R2 = 0.60$; $p < 0.01$) because in several of the lakes in group A (Brome, Echo) without thermal stratification, other large invertebrates (Pisidiidae bivalves, ostracods, and worms) produced the bulk of the meiobenthos. In comparison to meiobenthos and SDC, zooplankton biomass and active cyclopoids in the water column showed less variation among lakes and groups (Table 14.2). Zooplankton biomass varied from 157 to 1440 mg/m^3 and active cyclopoids biomass from 10 to 289 mg/m^3. However, the abundance of the invertebrate predator *Chaoborus* was highly variable among lakes and groups (Table 14.2, Fig. 14.4). The density of chaoborids varied from 0 to 1749 ind./m^2 and their biomass from 0 to 20614 mg/m^2. In the A group, the majority of lakes were free from *Chaoborus*, except for few individuals in Lakes Lovering and Waterloo. The highest abundances of *Chaoborus* were found in lakes of the C group (mean density and biomass, 543 ind./m^2 and 7241 mg/m^2), especially in three lakes that were potentially meromictic and fishless or devoid of planktivorous fish (Geai, Tracy, and Brassard). Both the density and biomass of chaoborids were lower in lakes of the B group (mean density and biomass, 235 ind./m^2 and 758 mg/m^2). Significant chaoborid populations were only found in three lakes (Desmarais,

Croche, and Cromwell). Finally, fish communities also differed among lake groups. Masson et al. (2004) listed fish species found in our study sites and defined three categories: (1) lakes without piscivorous fish and with only one planktivorous species (Lakes Cromwell and Desjardins in group B; Lakes Brassard and Triton in group C); (2) lakes with only 1–2 piscivorous fish and several (2–8) planktivorous species including cyprinids, yellow perch, and white fish (mainly from group B); and (3) lakes with more than 3 piscivorous fish (3–8) and several (5–9) planktivorous species (all lakes in group A).

Meiobenthos and diapausing cyclopoids presented similar patterns of variation among lake groups (Fig. 14.4). The mean abundance of meiobenthos in the sediments of lakes of A, B, and C groups was 20421, 71200, and 2810 ind./m^2 in density and 1796, 2118, and 1484 mg/m^2 in biomass. In comparison, the mean abundance of summer diapausing cyclopoids (SDC) in the sediments of lakes of A, B, and C groups was 10280, 64432, and 2933 ind./m^2 in density and 218, 1365, and 83 mg/m^2 in biomass. In Quebec lakes, SDC abundances reached similar values as reported in lakes and ponds of Northern European lakes (up to 10^6 ind./m^{-2}) (Elgmork et al. 1990; Fryer 1996; Rivier 1996; Frisch 2002; Seebens et al. 2009). As stated in our conceptual model (Fig. 14.1), the highest abundances of meiobenthos and SDC occurred in the B group of lakes with moderate wind-mixing ($1 < Ko < 10$), intermediate abundances were observed in the A group of lakes with high wind-mixing index ($Ko > 20$), and the lowest abundances were found in the C group of lakes with low wind-mixing index ($Ko < 1$).

In lakes from group B with the highest density and biomass of summer diapausing cyclopoids, *Diacyclops thomasi* was the most abundant species, coexisting with *Cyclops scutifer* and *Acanthocyclops vernalis*. *D. thomasi* was dominant in four lakes (Beaulac, Connelly, d'Argent, Desmarais, Ouimet). This species reached density and biomass as high as 287000 ind./m^2 and 5000 mg/m^2 in July 2000 in the most transparent Lake Beaulac (SD = 6 m) (Table 14.2). Lower abundance (820 ind./m^{-2}, 16 mg/m^{-2}) was recorded in Lake Cromwell, a dystrophic and coloured lake. This cold-water species is known to be the only cyclopoid producing a protective cyst during summer diapause (Alekseev 2007). We found all of the four stages of the encysted diapausing copepodites of *D. thomasi* in the sediments of Lake Cromwell (Fig. 14.5). The cyst developed around the dormant copepodite stage IV accumulating in sediments; the cyst was formed from different particles attached, including bacteria, diatom frustules, and organic mud. The cyst could cover all of the body, increasing the size of encysted diapausing stages enough to avoid *Chaoborus* predation (Alekseev 2007). A masking effect might also be important, as it was often difficult to recognize dormant cysts of cyclopoids in sediments, with only the long setae of the furcal ramus standing out (see stage IV of the cyst in Fig. 14.5). Our observations of diapausing patterns of *D. thomasi* are consistent with previous studies. In a temporary pond in South Carolina, *Diacyclops* species showed a strong seasonality in dormancy patterns, emerging from diapause when photoperiod is short during fall, winter, or early spring, and not emerging during summer (Medland and Taylor 2001). High densities of encysted CIV copepodites of *D. thomasi* were

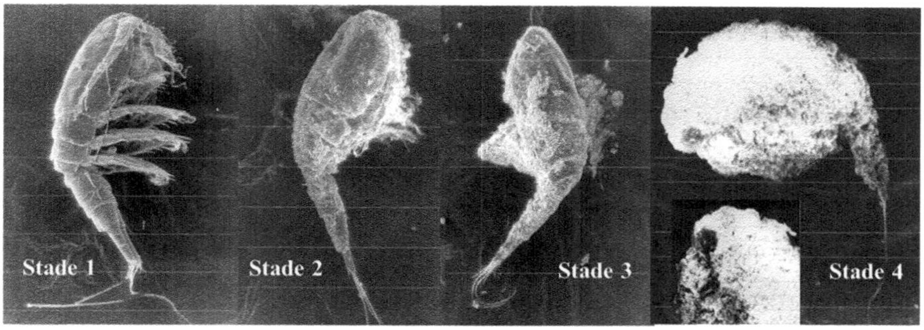

Fig. 14.5 Scanning electronic microscopic (SEM) photographs of cyst formation of diapausing stages IV of *Diacyclops thomasi* found in superficial sediment (0–5 cm) of Lake Cromwell

found in a nonmoving state and encased in a layer of mucilage covered with sediment particles at depths of 12 m in sediments of Lake Oneida (NY) (Hansen and Hairston 1998). This is similar to what we observed in Lake Cromwell. In our study, *C. scutifer* was found to be a dominant SDC only in Lake Croche (132000 ind./m^2 and 4500 mg/m^2), but never as encysted dormant stages (Table 14.2), as was reported for boreal lakes of south-central Norway (245000 ind./m^2) (Elgmork et al. 1990). *C. scutifer* coexisted with *D. thomasi* as a subdominant species in several lakes (Marois, Ouimet, Cristal, and Cromwell). Elgmork (2004) found striking similarity in life cycles of *C. scutifer* between lakes in Scandinavia and North America. In Northern European lakes, this common cold-water species exhibited complex and sometime multiyear life cycles, including summer diapause (Halvorsen and Elgmork 1976; Elgmork 1996) as in Quebec lakes. Diapause occurred in four copepodite instars (CII to CV) and at the adult stage, with CIV as the most common stage (Naess and Nilssen 1991). The induction phase was marked by an increased torpidity of cyclopoids entering diapause. The deepest torpor was observed during the summer refractory phase when a low metabolic rate enabled diapausing stages to survive anaerobiosis in deep waters and anoxic sediments (Elgmork and Nilssen 1978; Elgmork 1996). The activated phase started when half of the dormancy period was completed, and the termination phase was characterized by the emergence of diapausing stages from the sediments at spring, sometimes before ice-break. As *C. scutifer* diapause was induced at all photoperiods, this species probably used water temperature increases (> 12–15 °C) as a trigger for inducing diapause or had genetically determined obligate diapause during summer like for European populations of *Cyclops kolensis* (Rivier 1996; Alekseev 2007). *Acanthocyclops vernalis* also induced diapause at copepodite stage IV in very low abundance in several lakes of the B group (Beaulac, Connelly, Ouimet, Patrick, Cristal).

In the A group of lakes, the density and biomass of SDC in sediments were intermediate (Fig. 14.4, Table 14.2). These lakes were well-mixed, weakly stratified, and productive and contained large fish populations but few invertebrate predators. *Diacyclops bicuspidatus*, *Mesocyclops edax*, and *Acanthocyclops vernalis* were found in relatively low abundances in this group of lakes. *D. thomasi* and *A. vernalis*, two cold-water species, were present in Lake Echo in the Laurentian region and in Lake Lovering in the Eastern Township region. *M. edax*, known as a warm-water species, was observed only in sediments of three lakes (Achigan, Brome, and Waterloo) reaching the highest abundances in the two most productive lakes of the Eastern Townships region (Brome and Waterloo) with density of 15000 and 8500 ind./m^{-2} and biomass of 447 and 280 mg/m^{-2}, respectively. The lowest abundance was recorded in Lake Achigan (330 ind./m^{-2} and 24 mg/m^{-2}), an oligotrophic lake of the Laurentian region. However, as diapause in this species is induced by short-day photoperiod (Alekseev 2007), we do not consider *M. edax* as a cyclopoid with summer diapause. The high densities of *M. edax* found in October 1998 in Lakes Brome and Waterloo were probably the result of recent sedimentation of copepodites starting winter diapause. One exception is Lake Magog where we found high concentrations of the copepodite stages V of *Diacyclops nanus* in sediments in July 2000. This species normally inhabits fishless bogs and forest pools (Dussart and Defaye 1985). We hypothesize that the formation of an anoxic zone in the deep waters and sediments of the eutrophic Lake Magog has provided a refuge to *D. nanus* and enabled it to increase its density and become a dominant form in the meiobenthos in this lake.

In the C group of lakes, the density and biomass of summer diapausing cyclopoids (SDC) were the lowest and never exceeded 10000 ind./m^2 (Fig. 14.4, Table 14.2). The highest SDC abundances were recorded in the deepest lakes (Pin Rouge and Tracy). Lake Tracy, potentially meromictic (Pm = 1.45), may have provided a refuge to SDC in deep waters. Lake Pin Rouge was also strongly stratified, showing a sharp thermocline at 5 m, and totally oxygen-depleted in hypolimnetic waters, thus limiting the abundance of both fish and *Chaoborus* in deep waters. The lowest abundances of SDC were observed in Lakes Geai and Brassard supporting populations of large chaoborids (*C. americanus*). The small-bodied *Diacyclops thomasi* was commonly found in all lakes but was dominant in only two lakes (Tracy and Brassard), while the large-bodied *Cyclops scutifer* was the dominant SDC in two other lakes (Geai, Pin Rouge). *D. thomasi* was found in the highest density in Lake Tracy (6800 ind./m^2 and 122 mg/m^2) in July 2000 and *C. scutifer* in Lake Pin Rouge (7333 ind./m^2 and 281 mg/m^2) in October 1998 and July 2000; these lakes were the deepest lakes of the group C, strongly stratified, and potentially meromictic. Similarly, diapausing stages of *C. scutifer* were found in high density, up to 30000 ind./m^2, at depth corresponding to the chemocline of meromictic lakes of Norway and Poland in Northern Europe (Szlauer 1963; Elgmork 1973; Elgmork and Langeland 1980; Elgmork and Lie 1998). We did not found any diapausing copepods in Lake Triton, included in the C group because of its low Ko (0.27), because this small well-mixed lake contained planktivorous cyprinids which suppress both *Chaoborus* and cyclopoids.

14.5 Environmental Controls of Summer Diapause Patterns in Cyclopoids

To test our conceptual model and determine the relationships between the dormancy patterns of cyclopoids and the limnological features of the study lakes, we analysed the spatial variation in SDC abundance (density and biomass) in relation to their morphometry, trophy, and predation pressure by fish and *Chaoborus*.

With all lakes pooled together, no correlation was found between SDC abundances and morphometric or trophic indices. However, SDC abundances in lakes of A, B, and C groups were distributed along the wind-mixing gradient (Ko) in accordance with the predicted order of our conceptual model: SDC abundance in lakes of the group B > groups A and C (Fig. 14.6). However, the distribution of SDC in the lakes of the group B was bimodal which could be explained by changes in trophy level and SDC species dominance patterns among lakes.

14.5.1 Role of Morphometry and Wind-Mixing on SDC Patterns

The distribution of SDC density and biomass in sediments among lake groups supports our hypothesis that morphometric features play an important role in environmental control of diapausing patterns, according to the wind-mixing index (Ko) (Fig. 14.6). Plotting the biomass of SDC in relation to increasing Ko showed a bimodal distribution with two maxima among lakes of the B group with intermediate wind-mixing (1 < Ko < 10). The two ends of the distribution, with lower SDC abundance, corresponded to lakes with high wind-mixed index (Ko > 20, A group), inhabited by large populations of fish and generally devoid of *Chaoborus* at the right side, and to lakes with low wind-mixing index (Ko < 1, C group), potentially meromictic, with abundant populations of *Chaoborus* on the left side.

In the B group of lakes, the average abundance of SDC in sediments was about 6 to 16 times higher than in A and C groups, respectively (Fig. 14.4). This supports our hypothesis which predicts that the highest biomasses for SCD should be found in lakes with intermediate wind-mixing index. The SDC were very abundant in five lakes (Croche, Ouimet, Connelly, d'Argent, Beaulac) reaching high density (74000–287000 ind./m^{-2}) and biomass (1400–5000 mg/m^{-2}), intermediate in three lakes (Patrick, Cristal, Desmarais, 7000–17000 ind./m^{-2}, 100–300 mg/m^{-2}), and low in three other lakes (Cromwell, Marois, Desjardins, 0–820 ind./m^{-2}, 0–16 mg/m^{-2}) (Fig. 14.6, Table 14.2). The reasons for the bimodal SDC abundances in the B group are unclear. Moreover, two different species were responsible for the two maxima. The first mode was produced by *Cyclops scutifer* in Lake Croche and *Diacyclops thomasi* in Lake Ouimet. The second mode was formed by *D. thomasi* in three lakes (Beaulac, Connelly, d'Argent). This discrepancy can be partly explained by the gap in our data, as we did not survey lakes in the upper intermediate range of Ko (10–

20). It is possible that high accumulation of SDC would be found in lakes with higher Ko around 15. Differences in trophic states among lakes could be another explanation for the bimodal distribution in SDC in lakes from group B. Indeed, the two modes corresponded to oligotrophic lakes (Croche, Ouimet; TP, 3–4 μg/L^{-1}) in one hand and to mesotrophic lakes (Beaulac, d'Argent, Connelly; TP, 8–16 μg/L^{-1}) in the other hand. However, the highest abundances of SDC in each mode were produced in two lakes (Croche, Beaulac) only and by two different species (*Cyclops scutifer*, *Diacyclops thomasi*).

In the A group of lakes, the higher biomasses (15100 and 34500 ind./m^{-2}, 447 and 493 mg/m^{-2}) in Lakes Brome and Magog were produced by diapausing copepodites of *Mesocyclops edax* which is a warm-water species entering in winter diapause in fall (Fig. 14.6, Table 14.2). As SDC of *Diacyclops thomasi* were found only in low abundance in some lakes of the group A (Brome, Echo, Lovering, < 12000 ind./m^{-2} and 234 mg/m^{-2}), this group of lakes does not provide suitable habitats for summer diapausing cyclopoids, probably due to foraging of benthivorous fish on SDC in warm and oxygenated deep waters.

Only the two deepest lakes (Pin Rouge, Tracy) of the C group supported abundant SDC (> 6800 ind.m^{-2} and 122 mg/m^{-2}) because they may offer a refuge for SDC in deep and anoxic waters devoid of fish, although SDC can be affected by different predation pressures of *Chaoborus* species. *D. thomasi* was the dominant SDC in Lake Tracy, whereas *D. thomasi* and *C. scutifer* were codominant species in Lake Pin Rouge. In this group, only the deeper lakes which were potentially meromictic and anoxic in the hypolimnion were suitable habitats for summer diapausing cyclopoids, as reported in Northern European lakes (Elgmork and Lie 1998).

14.5.2 Role of Trophy and Hypolimnetic Anoxia on SCD Patterns

Although morphometric features and wind-mixing are important factors influencing among-lake patterns of summer diapause in cyclopoids, there are other potential sources of variation affecting SCD patterns in Quebec lakes. One potential factor is the lake trophic status which is related to hypolimnetic anoxia, changes in water quality (TP, water transparency), and productivity (chlorophyll biomass). The presence of hypolimnetic anoxia depends on lake depth, external nutrient loading from catchments, and total phosphorus concentration (Reckhow 1977; Charlton 1980). Increasing algal production and subsequent decomposition of organic matter in the water and sediments lead to the formation of anoxic zones in the hypolimnion. These zones can offer temporary refuges to diapausing copepods such as *M. edax* and *D. nanus* even in relatively shallow lakes from the A group with high Ko, for example, Lake Brome (Ko = 118) and Lake Magog (Ko = 57). Increasing eutrophication in these lakes due to higher productivity and subsequent anoxia should favour the presence of SDC and increase their density (> 15000 ind.m^2) and biomass (>

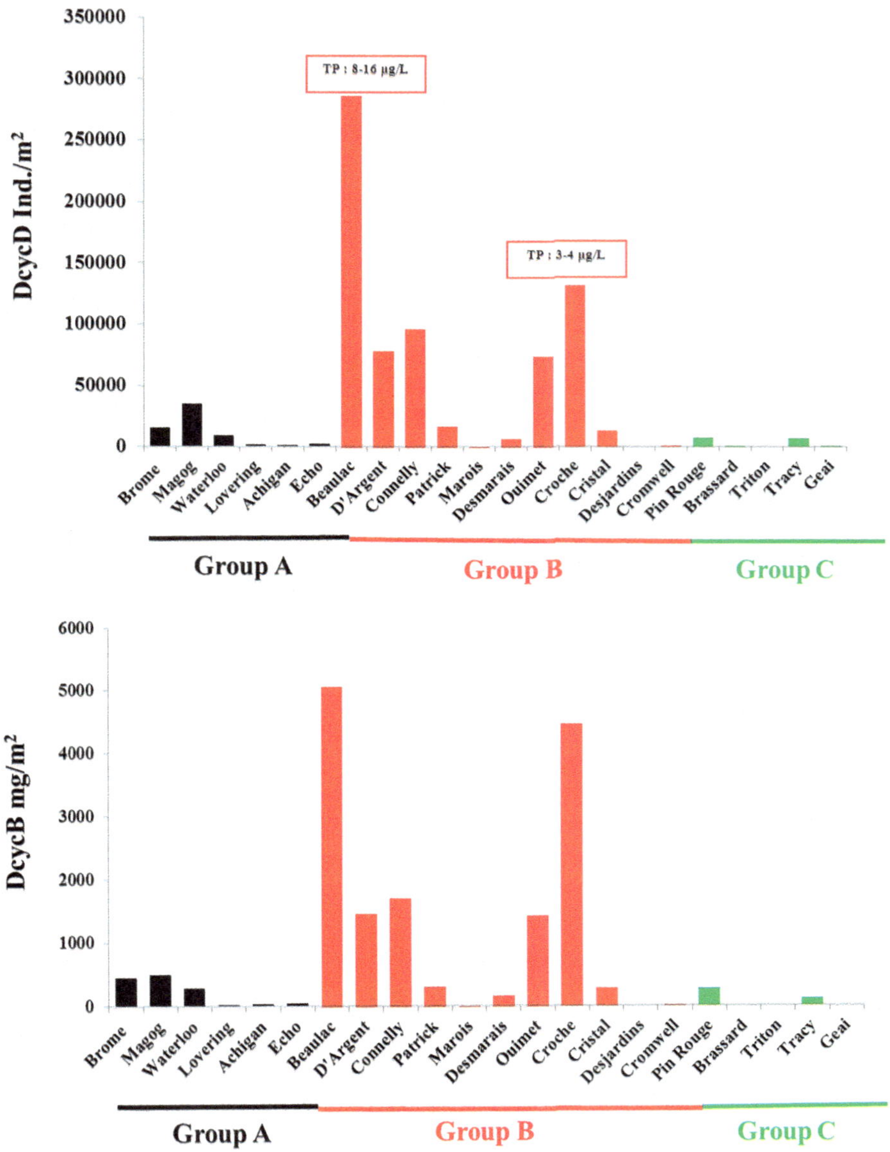

Fig. 14.6 Distribution of SDC abundance (density and biomass) in the studied lakes of each group according to increasing wind-mixing index (Ko). Difference in trophic conditions (TP) of lakes in Group B are reported. (Group C: Ko < 1, green; Group B: 1 < Ko < 10, red; Group A: Ko > 20, black)

400 mg/m²) like some lakes of the B group (Fig. 14.6). In the case of Lake Magog, fertilization of agricultural lands in the catchment has led to high levels of total phosphorus (43 $\mu g/L^{-1}$) in waters (Nürnberg 1988). In this lake, we found anoxia in waters above sediments in July and a high biomass of diapausing copepodites CV of a rare species *Diacyclops nanus*. This species normally inhabits fishless bogs and forest pools and is rarely found in lakes (Dussart and Defaye 1985). We hypothesize that the anoxic zone has provided this species a refuge in Lake Magog so that it has become the dominant meiobenthos species.

In the B group of deep lakes with relatively low wind-mixing index (Ko = 1–10 for our lakes), we expected to find increasing dominance of cyclopoid copepods (and consequently higher abundances of SDC) in the zooplankton community with increasing trophic level following a latitudinal gradient (Pinto-Coelho et al. 2005). Our data did not support a monotonic relationship between SDC abundance and trophic indices (TP and Chla), although the bimodal distribution of SDC for the lakes of group B indicated higher abundance of dormant stages of *D. thomasi* in the mesotrophic lakes (Beaulac, Connelly, d'Argent: TP, 8–16 $\mu g/L^{-1}$) compared to the oligotrophic lakes (Ouimet, Marois, Desmarais, Cristal: TP, 3–4 $\mu g/L^{-1}$). This trophic effect was mainly produced by higher abundances in Lake Beaulac only. Moreover, it was not observed in two other lakes (Cromwell, Patrick: TP, 10–12 $\mu g/LL^{-1}$). In these lakes, an interaction with hypolimnetic anoxia could have negatively affected the SDC and decreased their abundance due to increasing predation from *Chaoborus* larvae both on active and diapausing cyclopoid copepods. This interaction is supported by a decrease in the abundance of SDC in lakes with higher *Chaoborus* densities from the B group. For instance, Lake Desmarais and Lake Cromwell had dense populations of *Chaoborus* and low SDC biomasses. Therefore, the effect of trophy was not supported by this study as it would require a larger set of lakes to take into consideration all of the interactions with other abiotic (hypolimnetic anoxia) and biotic (chaoborid predation) factors.

14.5.3 Role of Fish and Invertebrate Predation on SCD Patterns

Invertebrate predation by *Chaoborus* is considered a strong biotic factor controlling the SDC abundance in our conceptual model (Fig. 14.1). Indeed, *Chaoborus* is well known as an active predator that reacts to prey movement and can easily suppress both the active planktonic cyclopoids and the diapausing cyclopoids, as well as most other invertebrates in deep waters and sediments (Kerfoot and Sih 1987; Hanazato and Yasuno 1989; Arnott and Vanni 1993; Pinel-Alloul 1995a, b). *Chaoborus* may recognize diapausing cyclopoids as prey in the anoxic zone of sediments due to their different torpidity during diapause and their reaction to tactile contact (Elgmork 1996). Daily feeding rates of large *Chaoborus* larvae could exceed 5% of their body weight, and they can eat 1–59 copepods per day in experimental tests (Anderson and Raasveldt 1974). To test the effect of *Chaoborus* predation in our lakes, we performed laboratory experiments and observed active predation of *Chaoborus* larvae on SDC of *C. scutifer*. We exposed 40–50 copepodites (CIII–CV) of *C. scutifer* during 2–3 days in plastic dishes filled with lake water in a fridge (4–6 °C, full darkness) in absence or presence of instars 3–4 of *Chaoborus* larvae. No mortality was observed in the absence of *Chaoborus*, while all copepodite stages were killed and eaten in 2–3 days in presence of *Chaoborus*. Important rate of predation (10–20 SDC per 1–2 *Chaoborus*) was also observed in the lake where we collected deep anoxic waters with SDC and exposed them for 24 h to *Chaoborus* larvae in 1 L bottles.

These results suggest a negative effect of *Chaoborus* on the abundance of diapausing cyclopoids in Quebec lakes. When present, *Chaoborus* larvae were the only invertebrate predator in sediment samples with SDC stages. No other large invertebrates were found in our lakes, including no amphipods or mysids, or planktonic predators such as *Epischura lacustris* or *Leptodora kindtii*. In two lakes from the C group, the biomass of diapausing cyclopoids of *D. thomasi*, which was encysted in sediments and less vulnerable to *Chaoborus* predation, was very low. This lake was without planktivorous fish (Geai and Brassard) and inhabited by large *Chaoborus* species (*C. americanus* in Lake Geai; Masson and Pinel-Alloul 1998; Le Jeune et al. 2012). This could be explained by previous predation of *Chaoborus* on copepodite stages in the water column before *D. thomasi* switched to summer diapause and produced protective cysts. When copepodite stages IV develop diapause, they reduce their metabolic processes, moving speed, and escape reactions, making them highly vulnerable to predation (Alekseev 1998). In contrast, in Lake Pin Rouge which was inhabited by planktivorous fish and with a low abundance of *Chaoborus* larvae, the diapausing stages of the large cyclopoid *C. scutifer* were abundant. Lake Tracy was a unique site, as it was a deep lake with some potential for meromixis, where we found high abundance of *D. thomasi* diapausing stages which may have survived as cysts in the anoxic deep waters unsuitable for *Chaoborus* larvae. Three lakes of the C group (Geai, Pin Rouge, and Tracy) presented some potential for intermittent meromixis as observed in Finland in small dark-coloured humic forest lakes (Hakala 2004) and might serve as refuge for SDC dormancy. Additionally, there were the trends found in lakes of B group with anoxic zones. The highest abundances of diapausing stages of *D. thomasi* were found in the sediments of Lake Beaulac without *Chaoborus* and in Lakes Connelly, d'Argent, and Ouimet with low biomass of *Chaoborus*. Lake Croche was a unique situation where *C. scutifer* diapausing stages were found in large density in presence of a small *Chaoborus* larvae (*C. punctipennis*), which migrate to deep anoxic zones and limit feeding rates during the daytime (Le Jeune et al. 2012). About 40% of the water column in this lake (especially below 5 m) became anoxic or depleted of oxygen ($\leq$ 2 mg/L^{-1}) during the summer. A seasonal survey carried out in Lake Croche from mid-June to mid-August showed a decrease in the abundance of diapausing stages of live copepodites from *C. scutifer* from 60000 to 9000 ind./m^2. Most of the SDC were found to be predated upon by *Chaoborus*. Finally, for the large and well-mixed lakes from the A group, fish predation was very effective in suppressing both invertebrate predators and active cyclopoid copepods. Therefore, invertebrate predation was not acting on diapausing cyclopoids in these large and well-mixed lakes, where summer dormancy was an inefficient strategy for the survival of the cyclopoids.

14.6 Conclusions

In this chapter, we proposed a model describing the abiotic and biotic controls on the diapause patterns of cyclopoid copepods in southern Quebec boreal lakes (Fig. 14.1). Our results support the hypothesis that morphometric features (lake size and wind-mixing) and predation from invertebrates (*Chaoborus*) and fish significantly affect diapausing patterns of cyclopoid copepods but that trophic effect does not show a clear effect on diapausing patterns of cyclopoids.

Morphometric features related to lake size (area and depth), wind-mixing effect (Ko), thermal stratification, and hypolimnetic anoxia were key abiotic factors regulating the accumulation of diapausing cyclopoids in lake sediments during summer and fall. In wind-mixed lakes (A group: right side in the model), with weak stratification and low anoxia in deep waters, there was low biomass and density of diapausing cyclopoids. Dormancy patterns were inhibited by abiotic morphological factors, mainly the wind-mixing effect and the lack of thermal stratification, which did not allow SDC to find a refuge in deep waters. Secondly, intense predation pressure by fish both on invertebrate predators (such as *Chaoborus*) and active cyclopoid copepods was the main biotic factors suppressing SDC and making dormancy an ineffective life strategy.

In very small, deep lakes with strong hypolimnetic anoxia and potential meromixis (C group: left side in the model), the abundance of SDC was found to be the lowest, even though deep hypolimnion waters could be a refuge for SDC. This is likely due to important predation pressure by *Chaoborus*, which was the main biotic driver controlling diapausing cyclopoids in these stratified anoxic lakes. Prey-predator interactions between *Chaoborus* species and diapausing cyclopoids were the main mechanism governing SDC diapause in this C group of lakes without planktivorous fish. When a large species such as *Chaoborus americanus* was present, the biomass of encysted stages of diapausing cyclopoids of *D. thomasi* was very low, likely because active stages were predated upon before entering dormancy. However, moderate abundance of SDC could be found in lakes offering strong potential for meromixis or in lakes with small species of *Chaoborus* (*C. punctipennis*, *C. trivittatus*, *C. flavicans*).

Finally, the highest biomass and density of diapausing copepods during summer and fall was found in small, deep lakes (B group: middle of the model) with intermediate wind-mixing, permanent thermal stratification, and moderate anoxia in deep waters. These lakes provided the optimal conditions for summer diapausing cyclopoids (SDC). However, the success of dormancy varied between lakes of varying trophic status, as well as between cyclopoid species. On average, SDC density was lower in oligotrophic lakes than in mesotrophic lakes; however there was no clear response to trophic features because of confounding changes in SDC species composition. In wind-mixed lakes (B group), high biomass of the diapausing stages of *Diacyclops thomasi* was found in mesotrophic lakes, while high biomass of diapausing stages of *Cyclops scutifer* was found in oligotrophic and humic lakes. Biotic factors related to the relative influence of invertebrate (*Chaoborus*) and fish predation were also important drivers of SDC success in the B group. In these stratified lakes with an anoxic hypolimnion, the highest abundances of SDC were found in lakes without *Chaoborus* or with low abundance of this predator.

In conclusion, modelling and field validation of the dormancy patterns of cyclopoid copepods in Quebec lakes have revealed that this life strategy varied greatly among boreal lakes, according to changes in SDC dominance, composition, and abundance between lakes. Cyclopoid dormancy is a complex response to environmental heterogeneity in morphometry, wind-mixing, lake size and depth, trophy, and hypolimnetic anoxia, as well as fish and invertebrate predation. However, the interactions between abiotic and biotic drivers of dormancy patterns remain difficult to disentangle, and further investigation into these interactions across morphometric and trophic limnological gradients is needed.

Acknowledgements This chapter is a contribution resulting from research collaborations between the Russian Academy of Sciences and the Ministry of Education of Quebec, Canada. Financial support for this first research on diapausing cyclopoid copepods in boreal lakes of Quebec was provided through grants from the Natural Sciences and Engineering Research Council of Canada (NSERC Discovery) and the Fond de Recherche du Québec (FRQ – Regroupement stratégique GRIL: Groupe de Recherche Interuniversitaire en Limnologie et Environnement Aquatique) to Bernadette Pinel-Alloul and from the Russian Foundation for Basic Researches to Victor R. Alekseev. Victor R. Alekseev was awarded by two grants "Excellence in Science" from the Ministry of Education, Quebec, Montreal, in 1998 and 2000. This chapter was written with the partial support from RFBR (grant 17-04-00027). The study was partly supported by the Russian Academy of Science, topics 65.4 and 65.5. Our thanks go to Ginette Méthot and Pierre Marcoux for helpful assistance during field sampling, to Louise Pelletier for SEM photos of encysted diapausing cyclopoids, to Louise Cloutier for valuable assistance in meiobenthos taxa identification, to David Lévesque and El-Amine Mimouni for statistical analysis and figure preparation, and to Antonia Cattaneo and Gwyneth McMillan for helpful comments and English editing on a previous version of the manuscript.

References

Alekseev VR (1991) Diapause in crustaceans: ecological and physiological aspects. Academia Publications "Nauka", Moscow, 143 [In Russian]

Alekseev VR (1998) Biochemical and physiological characteristics of crustaceans in diapause: the internal mechanism of reactivation. Archiv für Hydrobiologie 52:463–476

Alekseev VR (2007) Chapter 3: Diapause in crustaceans: peculiarities of induction. In: Diapause in aquatic invertebrates. Monographiae Biologicae 84. Springer, pp 29–63

Alekseev VR, Starobogatov YI (1996) Types of diapause in Crustacea: definitions, distribution, evolution. Hydrobiologia 320:15–26

Alekseev VR, Guissani G, Ravera O, Riccardi N (eds) (2004) Diapause in aquatic invertebrates. J Limnol 63(Suppl 1):97p

Alekseev VR, De Stasio BT, Gilbert J (2007) Diapause in Aquatic Invertebrates. Monographiae Biologicae 84. Springer, p 257

Anderson RS, Raasveldt LG (1974) Gammarus predation and the possible effects of Gammarus and Chaoborus feeding on the zooplankton composition in some small lakes and ponds in western Canada. Canadian Wildlife Service occasional paper no. 18

Arnott SE, Vanni MJ (1993) Zooplankton assemblages in fishless bog lakes: influence of biotic and abiotic factors. Ecology 74:2361–2380

Balushkina EV, Winberg GG (1979) Relation between body mass and length in planktonic animals. In: Winberg GG (ed) General bases of research into water ecosystems – Obshchiye Osnovy Izucheniya Vodnykh Ekosistem. Nauka, Leningrad, pp 169–172

Berger F (1971) Zur Morphometrie der Caranthia II, Special Issue 31:29–39

Birge EA, Juday C (1908) A summer resting stage in development of Cyclops bicuspidatus Claus. Trans Wis Acad Sci Arts Lett 16:1–9

Boers JJ, Carter JCH (1978) The life history of Cyclops scutifer Sars (Copepoda, Cyclopoida) in a small lake of the Matamek River System, Québec. Can J Zool 56:2603–2607

Boxshall GA, Defaye D (2008) Global diversity of copepods (Crustacea: Copepoda) in freshwater. Hydrobiologia 595:195–207

Brendonck L, De Meester L (2003) Egg banks in freshwater zooplankton: evolutionary and ecological archives in the sediment. Hydrobiologia 491:65–84

Brendonck L, De Meester L, Hairston N Jr (eds) (1998) Evolutionary and ecological aspects of crustacean diapause: proceedings of the Symposium "Diapause in the Crustacea – with Invited Contributions on Non-Crustacean Taxa" held in Gent, August 24–29, 1997. Archiv für Hydrobiologie, Special Issues: Advances in Limnology 52: 561p

Charlton MN (1980) Hypolimnion oxygen consumption in lakes: discussion of productivity and morphometry effects. Can J Fish Aquat Sci 37:1531–1539

Dahms H-U (1995) Dormancy in the copepods – an overview. Hydrobiologia 306:199–211

Dussart BH, Defaye D (1985) Répertoire mondial des Copépodes Cyclopoïdes. Centre National de la Recherche Scientifique (CNRS), Bordeaux/Paris. 236 pp

Elgmork K (1967) Ecological aspects of diapause in copepods. In: Proceedings of the Syrup. Crustacea. III. Marine Biological Association of India, symposium. series 2, pp 947–954

Elgmork K (1973) Bottom resting stages of planktonic cyclopoid copepods in meromictic lakes. Verhandlungen der Internationalen Vereinigung für Theoretische und Angewandte Limnologie 18:1474–1478

Elgmork K (1980) Evolutionary aspects of diapause in freshwater copepods. In: Kerfoot WC (ed) Evolution and ecology of zooplankton communities. The University Press of New England, Hanover/London, pp 411–417

Elgmork K (1981) Extraordinary prolongation on the life cycle in a freshwater planktonic copepod. Holarct Ecol 4:278–290

Elgmork K (1996) Variation in torpidity of diapause in freshwater cyclopoid copepods. Hydrobiologia 320:63–70

Elgmork K (2004) Life cycles of the freshwater, planktonic copepod Cyclops scutifer Sars on a north-south gradient in North America. Hydrobiologia 529:37–48

Elgmork K, Langeland A (1980) Cyclops scutifer Sars – one and two-year life cycles with diapause in the meromictic lake Blank vatn. Archiv für Hydrobiologie 88:178–201

Elgmork K, Lie S (1998) Diapause in the life cycle of Cyclops scutifer (Copepoda) in a meromictic lake and the problem of termination by an internal clock. Archiv für Hydrobiol, Special Issue Advances in Limnology 52: 371–381

Elgmork K, Nilssen JP (1978) Equivalence of copepod and insect diapause. Verhandlungen der Internationalen Vereinigung für Theoretische und Angewandte Limnologie 20:2511–2517

Elgmork K, Halvorsen G, Eie JA, Langeland A (1990) Coexistence with similar life cycles in two species of freshwater copepods (Crustacea). Hydrobiologia 208:187–199

Filion J-M, Chain P, Futter M (1993) Cantilevering vertical tow nets to reduce two-line induced zooplankton avoidance. J Plankton Res 15:581–587

Frisch D (2002) Dormancy, dispersal and the survival of cyclopoid copepods (Cyclopoida, Copepoda) in a lowland floodplain. Freshw Biol 47:1269–1281
Fryer G (1996) Diapause, a potent force in the evolution of freshwater crustaceans. Hydrobiologia 320:1–14
George DG (1973) Diapause in Cyclops vicinus. Oikos 24:136–142
Gliwicz ZM, Rowan MG (1984) Survival of Cyclops abyssorum tatricus (Copepoda, Crustacea) in alpine lakes stocked with planktivorous fish. Limnol Oceanogr 29:1290–1299
Gyllström M, Hansson L-A (2004) Dormancy in freshwater zooplankton: induction, termination and the importance of benthic-pelagic coupling. Aquat Sci 66:274–295
Hairston NG (1987) Diapause as a predator-avoidance adaptation: Chapter 18. In: Kerfoot WC, Sih A (eds) Predation: direct and indirect impacts on aquatic communities. University Press of New England, Hanover, pp 281–290
Hairston NG Jr, Hansen A-M, Schaffner WR (2000) The effect of diapause emergence on the seasonal dynamics of a zooplankton assemblage. Freshw Biol 45:133–145
Hakala A (2004) Meromixis as a part of lake evolution- observations and a revised classification of true meromictic lakes in Finland. Boreal Environ Res 9:37–53
Halvorsen G, Elgmork K (1976) Vertical distribution and seasonal cycle of Cyclops scutifer Sars (Crustacea, Copepoda) in two oligotrophic lakes in southern Norway. Nor J Zool 24:143–160
Hanazato T, Yasuno M (1989) Zooplankton community structure driven by vertebrate and invertebrate predators. Oecologia 81:450–458
Hansen A-M (1996) Variable life history of a cyclopoid copepod: the role of food availability. Hydrobiologia 320:223–227
Hansen A-M, Hairston NG (1998) Food limitation in a wild cyclopoid copepod population: direct and indirect life history responses. Oecologia 115:320–330
Husson F, Josse J, Le S, Mazet J (2011) FactoMineR: Multivariate exploratory data analysis and data mining with R. R. package version 1.16
Keller W, Conlon M (1994) Crustacean zooplankton communities and lake morphometry in Precambrian Shield lakes. Can J Fish Aquat Sci 51:2424–2434
Kerfoot WC, Sih A (1987) Predation: direct and indirect impacts on aquatic communities. University Press, Hanover
Kobari T, Ban S (1998) Life cycles of two limnetic cyclopoid copepods, Cyclops vicinus and Thermocyclops crassus, in two different habitats. J Plankton Res 20:1073–1086
Krylov PI, Alekseev VR, Frenkel OA (1996) Feeding and digestive activity of cyclopoid copepods in active diapause. Hydrobiologia 320:71–79
Le Jeune A-H, Bourdiol F, Aldamman L, Perron T, Amyot M, Pinel-Alloul B (2012) Factors affecting methylmercury biomagnification by a widespread aquatic invertebrate predator, the phantom midge larvae Chaoborus. Environ Pollut 165:100–108
Legendre P, Legendre L (2012) Numerical ecology. Elsevier Science, Amsterdam
Maier G (1994) Patterns of life history among cyclopoid copepods of Central Europe. Freshw Biol 31:77–86
Masson S, Pinel-Alloul B (1998) Spatial distribution of zooplankton biomass size fractions in a bog lake: abiotic and (or) biotic regulation. Can J Zool 76:805–823
Masson S, Pinel-Alloul B, Dutilleul P (2004) Spatial heterogeneity of zooplankton biomass and size structure in southern Québec lakes: variation among lakes and within lake among epi-, meta- and hypolimnion strata. J Plankton Res 26:1441–1458
Medland VL, Taylor BE (2001) Strategies of emergence from diapause for cyclopoid copepods in a temporary pond. Archiv für Hydrobiol 150:329–349
Naess T, Nilssen JP (1991) Diapausing fertilized adults: A new pattern of copepod life cycle. Oecologia 86:368–371
Naess T, Nilssen JP, Demmo R (1993) Individual lake characteristics modify the life cycle and diapause habitat of two neighbouring populations of the cyclopoid copepod Thermocyclops oithonoides. Can J Zool 71:1663–1672

Nilssen J, Elgmork K (1977) Cyclops abyssorum – life cycle dynamics and habitat selection. Memorie dell Istituto Italiano di Idrobiologia 34:197–238
Nürnberg GK (1988) A simple method for predicting the date of fall turnover in thermally stratified lakes. Limnol Oceanogr 32:1160–1164
Nürnberg GK (1995) Quantifying anoxia in lakes. Limnol Oceanogr 40:1100–1111
Paquette M, Pinel-Alloul B (1982) Cycles de développement de Skistodiaptomus oregonensis, Tropocyclops prasinus et Cyclops scutifer dans la zone limnétique du lac Cromwell, Saint-Hippolyte, Québec. Can J Zool 60:139–151
Pinel-Alloul B (1995a) Chapitre 18: Les invertébrés prédateurs du zooplancton. In: Pourriot R, Meybeck M (eds) Limnologie Générale. Collection d'écologie 25. Masson, Paris, pp 541–564
Pinel-Alloul B (1995b) Chapitre 21: Impacts des prédateurs invertébrés sur les communautés aquatiques. In: Pourriot R, Meybeck M (eds) Limnologie Générale. Collection d'écologie 25. Masson, Paris, pp 628–647
Pinto-Coelho R, Pinel-Alloul B, Méthot G, Havens KE (2005) Crustacean zooplankton in lakes and reservoirs of temperate and tropical regions: variation with trophic status. Can J Fish Aquat Sci 62:348–361
Prepas E (1978) Sugar frosted Daphnia: an improved fixation technique for Cladocera. Limnol Oceanogr 23:557–559
Radzikowski J (2013) Resistance of dormant stages of planktonic invertebrates to adverse environmental conditions. J Plankton Res 35:707–723
Reckhow KH (1977) Phosphorus model for lake management. PhD dissertation. Harvard University
Rivier IC (1996) Ecology of diapausing copepodids of Cyclops kolensis Lill. in reservoirs of the Upper Volga. Hydrobiologia 320:235–241
Santer B, Boldt E (1998) The seasonal patterns of lipids in the life cycle of the summer-diapausing freshwater copepod Cyclops kolensis (Lilljeborg). Archiv für Hydrobiol, Special Issues Advanced Limnology 52: 477–492
Santer B, Hansen A-M (2006) Diapause of Cyclops vicinus (Uljanin) in Lake Søbygård: indication of a risk-spreading strategy. Hydrobiologia 560:217–226
Santer B, Lampert W (1995) Summer diapause in cyclopoid copepods – adaptive response to a food bottleneck. J Anim Ecol 64:600–613
Seebens H, Einsle U, Straile D (2009) Copepod life cycle adaptations and success in response to phytoplankton spring bloom phenology. Glob Chang Biol 6:1394–1404
Szlauer L (1963) The resting stages of Cyclopidae in Stary Dwór Lake. Polka Archiv für Hydrobiologie 11:385–394
Watson NHF, Smallman BN (1971) The role of photoperiod and temperature in the induction and termination of an arrested development in two species of freshwater cyclopoid copepods. Can J Zool 49:855–862
Williams-Howze J (1997) Dormancy in the free-living copepod orders Cyclopoida, Calanoida, and Harpacticoida. Oceanogr Mar Biol Annu Rev 35:257–321
Wyngaard GA (1988) Geographical variation in dormancy in a copepod: evidence from population crosses. Hydrobiologia 167/168:367–374

Chapter 15
The Model of Dynamics of Population Number of the Plankton Cyclop *Mesocyclops leuckarti* (Copepoda), Based on Photoperiodism

Victor R. Alekseev, Boris V. Fiks, Karen G. Jenderedjian, and Susanna H. Hakobyan

Abstract A model of *Mesocyclops leuckarti* population dynamics in a forest lake in Nord-Western Russia is described. The model is based on detailed phonological data that included the following information on copepod dormancy: density of resting stages in the lake sediments, rate of dormant stage reactivation in spring, active stage growth rate after diapause, accumulation rate of dormant stages in fall, and their survivorship during wintertime. Other important modeling parameters, such as critical photoperiodic response, percentage of diapausing stages versus day length, growth, and feeding rates for each development stage, and sensitivity to photoperiod stages, were estimated experimentally both in field and in lab. The model was verified with natural population of the species in Lake Kalishevskoe, Leningradskaya district, Nord-West of Russia. After verification, the model was used to test several hypotheses including effects of climate warming on copepod population density and accumulation of dormant stages in the end of summer. Increasing of diapausing copepodite stage density in the end of season was related to population density in the next year. The last amount was evaluated as population success obtained in competition with other zooplankton species.

V. R. Alekseev (✉)
Zoological Institute of Russian Academy of Sciences, St. Petersburg, Russia
e-mail: alekseev@zin.ru

B. V. Fiks
Medicine University of Lubeck, Lübeck, Germany

K. Jenderedjian
Project "Promotion of Eco-corridors in the Southern Caucasus", Sevan Natural Reservation, Yerevan, Armenia

S. Hakobyan
Institute of Hydroecology and Ichthyology of Scientific Center of Zoology and Hydroecology of NAS of RA, Yerevan, Armenia

V. R. Alekseev, B. Pinel-Alloul (eds.), *Dormancy in Aquatic Organisms. Theory, Human Use and Modeling*, Monographiae Biologicae 92,
https://doi.org/10.1007/978-3-030-21213-1_15

Keywords Copepod dormancy · Photoperiodism · Population dynamics · Model experiments

15.1 Introduction

The widespread use of modeling in limnology is due to the systemic nature of the processes occurring in water bodies and the need to integrate diverse data to obtain a holistic picture of the functioning of a multilevel community. An indispensable condition for modeling is the formalization of connections between the elements of a model. As a consequence, a less detailed consideration of the dependencies occured that resulted in the inevitable discrepancy between the results of the machine experiment and the forecast with the natural course of processes and events (Puzachenko 1986). An important source of this discrepancy may be the underestimation of the role of intrasystemic rhythms, manifested both at the level of individuals or populations, as well as communities, and also ecosystems (Putilov 1987). Accounting for biorhythms in modeling of transformation processes of matter and energy in aquatic ecosystems is either not carried out at all or is made in the form of undifferentiated perception by the model of the seasonal rhythm of the development process. In the latter case, a fact of stability of rhythmic processes in the ecosystem is a priori accepted, which is based on the frequency of many biological phenomena in stable aquatic ecosystems. It is understandable that with this approach, the predictive capabilities of modeling are limited to a fairly narrow range of natural variability of rhythms within the stable (in the interannual aspect) changes in biotic and abiotic environmental factors.

Popov (1985) and Alekseev (1985, 1987) recorded the changes in the seasonal rhythms of the development of mass zooplankton species and, accordingly, the disruption of the natural course of production processes under the influence of a number of destabilizing factors. This puts us before the necessity of studying nature and modeling the seasonal rhythm of the development of hydrobionts and, in particular, crustaceans.

15.2 Description of Model and Modeling Population

Seasonal rhythm in the development of populations of crustaceans is determined by the alternation of diapause and active development. As an example, we consider the population dynamics model of *Mesocyclops leuckarti*. The periodicity of its development is determined by the conditions of reactivation and induction of diapause. The model is based on laboratory experiments of studying its growth rate (Alekseev 1983), the conditions for the formation of diapause (Alekseev 1984), and 2-year field samplings in Lake Kalishchevskoye of Leningrad region (Alekseev 1987).

The model simulated the dynamics of population size of the following age composition:

1. Young individuals capable of perceiving a photoperiodic signal (nauplia of the 5th stage and copepodites of the 1st stage). The maximum age of this group was 240 degrees-days.
2. Young individuals unable to estimate the length of the day, but as they reach the age limit (400 degrees-days), they pass to diapause if they have already received the signal information in the previous stage. Otherwise they completed the metamorphosis and switched to the next age group (copepodites of 2–5 stages).
3. Young sexually mature individuals before the appearance of the first offspring; the limiting age was 580 degrees-days. Fertility was 0.
4. Mature individuals, the age of up to 750 degrees-days. Fertility – 20 eggs per individual (including bisexual structure) for the first generation of females that emerged from the overwintered copepodites and 10 eggs per individual for subsequent generations.
5. Old females with age limit 1055 degrees-days, after which they perished. Fertility – five eggs per individual for the first generation and three eggs per individual for the others.

The model consisted of five time intervals, generally corresponding to the main stages of the seasonal cycle of *M. leuckarti* (Alekseev 1987). The life cycle began with the release of the diapause of the crustaceans in the stages 4 and 5 of copepodites. According to field observations, the duration of the activation phase was 20 days. During this time, obeying the law of normal distribution, about 60,000 specimens/m^2 of older copepodite stages emerged from the diapause. After the melting of ice, the temperature of the water, in accordance with the weather conditions, naturally increased (Fig. 15.1). The fertility of females was the highest due to the reserves accumulated in the autumn.

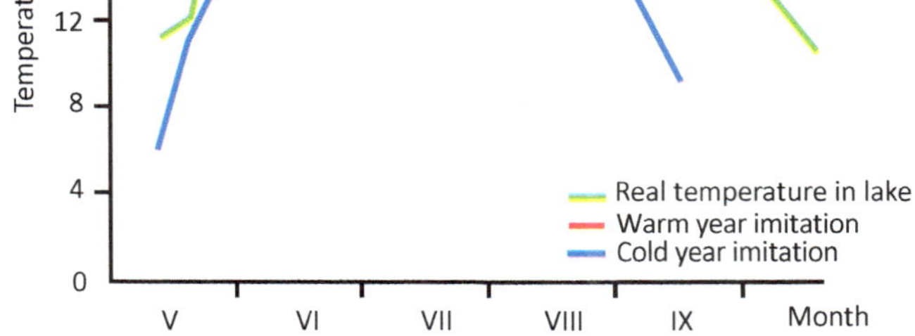

Fig. 15.1 Seasonal temperature graphs in the lake Kalishchevskoe in 1982 and imitations of warm and cold years

With the advent of females of a new generation, the model established a lower fertility. The stage most crucial for the functioning of the model began with the onset of the critical day length (day 90), which led to the division of the population into two groups, accepting and not accepting signal to dormancy. Having accumulated 300 degrees-days in excess of the limit established for diapausing copepodite stage, the crustaceans obtained a survival rate of 100%, which was simulated by the transition to the bottom, in a zone inaccessible to predators. The remaining individuals completed the metamorphosis and gave rise to the last cycle of development. The time of stopping the program is 150 days (September 15) – the time when *M. leuckarti* drops out of the plankton in most reservoirs of the northwest.

The peculiarity of the model, which in some way limited its capabilities, was the lack of a definite relationship between the survival of individuals with a trophic situation and the press of predator in a waterbody. Survival was chosen so that the model best describes the dynamics of the number of natural populations in the Lake Kalishchevskoye in 1982. Analysis of the samples showed that the survival rate of different ages of *M. leuckarti* did not change throughout the season, which was probably due to the indiscriminate press of predators and the absence of pronounced trophic restriction. A similar type of decline in numbers (with a constant relative velocity) was noted by M. M. Kamshilov (1958) in the copepod *Calanus finmarchicus* (Gunner) in the coastal zone of East Murman.

We took into account the limitation due to the constant survival rate (in most cases equal to 0.9 when analyzing the results of modeling). In addition, in one of the variants of the model, the survival rate changed according to a certain plan, which was simulated as a predator press. Our study showed that in a stabile (in the long-term aspect) developing copepod population, the survival is unlikely to experience major changes. It is possible that this is due to the peculiarity of reproduction of copepods, where the population consists mainly of immature individuals for a long time and which cannot quickly respond to increased mortality by increasing fertility and fecundity, as occurs in branching crayfish.

The aim of the study is to conduct model experiments at the population level of *M. leuckarti*. The model made it possible to vary the dates of exit from the diapause, to change the position of the photoperiodic reaction threshold in accordance with the geographic and temperature shift, to simulate the temperature features of the year, to trace the dynamics of the population size of *M. leuckarti* for a number of consecutive years, and to assess the influence of photoperiodism and some other factors. The main results of the model operation are presented below.

15.3 The Significance of Photoperiodism

To estimate the length of the day to prepare the population for unfavorable conditions, the dynamics of the *M. leuckarti* population in two scenarios of the model was analyzed. The first option fully corresponded to the abovementioned conditions: on 90th day nauplia accepted critical photoperiod and switched to dormant stage preparation. The second differed from the first model only in that the division of the population on day 90 did not occur, and all individuals of the second group continued metamorphosis.

In both cases, first of all, we should note the pronounced unevenness of the population dynamics (Fig. 15.2). These fluctuations with a period of about a month (the duration of the development cycle) represent a trace of a practically simultaneous release from the diapause of a large number of copepodites, which initiated the development of the population in the spring. During the "vegetation period," the model population in both variants switched to multiplication three times; the third largest peak was the youngest at the end of July, after which, according to the model conditions, either inhibition caused by photoperiodism occurred or development continued during the seasonal temperature decrease (see Fig. 15.2).

The number of model populations in the second scenario naturally decreased, and by the end of the season, the total number of older copepodite stages, capable of passing to a diapause with a sharp deterioration in conditions, amounted to about 20,000 specimens/m^2, i. e., almost three times less than the diapause in the spring. It is characteristic that the number of sexually mature individuals reproducing the young in vain at the end of the season even exceeded the number of diapausing stages. It is obvious that under the absence of preliminary synchronization, the number of population in a few years will fall sharply.

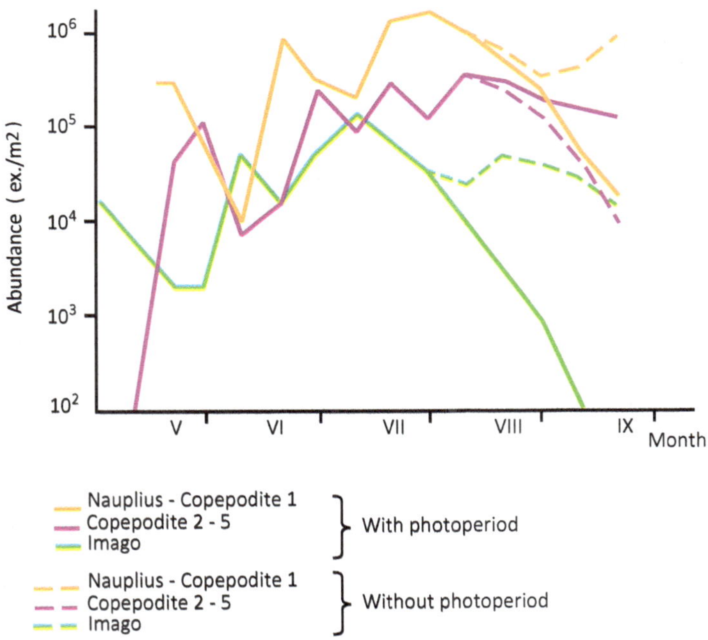

Fig. 15.2 Model population dynamics of *M.leucarti* based and non-based upon the photoperiodism

In the next scenario, the crustaceans were responding to photoperiodic conditions. In this situation, the differences in population dynamics began shortly after passing the critical length of the day. Firstly, they included the sexually mature stages and the older copepod stages. The number of adults decreased at a higher rate than in the previous scenario, and by the end of the season, sexually mature individuals were represented by a small number of old, nonreproductive females. The basis of the population was (approximately 70,000 specimens/m^2) the older copepodite stages (see Fig. 15.2). Considering that about 10% of diapausing crustaceans perish during the wintering season, we conclude that in the next year, the beginning of the population will be practically the same number of specimen.

The simulation results were compared with the phenological data on the development of this species in the Lake Kalishchevskoye in 1982 (Fig. 15.3). It turned out that by the end of July, the data of plankton samples correspond well to the model. In the future (almost all August), the relative number of diapausing stages in plankton was significantly lower. At this time, the active migration of older copepodites to the bottom has already begun in the reservoir. This was evidenced by an increase in the proportion of these crustaceans in the meiobenthos samples (see Fig. 15.3). In September, the basis of the population in the plankton was copepodites of the 4–5 stages, as in the model based on photoperiodism. It is very indicative that the second scenario of the model corresponds to the zooplankton data in August. This similarity reflects the underestimation (undernutrition of the plankton net) of older copepodites and the associated error in assessing the age structure of the population. Thus, a model based on photoperiodism made it possible to discover the methodological imperfection of the collection of phenological data, which at that time must necessarily be carried out both by the plankton network and by the tool for collecting meiobenthos.

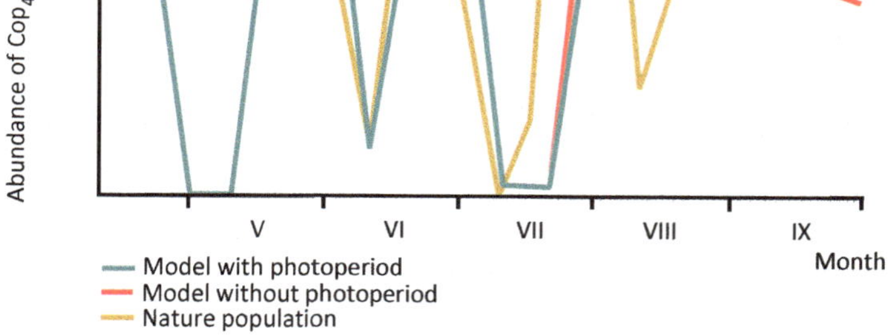

Fig. 15.3 Abundance of major copepodite stages in model populations and in nature populations

As a result, for all the relativity of specific figures, the model of population dynamics of *M. leuckarti*, taking into account the possibility of organisms to focus on the length of the day, made it possible to get a picture closer to the real state of the population. In the following study of the properties of the model, a model based on photoperiodism was used.

15.4 The Role of Meteorological Conditions in the Population Dynamics of *M. leuckarti*

It is rather difficult to estimate the influence of meteorological conditions on the development of populations of hydrobionts from field observations, since the change in the number of organisms is the result of the simultaneous action of many factors – trophic conditions, predator press, long-term cycling, etc. Modeling the reaction of a population under the influence of a temperature course makes it possible to isolate the effect of one factor and to assess its capabilities. On the other hand, the changes occurring in the model population can be used already for the analysis of natural populations.

In order to model the meteorological conditions, we used the seasonal temperature curves approaching the natural curves simulating the warm year (early spring–late autumn), the cold year (late spring–early autumn), as well as the intermediate variants (late spring–late autumn and early spring–early autumn) (see Fig. 15.1).

The survival indices, the dates of the onset of reactivation (melting of ice), and the inclusion of the photoperiodic signal remained unchanged.

1. Warm Year (Fig. 15.4a)

Early warming up of water contributed to the rapid development of the first, most effectively breeding generation of individuals emerging from the diapause. The high density of this generation determined the level of development of the species and subsequent increase in quantity of its population. As a result, the maximum number of youngsters exceeded the corresponding figure of the original year by more than an order of magnitude. The number of diapausing individuals left for the wintering has exceeded the original level by 50 times.

It is understandable that in real conditions, the population growth would not occur at such a high rate. In limiting the growth of numbers, both the predator press and trophic conditions will interfere. We observed coincidence of the periods of mass development of juveniles and sexually mature (predator) individuals at the end of the season (see Fig. 15.4). The results suggest the deterrent role of cannibalism. It is significant that at the beginning of the season (the first two generations), the peaks in the number of adults and juveniles do not coincide. A more obvious result is an increase in the number of complete cycles of the population development – 4 instead of 3 – in the original year. Beginning at the end of July, the reproductive

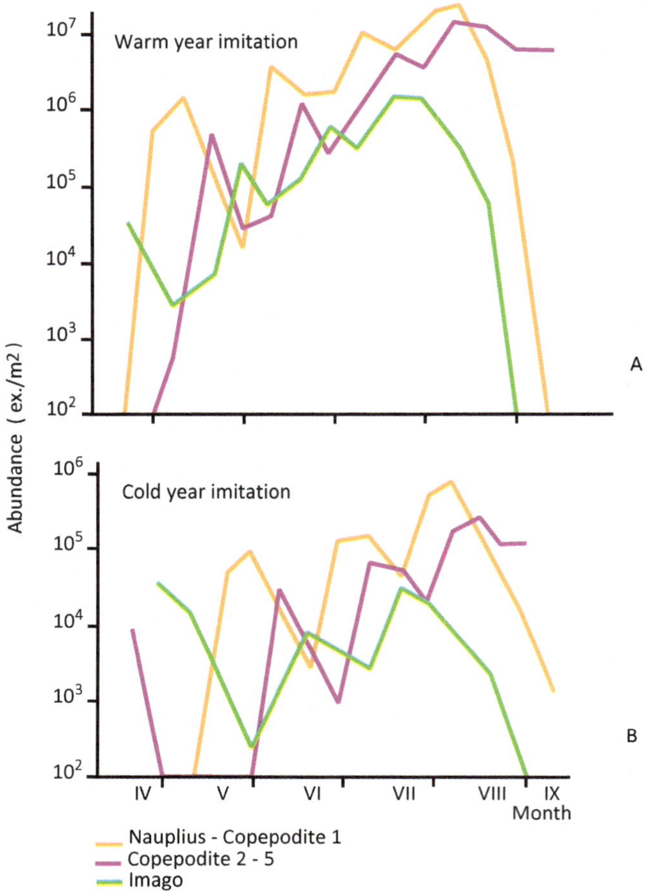

Fig. 15.4 Model population dynamics of *M.leucarti* when the year was "warm" and "cold" (for temperatures see Fig. 15.1)

processes were curtailed under the influence of the photoperiod, and on the 145th day, the model population consisted entirely of diapausing individuals.

2. Cold year (Fig. 15.4b). The slow development of cyclops in the early summer led to a significant drop in the number of sexually mature specimens of the overwintered generation. By the beginning of the breeding season, the population had almost ten times less females than in the warm year, which affected the numbers of subsequent generations. The largest population was almost an order of magnitude lower than in the original year and more than two orders of magnitude than in the warm year (see Figs. 15.2 and 15.4a). In addition, due to lengthening the development time, the population was able to complete only three complete cycles. This also affected the accumulation of diapausing stages of the cyclops. It is important to note that the number of crustaceans that left for wintering was

significantly different from the warm year and was only 1.3 times lower than in the original year.

3. Intermediate versions (early spring–early autumn and late spring–late autumn)

The dynamics of the number of all the age groups of the *Cyclops* in these variants was similar to that of the years considered above, and the beginning of population density (in spring) was the same as in warm year. Small deviations referred to the end of the periods of active development of the population. Thus, the number of sexually mature specimens at this time in the variant "early spring–early autumn" was even higher than in the warm year, due to the slow death of old females. For the same reason, the old imago in the version "late spring–late autumn" ended earlier than in the cold year.

15.5 Long-Term Dynamics of the Model Population

Considering the significant discrepancies in the number of *M. leuckarti* in the warm and cold years, it was of particular interest to trace the long-term dynamics of the population under conditions of a combination of a number of favorable and unfavorable years in meteorology. The 3-year repetition of the "warm year" cycle led to an increase in the population size by many orders of magnitude in comparison with the initial one. There is no doubt that in real conditions in such a situation, intraecosystemic and intraspecific homeostatic mechanisms come into force to limit the development of cyclops to a lower level.

More plausible are the results obtained for three consecutive "cold" years (Fig. 15.5). The number of all age stages has regularly decreased, but not so much

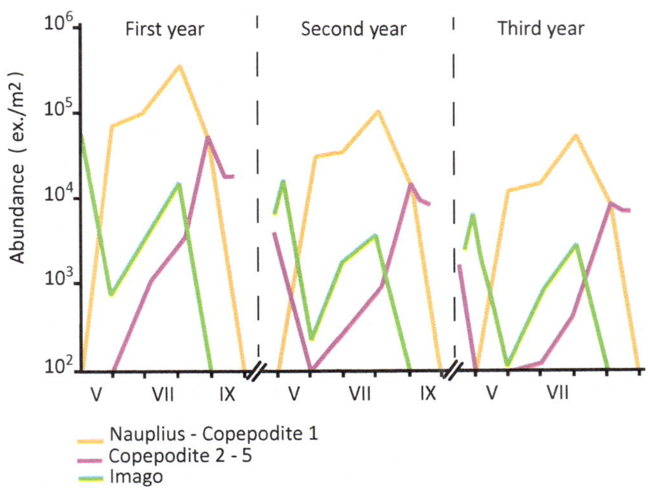

Fig. 15.5 Three-years model population dynamics of *M.leucarti* when the years were "cold"

as to call into question the existence of the population. The accumulation of diapausing stages at the end of the growing season allowed the population to successfully resist an unfavorable combination of climatic conditions. At the first "warm" year, the reserve of diapausing stages preserved by the end of the 3-year cold cycle was able to raise the population again to the initial and even higher levels.

During the acclimatization, works frequent are cases of transposition of hydrobionts in the meridional direction (Alekseev 1990). Frequent failures in such attempts, with seemingly complete conformity of temperature and trophic conditions (Ioffe 1974), allow assuming the influence of desynchronization of population rhythms and periodically advancing critical temperatures in new habitats (Alekseev 1986). The model of population dynamics of *M. leuckarti* allowed estimating the possible influence of latitudinal transpositions on the preparation of the population for wintering.

The following scenarios of moving the model population along the meridian were considered: 5 and 10° to the north and south of the site of research, which roughly corresponded to 46, 51, 61, and 66° n. l. The original model was set on latitude of 56° n. l. (Alekseev 1984). In order not to overload this paper with excessive illustrative material, let us consider the state of the structure of the population on the 150th day of its existence (Fig. 15.6).

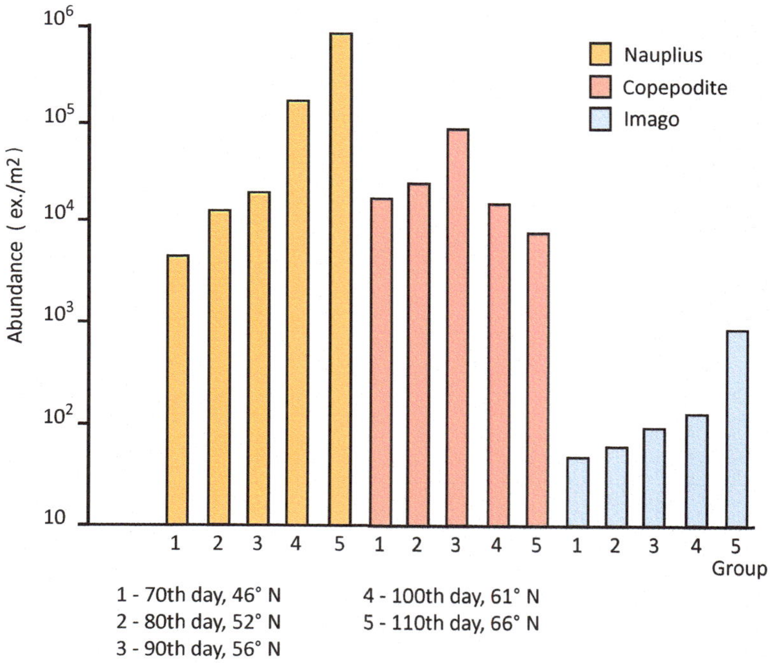

Fig. 15.6 Abundance of nauplius, copepodites and imago in model populations of *M.leucarti* on the 150th day of operation of the model, depending on the date of perception (geographical latitude) of the photoperiodic signal

The number of sexually mature individuals and juveniles to the south of the main habitat was significantly lower and to the north, respectively, more than in the original population. The main reason for such differences was the onset of the critical day length in the southern regions earlier than in the northern ones. In the southern regions, this led to unjustifiably early winding up of breeding processes and in the northern, on the contrary, contributed to the prolongation of reproductive activity. As a result, at the key date, the number of diapausing stages, both in the "southern" and "northern" versions, was significantly lower than the initial one. The model did not take into account the fact that in the real conditions of the south and north, the seasonal temperature variation turned out to be different from the one considered. An analysis of the possible impact of these differences suggests that the unfavorable situation with the accumulation of wintering stages was aggravated by a longer warm period in the south and an early cooling in the north. In the first case, higher temperatures would slow the transition of the diapausing stages to the bottom and thereby prolong the period of availability to predators; in the second case, a sharp cooling and minimizing production processes in the north would drastically reduce the survival of the young to the onset of diapause.

Accordingly, the transfer of the *M. leuckarti* population over a considerable distance in the meridional direction was accompanied by a decrease in the effectiveness of important adaptations such as the accumulation at the end of the season of diapausing stages.

We conclude that the model of the population dynamics of *M. leuckarti* based on photoperiodism in dormancy creation, when used in various regimes, made possible to obtain close to real descriptions of the state of populations. Special experiments with the model revealed significant effects of climate warming on population dynamics in this planktonic invertebrate predator. Significant increase of its density should play important role in possible transformation of zooplankton community as well as in phytoplankton and planktivorous fish dynamics.

Further improvement of the model is due to its introduction as a link in the model of the trophic chain of the aquatic ecosystem and an analysis of its response to changes in trophic status and the press of vertebrate and invertebrate predators.

Acknowledgments The chapter was written with support from the Russian Foundation for Basic Researches grant 17-04-00027. The study was partly supported by the Russian Academy of Science, topics 65.4 and 65.5. A special thanks for linguistic correction of the text to Anna Jenderedjian Columbus, Ohio, USA.

References

Alekseev VR (1983) Growth, development and production of Mesocyclops leuckarti (Claus) in different types of water bodies. Trans State Lakes Rivers Res Inst Leningrad 204: 48–56 (in Russian)

Alekseev VR (1984) Role of photoperiodic responses in formation of diapause in freshwater Cyclopoids. Trans State Lakes Rivers Res Inst Leningrad 223: 79–82 (in Russian)

Alekseev VR (1985) Non-specific resistance of diapausing crustaceans to toxic matter treatments. Trans State Lakes Rivers Res Inst Leningrad 234: 122–130 (in Russian)
Alekseev VR (1986) Role of diapause in acclimatization of crustaceans. Trans State Lakes Rivers Res Inst Leningrad 252: 61–68 (in Russian)
Alekseev VR (1987) Influence of intensification measures on the development of zooplankton and seasonal cycles of crustaceans in whitefish lake fish hatcheries. Trans State Lakes Rivers Res Inst Leningrad 264:20–32 (in Russian)
Alekseev VR (1990) Diapause in crustaceans: ecological and physiological aspects. Nauka Academic Publishers, Moscow, 144 pp. (in Russian)
Ioffe TSI (1974) Enrichment of the fodder base for fish in reservoirs of the USSR by acclimatization of invertebrates. Trans State Lakesand Rivers Res Inst Leningrad 226: 61–68 (in Russian)
Kamshilov MM (1958) Production of Calanus finmarchiciis (Gunner) in the coastal zone of eastern Murman. Proc Murmansk Biol Station 4: 45–55 (in Russian, English sum)
Popov NYA (1985) Development and production of zooplankton in intensively exploited lakes of the nursery of the south of the Tyumen region: Author's abstract. Ph.D. dissertation Leningrad, 1–23 p. (in Russian)
Putilov AA (1987) System-forming function of synchronization in living nature. Nauka Publishers, Novosibirsk, 144 p. (in Russian, English sum)
Puzachenko YuG (1986) Formal systems in ecology. Abstract of the second. All-Union conference. Common problems of biogeocenology. Moscow, pp 16–18 (in Russian)

Chapter 16
Use of the Simulation Model for the Analysis of the Impact of Parasites on the Dynamics of Abundance, Reproduction, and the Transition to Diapause in the Mass Rotifers *Brachionus calyciflorus* Pallas of the Volga Delta

Alexander K. Gorbunov, Boris V. Fiks, and Victor R. Alekseev

Abstract Rotifers are one of the most important groups of zooplankton, and they play an important role in aquatic ecosystems as initial energy converters of primary and bacterial products. In the course of monitoring studies of the state of ecosystems, many questions arise, the solution of which is necessary for correct interpretation of the data obtained. When considering the dynamics of the abundance of zooplankton and its constituent groups, the main focus is on the state of temperature and oxygen regimes, feed conditions, competitive relationships, and predator consumption, and too little attention is paid to the effects of parasitic organisms on plankton. The question of rotifer infection and the possible influence of this factor on their population dynamics is still poorly studied. Simulating of these processes with the help of mathematical model is described.

Keywords Dormancy, Zooplankton, Population dynamics, Resting egg production, Disease in aquatic invertebrates

A. K. Gorbunov (✉)
Astrakhan State Nature Biosphere Reserve, Astrakhan, Russia

B. V. Fiks
Medicine University of Lubeck, Lübeck, Germany

V. R. Alekseev
Zoological Institute of Russian Academy of Sciences, St. Petersburg, Russia
e-mail: alekseev@zin.ru

V. R. Alekseev, B. Pinel-Alloul (eds.), *Dormancy in Aquatic Organisms. Theory, Human Use and Modeling*, Monographiae Biologicae 92,
https://doi.org/10.1007/978-3-030-21213-1_16

16.1 Introduction

In Russia, N. V. Voronkov was the first to pay attention to the infection of rotifers (rotifer as host) by parasitic organisms; the work examined small temporary water bodies in the vicinity of Lake Glubokoe (Voronkov 1910).

Later, N. K. Decksbach noted the infection of four species of rotifers with the same type of parasite in the waters of Moscow neighborhoods (Decksbach 1928). In reservoirs outside of Russia, a number of researchers have indicated infection of rotifers by parasitic organisms. A. Ruttner-Kolisko (1977) noted a strong infection of the *Conochilus unicornis* Rousselet 1892 population by microsporidia in Lake Lunts (Austria) and its influence on rotifer population dynamics. In another body (Lake Banyoles, Spain), M. R. Miracle (1977) showed a strong influence of parasites on the development of two populations of rotifer species.

Infection of rotifers has not been previously studied in Volga delta-associated water bodies. The Volga delta, however, is a unique natural complex of great economic importance. Effective management of its resources and preservation of its ecosystems require an understanding of all of the natural components and their functions, including rotifers. The available data on infection of rotifers raises questions regarding parasite impacts on this important group of zooplankton. Abundance and other indicators of population dynamics including diapause (the most important seasonal and biotopic adaptation of aquatic invertebrates) are worth consideration (Alekseev et al. 2007).

The purpose of the research was to study the infection dynamics of predominant rotifer species in natural ecosystems and to quantify the impact of infection on population dynamics, productivity, and seasonal adaptations in *Brachionus calyciflorus* Pallas rotifers based on the data obtained. Direct assessment of infection's impact on rotifer population dynamics by field observations alone is quite difficult due to short rotifer life cycles. Assessments of this kind performed under controlled laboratory conditions, however, raise fair questions about the validity of generalizing such experimental infection data to natural populations. In this situation, an ideal compromise is one in which an experimental model is used that takes into consideration knowledge of a species' life cycle parameters and verifies any data obtained by examining (measuring, estimating, extrapolating, etc.) data regarding the states of real populations under natural conditions.

Modern simulation methods allow researchers to reproduce and trace the seasonal course of selected population parameters in model "populations" with a high level of approximation to reality. They make it possible to quantify indicators of population dynamics and seasonal adaptation which are difficult to obtain by direct observation. These include mortality, productivity, and population growth rate (Mnatsakanova and Polishuk 1996). Important indicators of the effectiveness of rotifer seasonal adaptations are the time of appearance and the proportion in the population of mictic females (which have transitioned from parthenogenesis to bisexual reproduction) and the production of diapausing (resting) eggs sufficient to maintain the population. Considering the multifactorial nature and complexity of parasite impacts on host population dynamics, a simulation model was developed to handle the problem, and this study is devoted to analysis of the model's behavior.

16.2 Materials and Methods

In order to quantify the impact of parasites on rotifer productivity and life cycle parameters, a population dynamics model featuring *Brachionus calyciflorus* Pallas was developed in collaboration with researchers from Germany. To create and verify this model, experiments were performed to evaluate the duration of embryonic development, postembryonic development, and other life cycle parameters of the species, and field studies were carried out. The main study materials were standard "quantity samples" (more than 700), as well as live and fixed "quality samples" of zooplankton collected in water bodies of the Volga delta at the Damchiksky site of the Astrakhan State Nature Biosphere Reserve in 2001–2005.

Samples were collected by filtration of 100 liters of water through Apstein plankton nets made from number 73 mill screen. Immediately after collection, samples were adjusted to a volume of 100 ml and fixed with 4% formalin. Collection of quality materials was carried out by hand towing of plankton nets. A 20 μm mesh plankton net was used for more effective collection for small rotifer species. When taking samples, water depths and temperatures were measured simultaneously. To characterize water temperature changes in the Bystraya channel, data from the "Damchik" pheno-hydrometeorological station were used (Fig. 16.1). Counting, viewing, and identification of rotifers were carried out using light microscopy

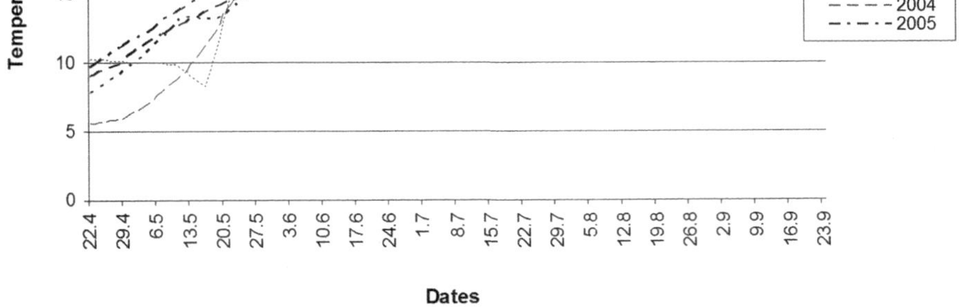

Fig. 16.1 Seasonal water temperature variations in the Volga delta Bystraya channel, by year and model reproduction, based on field measurements

(MBS-9 and MBI-1). To photograph infected rotifers in a living state, a custom microphotography apparatus was constructed; it is a refinement of earlier flash microphotography methods (Gorbunov 1979). The latest version uses a digital camera with a resolution of 8 million pixels, and the arrangement permitted us to effectively process both live and fixed materials.

For proper modeling and study of parasite involvement, it is necessary to establish baseline biological parameters of the living host rotifer (*Brachionus calyciflorus*). Therefore, laboratory manipulation and monitoring of samples were as follows: live rotifers from natural populations were placed in micro-aquariums with volumes of 1–50 ml and filled with water from the sampling site. Micro-aquarium water was changed several times a day. For visualization and photography, rotifers were pipetted onto slides and covered with coverslips featuring plasticine or wax legs. After photographing, rotifers were returned to micro-aquariums, where they continued their development.

16.3 Model Description

The model was developed in the C++ software environment and includes the following elements: a numerical model of *B. calyciflorus* population dynamics with step-by-step display of results (1 step = 1 day); and the ability to change infection levels, and other indicators, via macros. This makes it very convenient, both for simulation of natural conditions and also for quantifying the possible impact of parasites at specified infectiousness values.

The dynamics of the total number of rotifers, by age, on a specific date *N[d,k]* was approximated by the formula:

$$N[d,k] = N[d-1,k-1] * (1-M) * (1-I) \tag{16.1}$$

where

- N is the initial number of rotifers at the beginning of the population's development
- d is the day number from the beginning of observations (discrete time)
- k is the age group of rotifers, in days
- M is the coefficient of death
- I is the coefficient of infection

The dynamics of the number of infected rotifers specimens *(Ninf [d, k])* was estimated by the formula:

$$Ninf[d,k] = Ninf[d-1,k-1] * (1-M) + N[d,k] * I \tag{16.2}$$

If (degree-days [d, k > 0]) and (degree-days [d, k]) ≤ 60), then:

$$Ninf[d] = Ninf[d] + Ninf[d,k]$$

Rotifer age was determined in degree-days, with the following groups identified:

1. Parthenogenetic eggs: 20 degree-days (development time 24 h uniformly at 20 °C with the possibility of correction for the actual temperature)
1. Young females before the start of reproduction: 40 degree-days
2. Mature parthenogenetic females: life span up to 140 degree-days
3. Adult parasite-infected females: life span up to 60 degree-days

The dynamics of degree-day accrual (degree-days [d, k]), by each age stage, was carried out at a frequency of once per day according to the formula:

$$\text{degree days}[d,k] = \text{degree days}[d-1,k-1] + t^{\circ}C[d] \tag{16.3}$$

$t^{\circ}C[d]$ actual measured water temperature

The transition to the formation of males and diapausing eggs (seasonal adaptations) corresponded with conditions in which increasing numbers of rotifers were under the influence of a combination of two signal factors: temperature (natural data) and photoperiod (calendar data). Collectively, these determine the value of the diapause coefficient (K_D), which ranges from 0 to 1.

The number of mictic eggs (for males and females capable of laying diapausing eggs) was determined in this case by the formula:

$$N_{\text{stad}_{3[d,k]}} = N_{\text{stad}_{3[d,k]}} * (1 - K_D) \tag{16.4}$$

The model took into account the possibility of diapause reversibility (the return of some rotifers to parthenogenetic reproduction) in the case of high floods; this has been noted in cases in which water temperatures have decreased by no less than 5 degrees during 3–10 days (Alekseev 1978).

$$N_{\text{stad}_{3[d,k]}} = N_{\text{stad}_{3[d,k]}} + N_{\text{diap}[d,k]} * E_D \tag{16.5}$$

E_D fraction in diapause exit

The process of diapause exit lasted no more than 10 days during such temperature drops.

Population fecundity (Eggs[d]) was determined by the formula:

$$\text{Eggs}[d] = N_{\text{stad}_{3[d,k]}} * F \tag{16.6}$$

F average number of eggs laid by the female (determined by sampling).

The model reproduces *B. calyciflorus* seasonal dynamics and includes the following elements:

1. Parthenogenetic eggs, development time 24 h (uniformly at 20 °C with the possibility of correction for actual temperature)
1. Mature parthenogenetic females

2. Mictic eggs for males (mictic eggs from which males hatch)
3. Sexually reproducing mictic females
4. Diapausing (resting) eggs (as a result of fertilization)

Key population parameters used for simulation, including signal temperature and photoperiod, were based on empirical data. Specifically, data on the conditions surrounding development of natural Volga delta *B. calyciflorus* populations (2001–2005) was used. *Brachionus calyciflorus* production was estimated from the number of eggs produced (Edmondson 1960; Galkovskaya 1963). The level of infection of the *B. calyciflorus* population was reproduced in accordance with the results of real estimates following years of observations. In the model, it is taken into account with the help of the infection coefficient, which ranges from 0.01 to 0.05.

A few trends can be noted regarding the coefficient of infection in the model. Of the estimated infection parameters, the infection coefficient correlated best with the percentage of infection at peak population size. The average level of infection was less correlated with the infection coefficient over the population's entire existence. The maximum level of infection had almost no correlation to the coefficient of infection. The latter (max. inf.) usually occurred at the end of a population's existence (when their abundance is low). The coefficient of natural mortality during the years studied ranged from 0.12 to 0.26, and this range was determined by adjustment of model parameters until computed estimates matched observations.

16.4 Results of Model Analysis

To assess the impact of infection (infection of rotifer by bacteria, fungi, protozoan parasites, etc.) in a model that reproduces the population observed under natural conditions, population dynamic characteristics under conditions of natural infection were restored (0-variant). It was assumed that food was not a limiting factor during population growth. Accordingly, the increasing values of quantity, productivity, fertility, etc., seen in the model, after taking into account infection, reflect, in our opinion, the real potential of the population, although it is somewhat hindered by the negative impact of parasitism/disease (Fig. 16.2). As such, comparison of population development parameters without infection (the 0-variant) and with infection (the I-variant) allowed us to obtain a quantitative assessment of parasite/disease influence on *B. calyciflorus.* Infection is an external environmental factor, and its influence on population parameters is shown (Fig. 16.3). The parasite factor's role in changing *B. calyciflorus* seasonal adaptations was assessed by calculation of the accumulation of diapausing eggs by the population during the season. It is known that the outcome of competition between species, especially in temporary reservoirs, depends on the sizes of their diapausing egg banks (Mnatsakanova and Polishuk 1996). Therefore, the impact of parasites and diseases on the accumulation of diapausing eggs can also be considered as an environmental factor that changes the competitive relationship between species in subsequent years. As such, in order

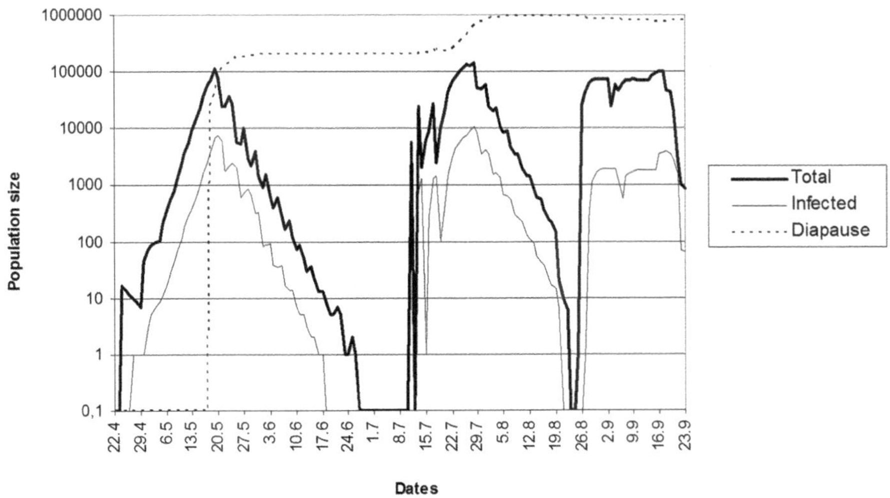

Fig. 16.2 Population dynamics (total number, number of infected individuals, and stock of diapausing eggs) of *Brachionus calyciflorus* Pallas, featuring an average infection level of 3.86%, by model for 2005 (thick line = total number; thin line = number of infected individuals; dotted line = stock of diapausing eggs)

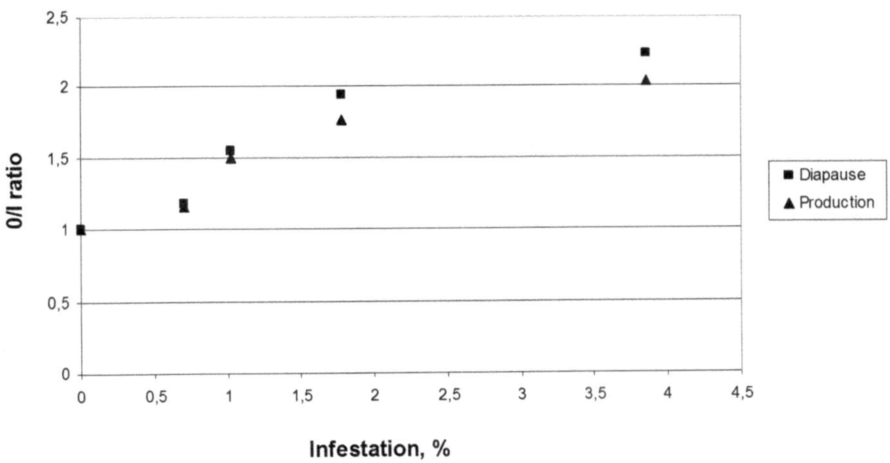

Fig. 16.3 0/I ratio of egg production and diapausing eggs in "clean" and infected *Brachionus calyciflorus* populations at different levels of infection close to those observed in nature, by model (triangles = 0/I ratio of egg production in "clean" and infected populations; squares = 0/I ratio of diapausing eggs in "clean" and infected populations)

to quantify the impact of infection on *B. calyciflorus* population dynamics and seasonal adaptation, we selected the following rotifer life cycle parameters: maximum total population size at their first and second peaks, average daily fertility per season, egg production per season, and total number of resting eggs by the end of the season. The results of these comparisons are summarized in the following final tables, by year (Tables 16.1, 16.2, 16.3, 16.4 and 16.5).

In 2001, the average infection rate was 5.9% in the model population and 9% in the natural (field data) population. The difference in these estimates is primarily due to different sample sizes for these quantities. In the model population, the average value was obtained from 152 points, while in natural conditions, it was obtained from only 6. Closer data regarding infection in the model and under natural conditions were obtained by comparing them at the population maxima: 7% (model) versus 7.9% (nature) (Table 16.1).

The average infection value, as estimated by the model, for 2002 was 0.7%; this almost coincides with field observations (1%). Infection values at the first population peak in the two systems were also very close: 1.2% (model) and 1% (nature) (Table 16.2).

Thus, *Brachionus calyciflorus* infection levels in 2002 were minimal in relation to all of the variants considered. Nevertheless, in this year, all of the indices clearly indicated a negative effect of even a weak infection on population parameters. The O/I ratios at first and second pop. peaks were 1.16 and 1.11.

In comparison with the previous year, 2001s first population peak was not as significantly suppressed by the infection. Yet, average fertility in 2001 was even more sensitive to infection than before, and it was 1.3-fold lower in the infected

Table 16.1 Simulation modeled *B. calyciflorus* population dynamics parameters reproducing the 2001 natural population

Parameters	0-variant	I-variant	0/I ratio
Maximum quantities at			
1st pop. peak	105,792	36,243	2.919
2nd pop. peak	144,290	119,736	1.205
Average seasonal daily fertility, eggs/ind.	1.158	1.09	1.16
Number of eggs, eggs/season	3,325,128	2,166,443	1.53
Stock of diapausing eggs	2,085,654	1,250,747	1.668

0-variant, population size without the influence of the infectious factor; I-variant, population size under the influence of the infectious factor

Table 16.2 Simulation modeled *B. calyciflorus* population dynamics parameters reproducing the 2002 natural population

Parameters	0-variant	I-variant	0/I ratio
Maximum quantities at			
1st pop. peak	96,737	83,178	1.163
2nd pop. peak	1,142	1,027	1.112
Average seasonal daily fertility, eggs/ind.	1.349	1.037	1.3
Number of eggs, eggs/season	319,981	276,023	1.159
Stock of diapausing eggs	185,109	156,436	1.18

0-variant, population size without the influence of the infectious factor; I-variant, population size under the influence of the infectious factor

Table 16.3 Simulation modeled *B. calyciflorus* population dynamics parameters reproducing the 2003 natural population

Parameters	0-variant	I-variant	0/I ratio
Maximum quantities at			
1st pop. peak	78,074	51,465	1.52
2nd pop. peak	169	107	1.58
Average seasonal daily fertility, eggs/ind.	1.273	1.317	0.966
Number of eggs, eggs/season	257,723	172,253	1.5
Stock of diapausing eggs	128,421	83,050	1.55

0-variant, population size without the influence of the infectious factor; I-variant, population size under the influence of the infectious factor

Table 16.4 Simulation modeled *B. calyciflorus* population dynamics parameters reproducing the 2004 natural population

Parameters	0-variant	I-variant	0/I ratio
Maximum quantities at			
1st pop. peak	19,377	12,496	1.55
2nd pop. peak	2,145	1,142	1.878
Average seasonal daily fertility, eggs/ind.	1.076	0.921	1.17
Number of eggs, eggs/season	139,955	78,905	1.774
Stock of diapausing eggs	77,018	39,614	1.94

0-variant, population size without the influence of the infectious factor; I-variant, population size under the influence of the infectious factor

Table 16.5 Simulation modeled *B. calyciflorus* population dynamics parameters reproducing the 2005 natural population

Parameters	0-variant	I-variant	0/I ratio
Maximum quantities at			
1st pop. peak	123,793	62,961	1.966
2nd pop. peak	138,970	62,772	2.213
Average seasonal daily fertility, eggs/ind.	0.617	0.6099	1.01
Number of eggs, eggs/season	613,362	299,358	2.049
Stock of diapausing eggs	433,822	194,429	2.231

0-variant, population size without the influence of the infectious factor; I-variant, population size under the influence of the infectious factor

variant than in the uninfected one. Differences in egg production during the season and in the stocks of diapausing eggs turned out to be small, 16% lower and 18% lower, respectively, in the infected population.

The average fraction of the population infected, in 2003, was 1.02% in the model; this is markedly different from the 4.73% level of the natural population (Table 16.3). At the same time, estimates of infection at peak population size were almost identical between the model and natural populations, amounting to 1.7 and 2%, respectively.

The maximum total population sizes for the first and second 2003 quantity peaks were almost the same for both variants. Their O/I ratios at first and second pop. peaks were1.52 and 1.58, respectively. The average fertility, over the entire 2003 observation period, was found to be only slightly lower in the uninfected population. The total egg production was 1.5 times lower in the infected population, and the stock of resting eggs was almost the same (Table 16.3).

Brachionus calyciflorus population dynamics parameters, reproduced by the model simulating 2004s natural population, are presented in Table 16.4.

In 2004, the average size of infection, according to the model, was 1.78%, and this is almost threefold different from field observations (5.1%). In contrast, the infection values at the first population peak were significantly closer between model and nature variants: 3.1% and 2%, respectively. The reason behind the differences in some of these parameters is likely related to size differences between the modeled data and field data. Of the variants considered, infection of *B. calyciflorus* in 2004 was intermediate between the extreme values of 2001 and 2002. Nevertheless, a number of indicators for 2004 indicated a significant negative effect of this average infection on population parameters. The O/I ratios at first and second pop. peaks were 1.55 and 1.88, respectively.

In 2005, the (population max.) first and second O/I ratios (0-variant/infected variant) were 1.97 and 2.2, respectively. The overall infection maximum was 3.86% in the studied populations. Average fertility values were almost identical between variants. Total egg production was more than twofold lower in the infected population, and the stock of diapausing eggs was even lower (Table 16.5).

Analysis of the changes in production and accumulation of diapausing eggs showed a definite correlation between these important *B. calyciflorus* parameters and infection levels (see Fig. 16.3). It is easy to see that the point at which infection begins to have a significant *impact on the population is at 0.5% infection (*defined as declines of 20% or more, in comparison with uninfected). Therefore, even 1% infection of rotifers (a moderate seasonal infection level) can be considered an important environmental factor that significantly influences their population dynamics and seasonal adaptations.

16.5 Conclusion

Thus, by comparing data on the effects of infection in populations of *Brachionus calyciflorus* from several years, including mean infection levels, we were able to model population features. It should be pointed out that the 2002–2005 model variants are in good agreement with field (nature) observations. The 2001 variant falls significantly outside of the general pattern. It is possible that in that year some unaccounted for external factor exerted an additional influence on population dynamics. It may appear as though the model produces a relatively high number of dormant eggs for the conditions, but it should be noted that the model focuses on relatively short time frames in which *Brachionus calyciflorus*' main survival/adaptive behavior emphasizes production of parthenogenetic eggs over dormant. In habitable, longer-term water bodies, different mechanisms regulate the balance between production of parthenogenetic eggs and dormant; the latter situation is also analogous to that which is seen in crustaceans (Alekseev 1978).

Examination of the simulation model shows that even an insignificant seasonal infection of the *Brachionus calyciflorus* population with parasitic and/or bacterial infections (starting at 1%) leads to a twofold or greater reduction in the population's production and diapausing (resting) egg stock values. This significantly impacts populations' adaptive and competitive capabilities. Collectively, these results indicate that infection is an important environmental factor, and work related to rotifers and other hydrobionts, such as monitoring and general environmental studies, must take it into account.

Acknowledgments Victor R. Alekseev's participation in this work was supported by the Russian Foundation for Basic Research (grants 07-04-00006a and 17-04-00027).

The study was partly supported by the Russian Academy of Science, topics 65.4 and 65.5.

References

Alekseev VR (1978) *Microcyclops gracilis* (Lill.) and *M. varicans* (Sars) in temporary water bodies of the Volga delta. Hydrobiol Mag 14b(4):29–34

Alekseev VR. De Stasio B, Gilbert JJ (2007) Diapause in aquatic invertebrates. theory and human use, Springer, 257 pp.

Decksbach NK (1928) Zur Kenntnis der Parasiten der Radertiere. Zoolog Anz 78:274–278

Edmondson WT (1960) Reproductive rates of rotifers in natural populations. Mem Ist ital Idrobiol 12:21–77

Galkovskaya GA (1963) About the use of food for growth and the conditions for the maximum output production of rotifers *Brachionus calyciflorus* Pallas. Zool Mag 43(4):506–512

Gorbunov AK (1979) A method for rapid intravital measurement of microscopic animals using flash backlight microphotography. Hydrobiol Mag 6:114–115

Miracle MR (1977) Epidemiology in rotifers. Arch Hydrobiol Beih Ergebn Limnol H 8:138–141

Mnatsakanova EA, Polishuk LV (1996) Role of parthenogenetic natality and emergence from diapausing eggs in the dynamics of some rotifer populations. Diapause in the Crustacea. Dev Hydrobiol 114:169–178

Ruttner-Kolisko A (1977) The effect of the microsporid *Plistophora asperospora* on *Conochilus unicornis* in Lunzer Untersee (LUS). Arch Hydrobiol Beih Ergebn Limnob Stuttgart 8:135–137

Voronkov NV (1910) On the question of the development cycle of the sporozoa *Ascosporidium asperosporum*. Transactions of the Hydrobiological station on the Glubokoe lake 3:219–222

Index

V. R. Alekseev, B. Pinel-Alloul (eds.), *Dormancy in Aquatic Organisms. Theory, Human Use and Modeling*, Monographiae Biologicae 92,
https://doi.org/10.1007/978-3-030-21213-1

Printed by Printforce, the Netherlands